APPLIED STRUCTURAL STEEL DESIGN

Steel framing for the new international terminal. Los Angeles International Airport, Los Angeles, California.

APPLIED STRUCTURAL STEEL DESIGN

Leonard Spiegel, P.E.

Consulting Engineer

George F. Limbrunner, P.E.

Hudson Valley Community College

PRENTICE-HALL, INC., *Englewood Cliffs, NJ 07632*

Library of Congress Cataloging in Publication Data

Spiegel, Leonard.
   Applied structural steel design.

   Bibliography: p.
   Includes index.
   1. Building, Iron and steel. 2. Structural
design. I. Limbrunner, George F. II. Title.
TA684.S656 1986   624.1′821   85-6451
ISBN 0-13-041567-7

Editorial/production supervision and
   interior design: *Eileen M. O'Sullivan*
Cover design: *Debra Watson*
Manufacturing buyer: *John Hall*

Printed in the United States of America

10 9 8 7 6

ISBN 0-13-041567-7 01

PRENTICE-HALL INTERNATIONAL (UK) LIMITED, *London*
PRENTICE-HALL OF AUSTRALIA PTY. LIMITED, *Sydney*
PRENTICE-HALL OF CANADA INC., *Toronto*
PRENTICE-HALL HISPANOAMERICANA, S.A., *Mexico*
PRENTICE-HALL OF INDIA PRIVATE LIMITED, *New Delhi*
PRENTICE-HALL OF JAPAN, INC., *Tokyo*
PRENTICE-HALL OF SOUTHEAST ASIA PTE. LTD., *Singapore*
EDITORA PRENTICE-HALL DO BRASIL, LTDA., *Rio de Janeiro*
WHITEHALL BOOKS LIMITED, *Wellington, New Zealand*

Contents

Preface

The primary objective of this book is to furnish the reader with a basic understanding of the strength and behavior of structural steel members and their interrelationships in simple structural systems.

There is bountiful technical literature available which reflects recent research and advances in the trade. It is the intent of this book to set forth the most current relevant information in the context of traditional structural steel design. It is not intended to be a comprehensive treatise on the subject because it is our contention that such a document could easily obscure the fundamentals that we strive to emphasize in the engineering technology programs. In addition, we are of the opinion that adequate comprehensive books on structural steel do exist for those who seek the theoretical background, the research studies, and more rigorous applications.

Throughout the book we have referred continuously to the *Manual of Steel Construction* which contains the Specification for the Design, Fabrication and Erection of Structural Steel for Buildings. It is published by the American Institute of Steel Construction (AISC). Its use as a ready reference and companion publication with this book is recommended.

The text content is primarily an elementary, non-calculus, practical approach to the design and analysis of structural steel. We have utilized numerous example problems and a step-by-step solution format. In addition, chapters on structural steel detailing of beams and columns are included in an effort to convey to the reader a feeling for the design-detailing sequence.

The book has been classroom-tested in our engineering technology programs and should serve as a useful design guide and resource for technologists, technicians, and engineering and architectural students. Additionally, it will aid architects and engineers in preparing for state licensing examinations for professional registration.

We wish to express our appreciation to our colleagues, associates, and students who assisted in various ways in the preparation of this book. We are particularly indebted to Victoria R. Lee who did a masterful job in typing the entire manuscript. We are also especially indebted to our wives, Beverly Spiegel and Jane Limbrunner for their continuing patience, encouragement, and support during this time of preoccupation and it is to them that we affectionately dedicate this book.

Leonard Spiegel
George F. Limbrunner

1 Introduction to Steel Structures

1·1 STEEL STRUCTURES

The material steel, as we know it today, is a relatively modern human creation. Its forerunners, cast iron (which may have been invented in China as early as the fourth century B.C.) and wrought iron, were used in building and bridge construction in the period from the mid-eighteenth century to the mid-nineteenth century. However, in the United States, the age of steel began when it was first manufactured in 1856. The first important use of steel in any major construction project was in the still-existing Eads Bridge at St. Louis, Missouri, which was begun in 1868 and completed in 1874. This was followed in 1884 by the construction of the first high-rise steel-framed building, the 10-story (later, 12-story) Home Insurance Company Building in Chicago. The rapid development of steel-framed buildings in the Chicago area at that time seems to have resulted from that city's position as the commercial center for the booming expansion of the midwest economy. The rapid expansion caused an increased demand for commercial building space. This demand resulted in soaring land prices, which, in turn, made high-rise buildings more cost-effective.

In the century that has passed since those beginnings, steel has been vastly improved in both material properties and in methods and types of applications. Steel structures of note at present include the Humber Estuary suspension bridge in England, the main span of which stretches 4626 ft; a guyed radio mast in Poland with a height of 2120 ft; and the Sears Tower in Chicago, with 109 stories, which rises to 1454 ft. Each of these structures owes its notability to the strength and quality of the steel that supports it.

This is not to say that steel offers the builder an answer to all structural problems. The other major common building materials (concrete, masonry, and wood) all have their place and in many situations will offer economies which will dictate their use. But for building applications where the ratio of strength to weight (or the strength per unit weight) must be kept high, steel offers feasible options.

Steels for construction are **alloy steels,** which generally contain over 98% iron and usually less than 1% carbon. Although the actual chemical composition varies according to the desired properties, such as strength and corrosion resistance, steel may also contain alloying elements, such as silicon, manganese, sulfur, phosphorus, copper, chromium, and nickel, in varying amounts. Steel is not a renewable resource, but it can be recycled, and its primary component, iron, is plentiful.

Among the advantages of steel are uniformity of material and predictability of properties. Dimensional stability, ease of fabrication, and speed of erection are also beneficial characteristics of this building material. One may also list some disadvantages, such as susceptibility to corrosion (in most but not all steels) and loss of strength at elevated temperatures. Steel is not combustible, but it should be fireproofed to have any appreciable fire rating.

Some of the common types of steel structures are shown in Fig. 1-1.

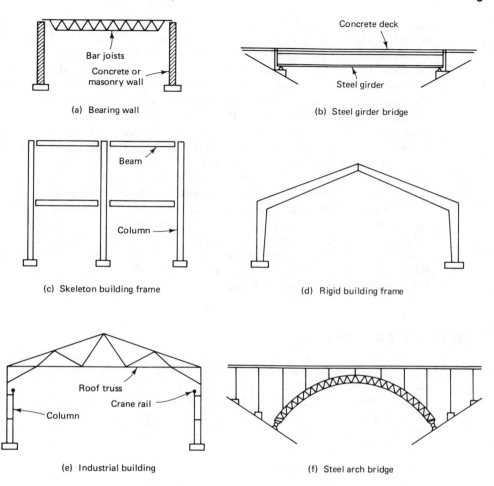

Figure 1-1   Types of steel structures.

1-2 HANDBOOKS AND SPECIFICATIONS

Structural steel is a manufactured product and is available in various grades, sizes, and shapes. The use of standard handbooks is absolutely essential to anyone working in any phase of steel construction. The most complete, comprehensive, and widely used handbook in the steel construction industry is the *Manual of Steel Construction,* published by the American Institute of Steel Construction (AISC).[1] This handbook, commonly called the *AISC Manual* or simply the *Manual,* is presently in its 8th edition. For simplicity we will refer to this manual as "AISCM." It will be referred to continually. It is meant to be a companion reference source for users of this textbook. It contains a wealth of information on products available, design aids, and erection guidelines, as well as useful specifications. The American Institute of Steel Construction is a trade association of steel producers, fabricators, and erectors in the United States and has as its objective to improve, advance, and promote the use of structural steel. The *AISC Specification for the Design, Fabrication and Erection of Structural Steel for Buildings,* contained in Part 5 of the AISCM, is a set of guidelines covering the various facets of steel construction. The first AISC Specification for buildings was issued in 1923. Over the years it has been revised many times to reflect new product developments, new philosophies, improved analysis methods, and research results. Although sworn at as well as sworn by, it is generally accepted that the current specification reflects a reasonable set of requirements and recommendations on which to base steel design and construction. When the AISC Specification is incorporated into the building code of a state or municipality (as it usually is), it becomes a legal document and is part of the law governing steel construction in a particular area. For simplicity, we will refer to this specification as "AISCS."

1-3 STEEL PROPERTIES

A knowledge of the various properties of steel is a requirement if one is to make intelligent choices and decisions in the selection of particular members. The more apparent mechanical properties of steel in which the designer is interested may be determined by a tension test. The test involves the tensile loading of a steel sample and the simultaneous measuring of load and elongation from which stress and strain may be calculated using

$$\text{stress*} = f_t = \frac{P}{A}$$

$$\text{strain} = \epsilon = \frac{\Delta L_0}{L_0}$$

where f_t = computed tensile stress (ksi)
$\quad\quad P$ = applied tensile load (kips)
$\quad\quad A$ = cross-sectional area of the tensile specimen (in.^{2}); this value is assumed constant throughout the test; decrease in cross-sectional area is neglected
$\quad\quad \epsilon$ = unit strain, elongation (in./in.)

*Stress and strain, as referred to in this text, would be more precisely defined as unit stress and unit strain.

ΔL_0 = elongation or the change in length between two reference points on the tensile specimen (in.)

L_0 = original length between two reference points (may be two punch marks) placed longitudinally on the tensile specimen before loading (in.)

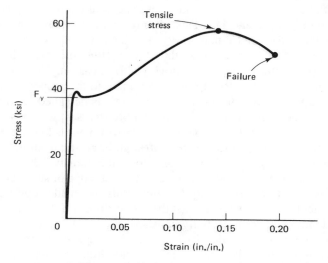

Figure 1-2 Typical stress-strain diagram for structural steel.

The sample is loaded to failure. The results of the test are displayed on a **stress-strain diagram.** Figure 1-2 is a typical diagram for commonly available structural steels. Upon loading, the tensile sample initially exhibits a linear relationship between stress and strain. The point at which the stress-strain relationship becomes nonlinear is called the **proportional limit.** This is noted in Fig. 1-3, which is the left part of Fig. 1-2 shown to a large strain scale. The steel remains elastic (meaning that

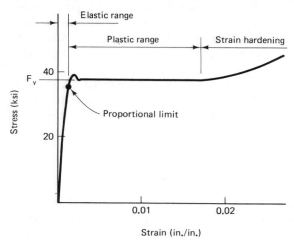

Figure 1-3 Partial stress-strain diagram for structural steel.

upon unloading it will return to its original length) as long as it is not stressed past a value slightly higher than the proportional limit called the **elastic limit.** The proportional limit and the elastic limit are so close that they are often considered to be the same value. Continuing the loading, a point will be reached at which the strain in the specimen increases rapidly at constant stress. The stress at which this occurs is called the **yield stress,** F_y, noted on Figs. 1-2 and 1-3. This striking characteristic is typical of the commonly used structural steels. Also note in Fig. 1-3 that F_y is the stress value of the horizontal plateau region of the curve. The slightly higher stress that exists just after the proportional limit (sometimes called the **upper yield**) exists only instantaneously and is unstable. That part of the curve from the origin up to the proportional limit is called the **elastic range.** At present, most structural steel is designed so that actual stresses in the structural members do not exceed **allowable stresses,** well below F_y. This method of design, which keeps stresses within the elastic range, is known by various names: **allowable stress design, working stress design,** or **elastic design.** As such, it is only the extreme left portion of the stress-strain diagram which is of immediate importance to the designer. However, there is still a vast range of stress and strain through which the steel will pass before it ultimately fails by tensile rupture.

In Fig. 1-3, once the steel has been stressed past the proportional limit, it passes into the **plastic range** and strains under essentially constant stress (F_y). As the steel continues to strain, it reaches a point at which its load-carrying capacity increases. This phenomenon of increasing strength is termed **strain hardening.** The maximum stress to which the test specimen is subjected is called the **tensile stress,** noted in Fig. 1-2. Although elastic design is still the most widely accepted design method, another design method exists which allows small but definite areas of members to be stressed to F_y and strained into the plastic range. This is called **plastic design** and will be discussed further in Chapter 11. For all practical purposes, in structural steel design, it is only the elastic range and the plastic range that need be of interest since the strains in the strain-hardening range are of such magnitude that the **deformation** of the structure would be unacceptable. Figure 1-4 shows an idealized diagram for structural steel which is sufficient for purposes of illustrating the steel stress-strain relationships. In Fig. 1-4, the strain at the upper limit of the plastic range, ϵ_p, is approximately 10 to 15 times the strain at the yield point, ϵ_y.

There are two other properties which become apparent from the stress-strain diagram. The first is **modulus of elasticity** E (or Young's modulus), which is the proportionality constant between stress and strain *in the elastic range*. It is also the **slope** of the stress-strain curve in the elastic range.

$$E = \frac{\text{stress}}{\text{strain}} = \frac{f_t}{\epsilon}$$

E is reasonably constant for structural steel. AISC recommends the value of E to be 29,000 kips/in.[2]. The second property of note is that of **ductility,** the ability to undergo large deformations before failure. Ductility is frequently the reason that a steel frame will remain standing after portions of its members have been stressed far

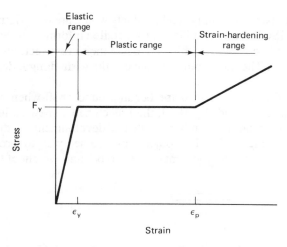

Figure 1-4 Idealized stress-strain diagram for structural steel.

beyond the design allowable stresses. The deformation of parts of the structure will serve to transfer loads to other less heavily loaded parts and thus will keep the structure from collapsing even though all or parts of it may have deformed to the point where the structure is no longer usable. This "forgiving nature," the ductility of steel, is important in the safeguarding of lives and property in unknown and uncertain loading situations such as earthquakes.

1-4 PRODUCTS AVAILABLE

In this text we consider only structural members made of hot-rolled steel. This process involves forming the steel to a desired shape by passing it between rolls while it is in the red-hot condition. Steel, as a material, is available in many different *types*. The primary property of interest is strength, in terms of F_y, since most allowable stresses are based on F_y. Also of interest may be the tensile strength, resistance to corrosion, and suitability for welding. Steels are usually specified according to ASTM (American Society for Testing and Materials) number. The AISCM, Part 1, Table 1*, contains a list of ASTM structural steels. Note that minimum F_y ranges from 32 to 100 ksi and tensile strength F_u ranges from 58 to 130 ksi.[†] Those interested in more detail concerning these steels are referred to the applicable ASTM literature.[2] Among these steels, the most widely used is the structural carbon steel designated A36. Its yield stress F_y is 36 ksi except for cross sections in excess of 8 in. thick. Grades A588 and A242 are **weathering steels,** the unpainted, maintenance-free steels which are popular for use in bridges and exposed building frames.

Various standardized structural products are available in the different types of steels. The AISCM, Part 1, Tables 1 and 2, groups these products into **shapes** and

---

*Note that all references to the AISCM in this text are to the *8th edition*.

[†]The determination of F_y is discussed further in Section 2-3 of this text.

plates and bars. Plates and bars are simply steel products which are of various widths and thicknesses. See the AISCM, Bars and Plates, for detailed information. Structural shapes, also called **sections,** are products which have had their cross sections tailored to one or more specific needs. The most common shape is the **wide flange,** designated as a **W shape.**

The development of the wide-flange shape began around 1830 when wrought-iron rails were being rolled in England. By 1849, the French had somewhat improved the cross section for use in bending members with the development of the rolled wrought-iron I section (see Fig. 1-5). The object, of course, was to provide an increased resistance to bending (greater moment of inertia) per unit weight of the cross section.

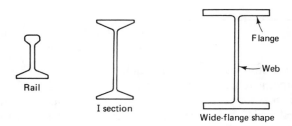

Figure 1-5 Development of the wide-flange shape.

The wide-flange shape, very efficient for bending applications, further improved on its forerunner by widening the flanges and thinning the web and thus obtaining an even better moment of inertia-to-weight ratio. The wide-flange beam is the invention of Henry Grey, a native of England who emigrated to the United States in 1870. He perfected the production method in 1897 but could find no company in the United States interested in manufacturing his new beam. However, his invention met with better acceptance in Europe. In 1902, in a German-owned steel mill in Differdange, in the Duchy of Luxembourg, the production of steel wide-flange shapes began. Shortly thereafter, Grey's beam captured the interest of Charles Schwab, president of Bethlehem Steel Company, and in 1908 the first wide-flange beam manufactured in the United States was rolled at Bethlehem, Pennsylvania.[3]

Figure 1-6 summarizes the structural shapes that are included in the AISCM, Part 1.

W shapes are used primarily as *beam and column members*. An example of their standard designation is W36 × 300. This indicates a W shape that is nominally (approximately) 36 in. deep from outside of flange to outside of flange and weighs 300 lb/ft. Note the range of W36 shapes in the AISCM, Part 1, Dimensions and Properties. Shapes in the set that runs from W36 × 135 to W36 × 210 are produced by the same set of rolls at the rolling mill. The weight difference is created by varying web thickness and flange width *and* thickness, as shown in Fig. 1-7. Interestingly, while the W36 is the deepest shape commonly rolled in the United States, nominal 40-in. (approximately 1-m)-deep

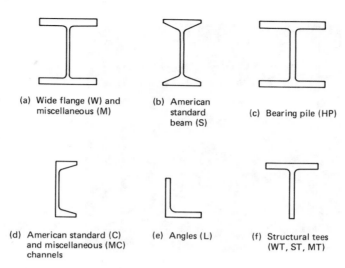

(a) Wide flange (W) and miscellaneous (M)

(b) American standard beam (S)

(c) Bearing pile (HP)

(d) American standard (C) and miscellaneous (MC) channels

(e) Angles (L)

(f) Structural tees (WT, ST, MT)

Figure 1-6   Hot-rolled steel shapes.

beam sections are rolled in Europe[7]. The deeper shapes will offer economies in some applications. These are discussed further in Chapter 3.

M shapes are *miscellaneous shapes* and have cross sections that appear to be exactly like W shapes, except that their depths are full inches. These shapes are rolled by fewer or smaller producers (see the AISCM, Part 1, Tables 3 and 5) and therefore may not be commonly available in certain areas. Availability should be checked before their use is specified. Their application is similar to that of the W shape, as is their designation (e.g., M4 × 13).

S shapes are *American standard beams*. They have sloping inner faces on the flanges, relatively thicker webs, and depths that are mostly full inches. They are infrequently used in construction, but do find some application where heavy point loads are applied to the flanges, such as in monorails for the support of hung cranes.

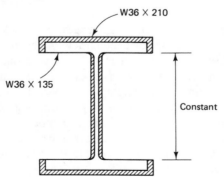

W36 × 210

W36 × 135

Constant

Figure 1-7   Cross-sectional variations.

World's largest hot-rolled wide-flange beam. Approximately $44\frac{1}{2}$ in. deep, 703 lb/ft. (Courtesy of Trade Arbed, Inc., New York.)

HP shapes are *bearing pile shapes* and are characterized by a rather square cross section with flanges and webs of nearly the same or equal thickness (in order that the web will withstand pile-driving hammer blows). They are generally used as driven piles for foundation support. They may also be used occasionally as beams and columns but are generally less efficient (more costly) for these applications.

C and **MC shapes** are *American Standard Channels* and *miscellaneous channels*. Examples of their designations are C15 × 50 and MC18 × 58, where the first number indicates the depth in inches and the second indicates the weight in pounds per foot. As with the M shapes, the MC shapes are generally less readily available and are produced by only a limited number of mills. The channel shapes are characterized by short flanges which have sloping (approximately $16\frac{2}{3}$% or 1:6) inner surfaces and depths to full inches. Their applications are usually as components of built-up cross sections, bracing and tie members, and members that frame openings.

Angles are designated by the letter L, leg length, and thickness. They may be equal leg or unequal leg angles. For unequal leg angles the longer leg is stated first. For example, L9 × 4 × $\frac{1}{2}$ indicates an angle with one leg 9 in. long and one leg 4 in. long, both having a thickness of $\frac{1}{2}$ in. Note that the designation *does not* provide the unit weight of the angle as has been the case with all shapes discussed thus far. The weight in pounds per foot is tabulated. Angles are commonly used singly and in pairs as bracing members and tension members. They are also used as brackets and connecting members between beams and their supports. Light trusses and open web steel joists may also utilize angles for component parts.

Structural tees are shapes that are produced by splitting the webs of W, M, or S shapes. The tees are then designated WT, MT, or ST, respectively. For example, a WT18 × 105 (18 in. deep, 105 lb/ft) is obtained from a W36 × 210. This can be verified by checking the dimensional properties of the tee with those of the wide-flange shape. Tees are used primarily for special beam applications and as components in connections and trusses.

In addition to the foregoing *shapes,* structural steel **tubing** is also available to the designer and may be observed in Fig. 1-8. Tubes are covered by different ASTM material specifications as reflected in the AISCM, Part 1, Table 4. One production method involves the forming of flat-rolled steel into a round profile. The edges are then welded together by the application of high pressure and an electric current. Square and rectangular profiles are then made by passing the round profile through another series of forming rolls.

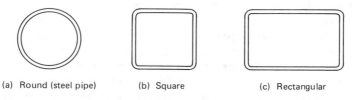

(a) Round (steel pipe)   (b) Square   (c) Rectangular

Figure 1-8   Structural steel tubing.

The AISCM does provide dimensions and properties for commonly used round, square, and rectangular tubes. Round tubes, or pipes, are available in three weights, with wall thickness being the influencing factor. With reference to the AISCM, Part 1, Pipe Dimensions, a standard-weight pipe section of 4 in. *nominal* outside diameter would be designated Pipe 4 Std. The same section in *extra strong* and *double extra strong* weights would be designated Pipe 4 X-Strong and Pipe 4 XX-Strong, respectively. Note the variation in wall thicknesses. Examples of structural steel tubing designations are TS5 × 5 × $\frac{1}{4}$ for a 5-in. square tube with $\frac{1}{4}$ in. wall thickness and TS12 × 8 × $\frac{1}{2}$ for a 12 in. × 8 in. rectangular tube with $\frac{1}{2}$ in. wall thickness. Tubes make excellent compression members although the connections usually involve some welding. The most common use of tubes is in compression and tension members, but they are also used as beams in some situations. Structural members made of tubes are easier to clean and maintain than are their wide-flange counterparts.

All of the foregoing product shapes and sections have advantages and disadvantages in particular structural applications. To make a reasonable decision concerning which to use, one should be familiar with the various properties of the available products. Availability will also play a role, and an alternative or substitute product may frequently be more economical. How the structure will be fabricated and erected must also be considered in order that every possible economy be realized. Minimum weight is not always the least expensive. Principal producers of the various structural shapes and tubes are designated in the AISCM, Part 1, Table 3. The reader should be aware that most of the producers (such as Bethlehem Steel) publish design aids and technical information concerning their products. This is in addition to the information in the AISCM, which does not contain information on *all* available products.

1-5 THE BUILDING PROJECT

The building of a structure involves many steps. The actual sequence of events, and the presence or absence of certain events, depends heavily on the size, the scope, the type of construction, and on the chosen method of managing the project. At the risk of oversimplifying the procedure, assuming that a low-rise steel frame building is to be constructed, a typical progression of events (primarily with respect to the structural aspect) may be broadly categorized into a design sequence and a construction sequence.

In the **design sequence,** a person or group (commonly called the **owner**) determines that there is a need, or a desire, for a structure. An architectural firm is contacted which studies the owner's requirements and the features of the site, and investigates various systems and layouts which offer solutions. Design calculations may be very rough and sketchy at this point, but the main building dimensions and the locations of the principal members are determined. Together with a structural engineer, the architect will recommend a design from the various possible solutions investigated. This phase may be called the **planning** or **preliminary design phase.** Upon the owner's approval, as well as a municipality approval, the **final design** begins. As the architectural, mechanical, and electrical designs are progressing, the structural design portion of the project proceeds through the stages of firming up the layout and selecting or designing all of the structural members. Contract drawings and specifications are also prepared.

In the **construction sequence,** the project is advertised and contractors are invited to submit bids. Based on bids submitted, a contractor is selected and a contract is signed. The contractor will then engage the services of a **steel fabricator** (a subcontractor). The steel fabricator will utilize the contract drawings and specifications to prepare **shop drawings** of all the steel members, showing precise dimensions and details and including all the details of the connections. Economy is gained in this way since the fabricator is well versed in structural practice and knows the strengths (and weaknesses) of his own shop. The shop drawings are submitted to the architect/engineer for approval before actual **fabrication** is begun. The fabrication shop work includes the actual cutting to length of the members, punching, drilling, shaping, grinding, welding, painting, and sometimes, preassembling parts of the structure. Following fabrication, the steel is transported to the site and **erection** of the steel commences.

In summary, the foregoing is only one way in which various events in a building project may take place. Any particular project may be completely different. One example is the current "fast-tracking" method, whereby construction is begun based on a sketchy and incomplete design. Design and construction then progress together. The economic benefits of early completion and occupancy outweigh the increased cost of construction brought about by the various unknown factors created by the incomplete design. However, traditionally, steel construction projects proceed generally as described.

1-6 DESIGN CONSIDERATIONS

Above all other considerations, the structural designer must be concerned with the safety and well-being of the general public, who will become the users of the structure. Economy, beauty, functionality, maintainability, permanence, and the like, are all secondary considerations when compared to safety and well-being of the users. The competence of the designer is of utmost importance. Not only must codes be followed, but their requirements (and the recommendations of the various specifications) must be tempered and applied with sound judgment. The blind following of a code will not

Rolled beams of nominal 40 in. depth and 102 ft span being placed for a bridge span. Note shear studs on the top flange for composite action between concrete deck and steel girders. Ditgesbaach Valley, Luxembourg. (Courtesy of Trade Arbed, Inc., New York.)

relieve the designer of ultimate responsibility. Codes and specifications, because of their bureaucratic nature, normally lag the state of the art. It therefore behooves the designers and practitioners to keep abreast of current developments and happenings through all the trade, technical, and professional channels available.

In the consideration of safety, a decision must be made as to just how safe a structure should be. The expression of safety is normally made in terms of a **factor of safety.** The factor of safety may be defined in many ways, but in a broad sense it is the ratio of (1) the load (or stress) that causes failure to (2) the maximum load (or stress) actually allowed in the structure. In **allowable stress design** the attainment of yield stress in a member is considered to be analogous to failure. Although the steel will not actually fail (rupture) at yield, significant and unacceptable deformations are on the verge of occurring, which may render the structure unusable. The maximum stress (allowable stress) to be used in the proportioning of the member is specified. This stress will exist, as a limit, in the member. The factor of safety is then a factor of safety *against yielding*. As an example, assume that a member composed of a steel having yield stress F_y has a specified allowable stress of $0.66F_y$. The factor of safety (F.S.) against yielding would then be

$$\text{F.S.} = \frac{\text{``failure stress''}}{\text{maximum stress}} = \frac{\text{yield stress}}{\text{maximum stress}} = \frac{F_y}{0.66F_y} = 1.5$$

Another way of considering this case is to think of the member as having a 50% **reserve of strength** against yielding in this particular application.

The factors of safety recommended by the various specifications and codes depend on many things. Danger to life and property as a result of the collapse of a particular type of structure, confidence in the analysis methods, confidence in the prediction of loads, variation in material properties, and possible deterioration during the design life of the structure are all possible considerations. Recommended factors of safety are the result of cumulative pooled experience and history, and are the minimum values that have been traditionally accepted as good practice.

The designer must also consider **loadings.** All the forces produced by loads that act on a structure must be transmitted through the structure to the underlying foundation. The designer must determine, based on the code requirements, judgment, and experience, just which loads are applicable and what their magnitudes will be. Codes and specifications again give guidelines in terms of minimum loads to be used based on general occupancy categories. The designer must decide if the minimum loads are satisfactory and, if not, make a better estimate.

Loads can be broadly grouped into dead loads and live loads. **Dead loads** are static loads that produce vertical forces due to gravity and include the weight of the steel framework and all materials permanently attached to it and supported by it. Reasonable estimates of the weight of the structure can usually be made based on the preliminary design work. **Live loads** include all vertical loads that may be either present on or absent from the structure. Generally, lateral loads are considered live loads whether they are permanent or not. Examples of live loads are:

Snow
People and furniture
Stored materials
Vehicles (on bridges and in warehouses)
Cranes
Wind
Lateral pressure due to earth or stored liquids
Earthquake

State, municipal, or other applicable building codes normally provide minimum loads for a designer's use in a particular area. In the absence of such requirements, the reader is referred to Reference 4. A detailed treatment of load calculations and theory is beyond the scope of this book. For an excellent discussion on theory and determination of loads and forces for structural design, the reader is referred to Reference 5.

1-7 NOTATION AND CALCULATIONS

For the many calculations that will be part of this text, notation of the AISC Manual will be used. The reader is referred to the "General Nomenclature" tables in the AISCM. In some cases, the AISC notation will be supplemented.

With regard to numerical accuracy of calculations, some things must be said in light of the widespread use of pocket electronic calculators. One should keep in mind the accuracy of the starting data in each calculation. Numbers resulting from the calculation need not be more accurate than the original data. The following guidelines, with respect to numerical precision in structural steel design calculations, are suggested. They are partially adopted from other sources.[6]

1. Loads to the nearest 1 psf; 10 lb/ft; 100 lb concentration
2. Span lengths and dimensions to 0.1 ft (for design)
3. Total loads and reactions to 0.1 kip or three-figure precision
4. Moments to the nearest 0.1 ft-kip or three-figure precision

The practice of presenting engineering numerical data with many more digits than warranted by the intrinsic certainty of the data (referred to by some as "super digitation") is common. To help guard against super digitation, several rules of thumb are advocated. One is based on the premise that engineering data are rarely known to an accuracy of greater than 0.2%. This would be equivalent to a possible error of 100 lb in a load of 50 kips. Therefore, loads in the range of 50 kips should be represented no more precisely than 0.1 kip.

Traditionally, three-significant-digit accuracy has been sufficient and acceptable for engineering calculations, although four digits may be used for numbers beginning with 1. (This tradition is rooted in the not-so-ancient widespread use of the 10-in. slide rule.) More accuracy is generally not warranted for civil engineering and structural design calculations. Therefore, the following representations would be common:

$$
\begin{array}{ll}
4.78 & 1.742 \\
32.1 & 0.00932 \\
728 & 0.1781 \\
88,300 &
\end{array}
$$

We will attempt to follow this tradition.

REFERENCES

1. *Manual of Steel Construction,* 8th ed., American Institute of Steel Construction, Inc., 400 North Michigan Avenue, Chicago, IL 60611.
2. American Society for Testing and Materials, 1916 Race Street, Philadelphia, PA 19103.
3. Robert Hessen, *Steel Titan, The Life of Charles M. Schwab* (New York: Oxford University Press, 1975), pp. 172–173.
4. *Building Code Requirements for Minimum Design Loads in Buildings and Other Structures,* ANSI A58.1, 1972, American National Standards Institute, 1430 Broadway, New York, NY 10018.
5. William McGuire, *Steel Structures* (Englewood Cliffs, N.J.: Prentice-Hall, Inc., 1968), Chapter 3.
6. *Concrete Reinforcing Steel Handbook,* Concrete Reinforcing Steel Institute, 933 North Plum Grove Road, Schaumburg, IL 60195
7. ARBED-Rolled Wide Flange Beams 40″ Standard and Tailor-Made Series, Third Edition, Trade Arbed, Inc., 825 Third Avenue, New York, NY 10022.

PROBLEMS

1-1. Sketch and completely dimension (use design dimensions) the cross sections of the following:
   (a) W36 × 210.
   (b) W14 × 500.

(c) HP13 × 60.
(d) MC13 × 50.
(e) L6 × 4 × $\frac{1}{2}$.
(f) WT12 × 52.
(g) TS7 × 5 × $\frac{1}{2}$.

1-2. Calculate the weight (in pounds) of a 1-ft length of a steel built-up member having a cross section as shown.

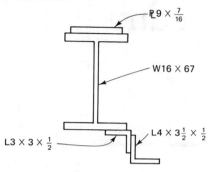

℞9 × $\frac{7}{16}$

W16 × 67

L4 × 3$\frac{1}{2}$ × $\frac{1}{2}$

L3 × 3 × $\frac{1}{2}$

Problem 1-2

1-3. How many of each of the following shapes are included in the AISCM?
(a) W33.
(b) W14.
(c) W8.
(d) Equal leg angles.

1-4. List the principal producers of
(a) W21 shapes.
(b) M14 × 18.
(c) TS10 × 6 × $\frac{1}{2}$.

2 Beams

18

2-1 INTRODUCTION

Beams are among the most common members that one will find in structures. They are structural members which carry loads that are applied at right angles to the longitudinal axis of the member. This causes the beam to **bend.** In this chapter we consider beams that carry no axial force. Figure 1-1(a), (b), and (c) illustrate some typical examples of beam applications.

When visualizing a beam (or any structural member) for the purposes of analysis or design, it is convenient to think of the member in some idealized form. The idealized form represents as closely as possible the actual structural member, but it has the advantage that it can be dealt with mathematically. For instance, in Fig. 2-1(a), the beam is shown with simple supports. These supports, a pin (knife-edge or hinge) on the left and a roller on the right, create conditions which are easily treated mathematically when it becomes necessary to find beam reactions, shears, moments, and deflections. Recall that the pin support will provide vertical and horizontal reactions (but no resistance to rotation) and the roller will provide only a vertical reaction. This is particularly significant for bridges, where provisions must be made for expansion and contraction due to temperature changes. In buildings, generally each support is capable of furnishing vertical and horizontal reactions. However, the beams are still considered to be simply supported since the requirement of a simple support is to permit freedom of rotation. In Fig. 2-1(b) the cantilever beam has a fixed support on the left side. This type of support provides vertical and horizontal reactions, as well as resistance (or a reaction) to rotation. The one fixed support is sufficient for static equilibrium of the beam. Although the idealized conditions generally will not exist in the actual structure, the actual conditions will approximate the ideal conditions and should be close enough to allow for a reasonable analysis or design.

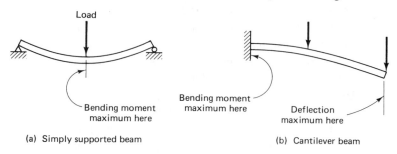

(a) Simply supported beam   (b) Cantilever beam

Figure 2-1   Beam types.

In the process of beam design, we will be concerned initially with the **bending moment** in the beam. The bending moment is produced in the beam by the loads it supports. Other effects, such as shear or deflection, may eventually control the design of the beam and will have to be checked. But usually moment is critical and it is, therefore, of initial concern.

Beams are sometimes called by other names, which are indicative of some specialized function(s):

Overhanging beams. Note the bearing plates and web stiffeners at the supports.

Girder: a major, or deep, beam which often provides support for other beams

Stringer: a main longitudinal beam, usually in bridge floors

Floor beam: a transverse beam in bridge floors

Joist: a light beam that supports a floor

Lintel: a beam spanning across an opening (a door or a window), usually in masonry construction

Spandrel: a beam on the outside perimeter of a building which supports, among other loads, the exterior wall

Purlin: a beam that supports a roof, and frames between or over supports, such as roof trusses or rigid frames

Girt: generally, a light beam that supports only the lightweight exterior sides of a building (typical in preengineered metal buildings)

2-2 THE MECHANICS OF BENDING

When the simply supported beam of Fig. 2-2(a) is subjected to two symmetrically placed loads, it bends as shown by the deflected shape. The diagrams of the induced shear (V) and moment (M) are as shown in Fig. 2-2(b) and (c). These diagrams neglect

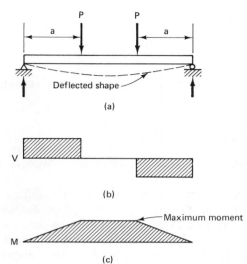

Figure 2-2 Load, shear, and moment diagrams for a simply supported beam.

the weight of the beam and consider the two concentrated loads (P) only. It is assumed that the reader is completely familiar with the development of shear and moment diagrams from strength of materials. We will consider a section at midspan (or anywhere between the two concentrated loads) where the moment is maximum as shown in Fig. 2-2(c). The maximum stress due to flexure (bending) in the beam may be determined by use of the **flexure formula:**

$$f_b = \frac{Mc}{I} = \frac{M}{S}$$

where f_b = computed bending stress (maximum at top and/or bottom)

$\quad M$ = maximum applied moment

$\quad c$ = distance from the neutral axis to the extreme outside of the cross section

$\quad I$ = moment of inertia of the cross section about the bending neutral axis

$\quad S$ = section modulus $(S = I/c)$ of the cross section about the bending neutral axis

The fundamental assumptions and the derivation of the flexure formula may be found in most textbooks on strength of materials.

The actual use of the flexure formula is straightforward, although the units must be carefully considered. Any conversion factors must be applied so that compatibility of units exist. For example, in the formula $f_b = M/S$ the *usual* units are

$\qquad f_b:$ kips/in.^{2} or ksi (stress)

$\qquad M:$ ft-kips (moment)

$\qquad S:$ in.^{3} (section modulus)

The *units* for the calculation of M/S may be written

$$f_b = \frac{M}{S} : \quad \frac{\text{ft-kips}}{\text{in.}^3}$$

For compatibility of units, with stress f_b resulting as ksi, a conversion factor of 12 in./ft must be used. For example:

$$\frac{50 \cancel{\text{ft}}\text{-kips}}{30 \text{ in.}^{\cancel{3}2}} \times 12 \frac{\cancel{\text{in.}}}{\cancel{\text{ft}}} = 20 \frac{\text{kips}}{\text{in.}^2}$$

For numerical problems worked out in this text, necessary conversion factors will be shown without further explanation.

It should be noted that for *empirical formulas,* such as those discussed in Section 2-3, units are sometimes *not* compatible (e.g., $\sqrt{F_y}$). In these formulas, numerical values with units precisely as defined in the AISCM must be used carefully.

Assuming that the beam of Fig. 2-2 is a typical wide flange (W shape), Fig. 2-3 shows the cross section and depicts the resulting bending stress diagram. The shape of the diagram is typical for bending stress at any point along the beam. Several points should be noted.

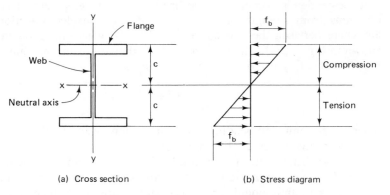

(a) Cross section   (b) Stress diagram

Figure 2-3   Simple beam bending.

1. For wide-flange beams, the moment of inertia about the x-x axis, I_x, is greater than the moment of inertia about the y-y axis. The beam is oriented so that bending occurs about the x-x axis. This is true except in very rare situations.

2. In this case, due to symmetry, the neutral axis is at the center of the cross section. The c distance is equal whether on the tension or compression side.

3. The maximum stress occurs at the top and the bottom of the cross section. The beam of Fig. 2-2 bends so that compression occurs above the neutral axis and tension occurs below the neutral axis (commonly, this is called **positive moment**).

4. Generally, only the maximum bending stress is of interest. Therefore, unless otherwise stated, f_b will be assumed to be the maximum stress. The flexure

formula may also be used to find the stress at any level in the cross section by substituting in place of c the appropriate distance to that level from the neutral axis.

Since f_b is bending stress that is induced in the beam by the applied loads (and the resulting moment), the steel of which the beam is composed must have sufficient strength to resist this moment without failure. (The reader is referred to Section 1-6 of this text for a discussion of *failure*.) The allowable bending stress F_b is specified by the AISCS based on the type of steel that is being used and other conditions that affect the strength of the beam in bending.

The stresses of Fig. 2-3(b) exist inside the beam. The summation of the stresses acting on their appropriate areas makes up an internal force system which creates what is frequently called an **internal couple** (or internal resisting moment). For equilibrium, the internal resisting moment at any point in the beam must be equal to the external applied moment at the same point [the external applied moment may be determined from the moment diagram of Fig. 2-2(c)].

In Fig. 2-4, if the external applied moment becomes the maximum allowed, the actual bending stresses at top and bottom of the beam will be equal to the allowable bending stress F_b. Additional moment should not be applied, since this would cause the actual bending stress to exceed F_b. The internal resisting moment (or, simply, resisting moment) that exists when the bending stress is F_b is termed M_R.

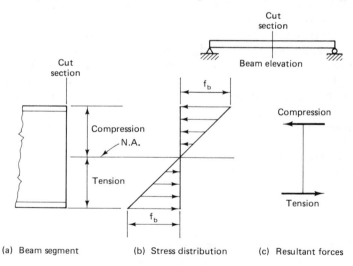

(a) Beam segment   (b) Stress distribution   (c) Resultant forces

Figure 2-4   Stress–moment relationships.

From the flexure formula, the resisting moment M_R can be calculated by substituting F_b for f_b, and M_R for M:

$$F_b = \frac{M_R c}{I} = \frac{M_R}{S}$$

Then

$$M_R = F_b S$$

M_R may be thought of as the bending strength (in terms of moment) of the beam cross section.

The use of the flexure formula, in any of its various forms, is basic to beam analysis and design.

2-3 ALLOWABLE BENDING STRESS

In dealing with beam problems it is necessary to have an understanding of the specified allowable bending stress F_b, the maximum bending stress to which a beam should be subjected. The AISC Specification (AISCS) treats this topic in Section 1.5.1.4. Neglecting later complications, the basic allowable bending stress (in both tension and compression) to be used for most rolled shapes is

$$F_b = 0.66F_y$$

where F_y is the material yield stress. For a member to qualify for an allowable bending stress F_b of $0.66F_y$, the member must have an axis of symmetry in, and be loaded in, the plane of the web. An important condition associated with the use of this value for F_b is the **lateral support** of the compression flange. The compression flange behaves somewhat like a column and it will tend to buckle to the side, or laterally, as the stress increases if it is not restrained in some way. Varying amounts and types of lateral support may be present. In Fig. 2-5(a), a concrete slab encases the top flange (assumed to be compression) and is mechanically anchored to it. The slab forms a horizontal diaphragm and effectively braces the top flange against any lateral movement. This may be termed **full lateral support.** In Fig. 2-5(c), *no* lateral support exists for the top flange. In Fig. 2-5(b), there are three points of lateral support. The *distance* between the points of lateral support (whatever it may be) *in inches* is denoted ℓ. For convenience, we will denote this distance L_b when it is *in feet.* To qualify for $F_b = 0.66F_y$, the compression flange of a beam must have *adequate* lateral support such that

$$\ell \leq \frac{76b_f}{\sqrt{F_y}} \quad \text{and} \quad \frac{20,000}{(d/A_f)F_y}$$

where b_f = flange width of the beam (in.)
F_y = yield stress of the steel (ksi)
d = depth of the beam (in.)
A_f = area of compression flange (in.^{2})

The smaller of the two values of ℓ is a tabulated property for each W shape (dependent on F_y) and is designated L_c (ft). See the AISCM, Part 2, Allowable Stress Design Selection Table.

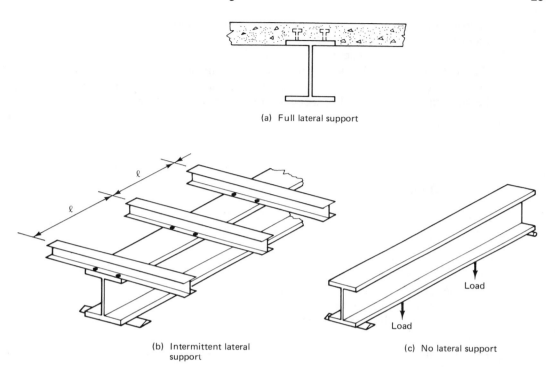

(a) Full lateral support

(b) Intermittent lateral support

(c) No lateral support

Figure 2-5 Lateral support conditions.

The amount of lateral support actually available may not be easy to determine. For instance, if a concrete slab rests on a beam but is not anchored to it, only the friction between the two will provide lateral support. Its adequacy is questionable and judgment must be used. A conservative estimate of no lateral support, in this case, would be prudent. If the compression flange of a beam has inadequate lateral support (ℓ is too large), the lateral buckling tendency will be counteracted by *reducing F_b*. This will be discussed in Section 2-6 of this chapter. We will, for now, assume beams to have adequate lateral support. For further discussion on lateral support and bracing, the reader is referred to Reference 1.

Another important condition that must be met if the beam cross section is to qualify for $F_b = 0.66F_y$ deals with the response of the beam in an overload situation. Elastic design assumes failure to occur when F_y is first reached. The beam will not fail at this point since it has a substantial reserve of strength. If the cross section continues to strain under increased moment, the outer fibers will further strain but the stress will remain at F_y (see Fig. 1-2). F_y will be reached by the fibers at levels progressively closer to the neutral axis until virtually the entire cross section is stressed to F_y. When this occurs the beam has achieved its **plastic moment capacity.** (This is the basis for *plastic design*.) However, the cross section must be proportioned so that no local buckling of the flange or web occurs before the plastic moment capacity is achieved.

A cross section that meets this criterion is said to be **compact.** Only compact sections qualify for $F_b = 0.66F_y$.

The test for compactness is found in the AISCS, Section 1.5.1.4.1. Assuming that there are no axial loads on the beam (and this assumption will be made for all beams until beam-columns are discussed in Chapter 6) the two formulas required may be simplified as follows. For a section to be considered compact:

Flange criterion:

$$\frac{b_f}{2t_f} \le \frac{65}{\sqrt{F_y}}$$

Web criterion:

$$\frac{d}{t_w} \le \frac{640}{\sqrt{F_y}}$$

where b_f = flange width of the beam (in.)
t_f = flange thickness of the beam (in.)
F_y = material yield stress (ksi)
d = depth of the beam (in.)
t_w = web thickness of the beam (in.).

Both the flange and the web criteria must be satisfied for a member to be considered compact.

Example 2-1

Determine if a W18 × 76 of A36 steel (F_y = 36 ksi) is compact.

Solution   W18 × 76 properties:

$$b_f = 11.035 \text{ in.}$$
$$t_f = 0.680 \text{ in.}$$
$$d = 18.21 \text{ in.}$$
$$t_w = 0.425 \text{ in.}$$

Flange criterion:

$$\frac{b_f}{2t_f} = \frac{11.035}{2(0.68)} = 8.11$$

$$\frac{65}{\sqrt{F_y}} = \frac{65}{\sqrt{36}} = 10.8$$

$$8.11 < 10.8 \qquad\qquad \textbf{O.K.}$$

Web criterion:

$$\frac{d}{t_w} = \frac{18.21}{0.425} = 42.8$$

$$\frac{640}{\sqrt{F_y}} = \frac{640}{\sqrt{36}} = 106.7$$

$$42.8 < 106.7 \qquad\qquad \textbf{O.K.}$$

Therefore, the W18 × 76 is compact. The preceding could be shortened by using tabulated quantities from the AISCM, Part 1, Properties of W Shapes, and Appendix A, Table 6. The tabulated quantities are rounded slightly in some cases.

A more rapid way to determine compactness of cross section for the rolled shapes is to calculate the value of a hypothetical yield stress F_y which would cause equality in each of the two criteria. For the flange criterion,

$$\frac{b_f}{2t_f} = \frac{65}{\sqrt{F_y}}$$

$$F_y = \left(\frac{65}{b_f/2t_f}\right)^2$$

For the W18 × 76,

$$F_y = \left(\frac{65}{8.11}\right)^2 = 64.2 \text{ ksi}$$

This shows that the flange criterion is satisfied provided that F_y *does not* exceed 64.2 ksi. This value is termed F_y' and is tabulated in the AISCM, Part 1, as a property of the W18 × 76. If the W18 × 76 is of a steel with F_y in *excess* of 64.2 ksi, it is *not* compact by the flange criterion. Therefore, *for compactness by the flange criterion,* the following condition should exist:

$$F_y \le F_y'$$

Should the calculated value of F_y' be in excess of the highest available F_y, this is reflected by the tabulation of a dash (–) for the F_y' value. For the web criterion,

$$\frac{d}{t_w} = \frac{640}{\sqrt{F_y}}$$

$$F_y = \left(\frac{640}{d/t_w}\right)^2$$

For the W18 × 76,

$$F_y = \left(\frac{640}{42.8}\right)^2 = 224 \text{ ksi}$$

This shows that the W18 × 76 is compact by the web criterion provided that F_y does not exceed 224 ksi. Reference to the AISCM, Part 1, Table 1, shows that shapes are currently not available which have F_y in excess of 65 ksi. Therefore, the W18 × 76 is compact by the web criterion *in all steels*. This is a general rule. All rolled, W, M, and S shapes tabulated in the AISCM are compact by the web criterion (when $f_a = 0$). This does not hold true for built-up sections and plate girders, which are discussed in Chapters 3 and 10. Web noncompactness will cause F_b to be reduced to $0.60\,F_y$ (assuming adequate lateral support).

If a shape does not satisfy the flange criterion, F_b must be reduced. For the range of $b_f/2t_f$ between $65/\sqrt{F_y}$ and $95/\sqrt{F_y}$, the AISCS provides for a linear reduction in F_b to $0.60F_y$ according to the following formula:

$$F_b = F_y\left[0.79 - 0.002\left(\frac{b_f}{2t_f}\right)\sqrt{F_y}\right] \qquad \text{AISCS formula 1.5-5a}$$

Shapes falling into this category are sometimes called **partially compact shapes.** The variation of F_b for rolled W shapes *which have adequate lateral support* is summarized graphically in Fig. 2-6.

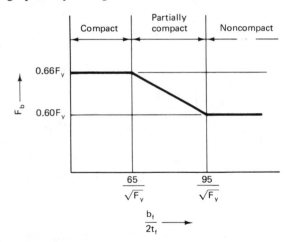

Figure 2-6 F_b for rolled W shapes with adequate lateral support.

In determining F_b, the yield stress F_y must be known. The best source for the value of F_y is the AISCM, Part 1, Tables 1 and 2. For a known shape, determine the appropriate *group* from Table 2. Then, knowing the steel type, utilize Table 1 to determine F_y.

Example 2-2

Find F_b for the following shapes. Assume adequate lateral support for the compression flange.

(a) W30 × 132 of A36 steel.

(b) W12 × 65 of A242 steel.

Solution

(a) All W shapes in A36 steel are compact except for the W6 × 15 since its F_y' is less than 36 ksi. From the AISCM, Part 1, Table 1, all shapes in A36 steel have $F_y = 36$ ksi.

$$F_b = 0.66F_y = 0.66(36)$$

where

$$F_b = 23.8 \text{ ksi (commonly rounded to 24.0 ksi).}$$

(b) From the AISCM, Part 1, Table 2, W12 × 65 is found in *Group 2*. From Table 1, $F_y = 50$ ksi. From the AISCM properties tables for the W12 × 65, $F_y' = 43.0$ ksi:

$$50.0 > 43.0$$

therefore, *not compact*. Check whether partially compact. (The following quantities can also be found in the AISCM, Part 1, Properties of W Shapes, and Appendix A, Table 6.)

$$\frac{b_f}{2t_f} = 9.9$$

$$\frac{65}{\sqrt{F_y}} = \frac{65}{\sqrt{50}} = 9.2$$

$$\frac{95}{\sqrt{F_y}} = \frac{95}{\sqrt{50}} = 13.4$$

Therefore, this shape *is* partially compact since

$$\frac{65}{\sqrt{F_y}} < \frac{b_f}{2t_f} < \frac{95}{\sqrt{F_y}}$$

Calculate F_b from AISCS formula 1.5-5a:

$$F_b = F_y\left[0.79 - 0.002\left(\frac{b_f}{2t_f}\right)\sqrt{F_y}\right]$$

$$= 50[0.79 - 0.002(9.9)\sqrt{50}] = 32.5 \text{ ksi}$$

2-4 ANALYSIS OF BEAMS FOR MOMENT

The analysis problem is generally considered to be the *investigation* of a beam whose *cross section* is known. One may be concerned with checking the adequacy of a given beam, determining an allowable load, or finding the maximum existing bending stress in the beam. All of these problems are related. All make use of the flexure formula and require an understanding of **allowable bending stress** F_b.

Example 2-3

A W21 × 44 beam is to span 24 ft on simple supports (as shown in Fig. 2-7). Assume full lateral support and A36 steel. The load shown is a **superimposed load,** meaning that it does *not* include the weight of the beam. Determine whether the beam is adequate by

(a) Comparing the actual bending stress with the allowable bending stress.

(b) Comparing the actual applied moment with the resisting moment M_R.

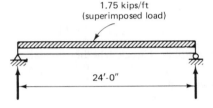

1.75 kips/ft
(superimposed load)

24'-0"

Figure 2-7 Load diagram.

Solution

(a) Determine the actual bending stress from the flexure formula

$$f_b = \frac{Mc}{I} = \frac{M}{S}$$

From the properties tables, for the W21 × 44, $S_x = 81.6$ in.^{3}. Moment may be determined by shear and moment diagram or by formula (see the AISCM, Part 2, Beam Diagrams and Formulas, for review). The total load should include the weight of the beam:

1.75	kips/ft	(applied load)
+0.044		(beam weight)
1.79	kips/ft	(total uniform load w)

$$\text{Applied moment } M = \frac{wL^2}{8} = \frac{1.79(24)^2}{8} = 129 \text{ ft-kips}$$

$$f_b = \frac{M}{S_x} = \frac{129(12)}{81.6} = 18.97 \text{ ksi}$$

With reference to the comments of Example 2-2(a): $F_y' = (\text{—})$ from the AISCM, Part 1. Therefore, since $F_y' > F_y$, the member is compact in A36 steel and the allowable bending stress F_b is 24.0 ksi. Therefore,

$$f_b < F_b \qquad\qquad \textbf{O.K.}$$

(b) The applied moment M has been determined to be 129 ft-kips. The resisting moment M_R may be calculated from the flexure formula:

$$M_R = F_b S_x = \frac{24.0(81.6)}{12} = 163 \text{ ft-kips}$$

Therefore,

$$M < M_R \qquad\qquad \textbf{O.K.}$$

Example 2-4

A W18 × 40 beam spans 20 ft on a simple span as shown in Fig. 2-8. Assume A36 steel. The compression flange is supported laterally at the quarter points where the concentrated loads P are applied. Therefore, $L_b = 5$ ft. Determine the allowable value for each load P (kips).

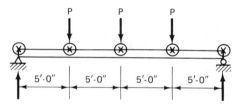

Ⓧ Indicates lateral support
for compression flange

5'-0"   5'-0"   5'-0"   5'-0"

Figure 2-8 Load diagram.

Solution   Determine F_b and M_R. W18 × 40 properties:

$$b_f = 6.015 \text{ in.}$$

$$\frac{d}{A_f} = 5.67$$

$$F_y = 36 \text{ ksi}$$

$$S_x = 68.4 \text{ in.}^3$$

$$F_y' = (\text{—})$$

Test for adequate lateral support by comparing L_b with L_c. From the AISCM, Part 2, Allowable Stress Design Selection Table, $L_c = 6.3$ ft. $L_b = 5$ ft. Since $L_b < L_c$, this beam has adequate lateral support. F_y' is (—), implying that F_y' is high enough so that this shape is always compact. Therefore,

$$F_b = 0.66F_y = 24.0 \text{ ksi}$$

$$M_R = F_b S_x = \frac{24.0(68.4)}{12} = 137 \text{ ft-kips}$$

The resisting moment is 137 ft-kips. The applied moment due to the beam's own weight and the moment due to the three equal loads P cannot exceed the resisting moment M_R. The applied moment due to beam weight is

$$M = \frac{wL^2}{8} = \frac{0.04(20)^2}{8} = 2.0 \text{ ft-kips}$$

The moment due to the concentrated loads may be determined using the shear and moment diagrams of Fig. 2-9 or aids such as those found in the AISCM, Part 2, Beam Diagrams and Formulas.

$$M = 10P \text{ ft-kips}$$

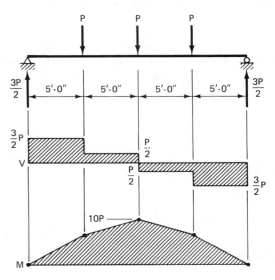

Figure 2-9   Load, shear, and moment diagrams.

The resisting moment remaining to support the concentrated loads is

$$137 - 2.0 = 135 \text{ ft-kips}$$

Equating, we have

$$10P = 135$$

$$P = \frac{135}{10} = 13.5 \text{ kips}$$

Therefore, the maximum allowable value of the concentrated load P on this beam is 13.5 kips.

It should be noted that the weight of the beam itself in both of the foregoing examples has been a very minor part of the total load carried. This is generally true.

Nevertheless, it should always be considered. As one becomes more experienced, the effect of beam weight is easier to estimate. There are various shortcuts used by designers to simplify the inclusion of weight of structure in design problems. In analysis problems, where the cross section is known, inclusion of the beam weight is a simple matter.

2-5 SUMMARY OF PROCEDURE: BEAM ANALYSIS FOR MOMENT ONLY

This procedure is general and typical for the various shapes that may be used for beams, primarily wide-flange sections and, to a lesser extent, other sections. The precise method of solution will depend on the nature of the particular problem, the known conditions, and the information sought.

1. Determine F_y. Use the AISCM, Part 1, Tables 1 and 2.
2. Check the adequacy of lateral support. See the AISCS, Section 1.5.1.4.1. If lateral support is inadequate, see Section 2-6 of this chapter.
3. Check the compactness of the cross section. Use F_y' from the table of properties in the AISCM, Part 1.
4. Using the preceding information, determine F_b.
5. If the applied loads are known, the applied moment can be found. Draw shear and moment diagrams or use beam formulas from the AISCM, Part 2.
6. If the magnitude of the applied loads is unknown, write an expression for the applied moment in terms of the unknown loads. This can then be equated to the resisting moment of the beam.
7. The flexure formula for use in analysis is

$$f_b = \frac{Mc}{I} = \frac{M}{S}$$

$$M_R = F_b S$$

2-6 INADEQUATE LATERAL SUPPORT

As the distance between points of lateral support on the compression flange (ℓ) becomes larger, there is a tendency for the beam to buckle laterally. There is no upper limit for ℓ. However, to guard against the buckling tendency as ℓ becomes larger, the AISCS provides that F_b be reduced. This, in effect, reserves some of the beam strength to resist the lateral buckling. Figure 2-10(b) shows a beam that has deflected vertically and buckled laterally. Note that the section has not only twisted in torsion, but the compression flange has moved laterally. This may be thought of as a lateral deflection in the horizontal plane. The compression flange will have some bending resistance to

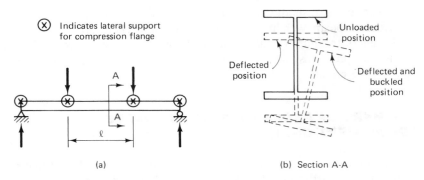

Figure 2-10   Beam deflection and lateral buckling.

this lateral deflection. These are the two general resistances available to counteract lateral buckling: torsional resistance of the cross section and lateral bending resistance of the compression flange. The total resistance to lateral buckling is the sum of the two. The AISCS conservatively considers only the *larger* of the two in the determination of a reduced F_b.

The AISCS, Section 1.5.1.4.5, establishes empirical expressions for F_b for the inadequate lateral support situation. Tension and compression allowable bending stresses are treated separately. The tension $F_b = 0.60F_y$ always. Only the compression F_b is reduced. For typical rolled shapes this is of no consequence since the shapes are symmetrical and the lower F_b of the two values will control. Note that the provisions of this section pertain to members having an axis of symmetry in, and are loaded in, the plane of their web. They also apply to compression on extreme fibers of channels bent about their major axis.

The AISCS provides three empirical formulas for the reduced compression F_b. The mathematical expressions that give an exact prediction of the buckling strength of beams are too complex for general use. Therefore, the AISCS formulas only *approximate* this strength for purposes of determining a reasonable F_b. The F_b that is finally used is the *larger* of the F_b values determined from the applicable formulas. The first two AISCS formulas (1.5-6a and 1.5-6b) give the F_b value when the lateral bending resistance of the compression flange provides the lateral buckling resistance. The third AISCS formula (1.5-7) gives F_b when the torsional resistance of the beam section provides the primary resistance to lateral buckling. In no case should F_b be greater than $0.60F_y$ for beams that have inadequate lateral support. The formulas that will be applicable will depend on the value of the ratio ℓ/r_T, where

ℓ = distance between points of lateral support for the compression flange (in.)

r_T = radius of gyration of a section comprising the compression flange plus one-third of the compression web area taken about an axis in the plane of the web (in.), as shown in Fig. 2-11

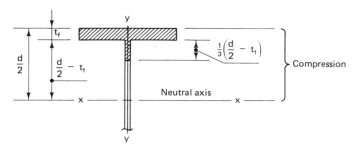

Figure 2-11 r_T determination.

r_T is a tabulated quantity for rolled shapes (see the AISCM, Part 1). ℓ/r_T may be considered a slenderness ratio of the compression portion of the beam with respect to the y-y axis. The formulas for F_b are as follows:

$$F_b = \left[\frac{2}{3} - \frac{F_y(\ell/r_T)^2}{1530 \times 10^3 C_b} \right] F_y \qquad \text{AISCS formula 1.5-6a}$$

$$F_b = \frac{170 \times 10^3 C_b}{(\ell/r_T)^2} \qquad \text{AISCS formula 1.5-6b}$$

$$F_b = \frac{12 \times 10^3 C_b}{\ell d/A_f} \qquad \text{AISCS formula 1.5-7}$$

where C_b = a liberalizing modifying factor whose value is between 1.0 and 2.3 which accounts for a moment gradient over the span and a decrease in the lateral buckling tendency; C_b may be conservatively taken as 1.0; see the AISCS, Section 1.5.1.4.5, for details

  d = depth of cross section (in.)

  A_f = area of compression flange (in.^{2})

Figure 2-12 depicts the decision-making process for the calculation of F_b. Note that one will use AISCS formulas (1.5-6a) and (1.5-7) *or* AISCS formulas (1.5-6b) and (1.5-7). The *larger* resulting F_b is used. Note that the AISCM, Appendix A, Table 6, provides the following numerical equivalents for A36 steel ($F_y = 36$ ksi):

$$\sqrt{\frac{102 \times 10^3 C_b}{36}} = 53\sqrt{C_b}$$

$$\sqrt{\frac{510 \times 10^3 C_b}{36}} = 119\sqrt{C_b}$$

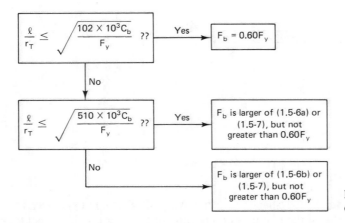

Figure 2-12 F_b for beams with inadequate lateral support.

Example 2-5

A W21 × 50, shown in Fig. 2-13, spans 36 ft on a simple span. The compression flange is laterally supported at the third points. A36 steel is used. Determine F_b for this beam.

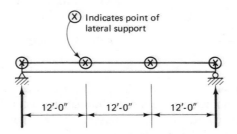

Figure 2-13 Beam diagram.

Solution   The shape is compact and $F_y = 36$ ksi. Check for adequacy of lateral support. ($\ell = 144$ in. or $L_b = 12$ ft.)

For the W21 × 50:

$$b_f = 6.53 \text{ in.}$$

$$\frac{d}{A_f} = 5.96$$

$$r_T = 1.60 \text{ in.}$$

To qualify for $F_b = 0.66F_y$, the actual unbraced length of the compression flange must be equal to, or less than,

$$\frac{76b_f}{\sqrt{F_y}} \qquad \text{and} \qquad \frac{20,000}{(d/A_f)F_y}$$

as required in the AISCS Section 1.5.4.1. For A36 steel, these expressions become

$$12.7b_f \quad \text{and} \quad \frac{556}{d/A_f}$$

As mentioned previously, the lesser of these two expressions is designated L_c (in feet). If $L_b \leq L_c$, the beam will qualify for $F_b = 0.66F_y$.

$$L_b = 12 \text{ ft}$$

$$L_c = 6.9 \text{ ft} \quad \text{(AISCM, Part 2)}$$

Since $L_b > L_c$, F_b must be reduced. The beam compression flange is therefore said to be **inadequately braced.**

The value of ℓ/r_T, which will determine the applicable AISCS formulas for F_b, will be compared with

$$53\sqrt{C_b} \quad \text{and} \quad 119\sqrt{C_b}$$

from the AISCS, Appendix A, Table 6.

$$\frac{\ell}{r_T} = \frac{144}{1.60} = 90.0$$

Conservatively assuming that $C_b = 1.0$, we have

$$53\sqrt{C_b} = 53$$
$$119\sqrt{C_b} = 119$$
$$53 < \frac{\ell}{r_T} < 119$$

Therefore, from Fig. 2-12, the applicable F_b formulas are:

AISCS formula 1.5-6a:

$$F_b = \left[\frac{2}{3} - \frac{F_y(\ell/r_T)^2}{1530 \times 10^3 C_b}\right]F_y = \left[\frac{2}{3} - \frac{36(90)^2}{1530(10^3)(1.0)}\right]36 = 17.1 \text{ ksi}$$

AISCS formula 1.5-7:

$$F_b = \frac{12 \times 10^3 C_b}{\ell(d/A_f)} = \frac{12(10^3)(1.0)}{144(5.96)} = 14.0 \text{ ksi}$$

The *larger* F_b is used:

$$F_b = 17.1 \text{ ksi}$$

As indicated, the tables of the AISCS, Appendix A, are useful in shortening the calculations somewhat.

This discussion has pertained primarily to the rolled shapes commonly used for beams and loaded for strong-axis bending. The AISCS formulas for F_b for beams that have inadequate lateral support are also applicable to built-up members and plate girders, provided that they have an axis of symmetry in the plane of the web.

Example 2-6

Determine the allowable superimposed uniformly distributed load that may be placed on the W21 × 50 of Example 2-5 (see Fig. 2-13).

Solution F_b was determined to be 17.1 ksi.

$$M_R = F_b S_x$$

$$= \frac{17.1(94.5)}{12} = 134.7 \text{ ft-kips}$$

Since the beam is a simple span member with a 36-ft span,

$$M = \frac{wL^2}{8} \qquad \text{and} \qquad w = \frac{8M}{L^2}$$

Since, as a limit, the applied moment M can equal M_R,

$$w = \frac{8M_R}{L^2} = \frac{8(134.7)}{36^2} = 0.831 \text{ kip/ft}$$

Subtracting the beam's own weight, the allowable superimposed load is

$$831 - 50 = 781 \text{ lb/ft} = 0.781 \text{ kip/ft}$$

2-7 DESIGN OF BEAMS FOR MOMENT

The basis for moment design is to provide a beam that has a moment capacity (M_R) equal to or greater than the anticipated maximum applied moment M. The flexure formula is used to determine a required section modulus S:

$$\text{required } S = \frac{M}{F_b}$$

The section modulus is assumed to be the strong-axis section modulus S_x. There is a convenient table set up for this selection in the AISCM, Part 2, the Allowable Stress Design Selection Table (S_x Table). It lists common beam shapes in order of decreasing section modulus.

Example 2-7

Select the lightest W shape for the beam shown in Fig. 2-14(a). Assume full lateral support ($L_b = 0$) and A36 steel. Consider moment only.

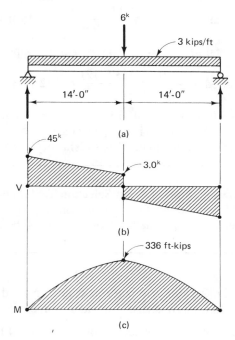

Figure 2-14 Load, shear, and moment diagrams.

Solution   The beam reactions are determined from the load diagram; the shear and moment diagrams are drawn. These are shown in Fig. 2-14(b) and (c). Note that no beam weight is included since the beam is unknown. An estimate of the beam weight could be made and its effect on the moment included. This is optional at this point. However, the beam weight must be included before the final selection of section is made. The maximum moment is determined to be 336 ft-kips, as shown in Fig. 2-14(c).

Select F_b, if possible. In this case, the A36 steel beam has full lateral support and F_b almost assuredly equals $0.66F_y$, since virtually all W shapes of A36 steel are compact. Therefore, assume that

$$F_b = 0.66F_y = 0.66(36) = 24 \text{ ksi}$$

Determine the required S_x:

$$\text{required } S_x = \frac{M}{F_b} = \frac{336(12)}{24} = 168 \text{ in.}^3$$

From the AISCM, Part 2 (Allowable Stress Design Selection Table), select a W24 × 76 with an $S_x = 176$ in.^{3}. It is the lightest W shape that will furnish the required section modulus. Note that the section selected weighs 76 lb/ft. Add in the effect of the beam weight.

$$\text{additional } M = \frac{wL^2}{8} = \frac{0.076(28)^2}{8} = 7.45 \text{ ft-kips}$$

$$\text{new total } M = 336 + 7.45 = 343.5 \text{ ft-kips}$$

$$\text{new required } S_x = \frac{M}{F_b} = \frac{343.5(12)}{24} = 172 \text{ in.}^3$$

The W24 × 76 is satisfactory since 176 in.^{3} > 172 in.^{3} required. Also, now check the assumed F_b: the W24 × 76 is compact since $F_y' > F_y$ and has adequate lateral support; therefore, the assumed F_b is satisfactory. **Use W24 × 76.**

In the use of the Allowable Stress Design Selection Table, note that any shape having at least the required S_x will be satisfactory (for moment). If a *shallower* section is required, the choice of one is a simple matter. However, it will be a heavier section.

Example 2-8

The beam shown in Fig. 2-15(a) is to be of A36 steel. Note the lateral support conditions. Select the lightest W shape. Consider moment only.

Solution   For this design, an estimated beam weight of 40 lb/ft (0.04 kip/ft) has been added to the given uniform load. (This estimate may be based on anything from an educated guess to a rough design worked quickly on scrap paper.) The shear and moment diagrams are shown in Fig. 2-15(b) and (c). The maximum moment is 129 ft-kips.

Establish F_b. F_y is 36 ksi and a compact shape that has adequate lateral support will be assumed. Therefore, the assumed allowable bending stress is

$$F_b = 0.66F_y = 0.66(36) = 24 \text{ ksi}$$

from which

$$\text{required } S_x = \frac{M}{F_b} = \frac{129(12)}{24} = 64.5 \text{ in.}^3$$

Select a W16 × 40 with $S_x = 64.7$ in.^{3}. This is a compact shape since $F_y' > F_y$. Check the adequacy of lateral support ($\ell = 60$ in. or $L_b = 5$ ft) by comparing L_b with L_c. From the AISCM, Part 2, $L_c = 7.4$ ft. Therefore, since $L_b < L_c$, the beam has adequate lateral support and the assumed F_b is satisfactory. The assumed beam weight is satisfactory. **Use W16 × 40.**

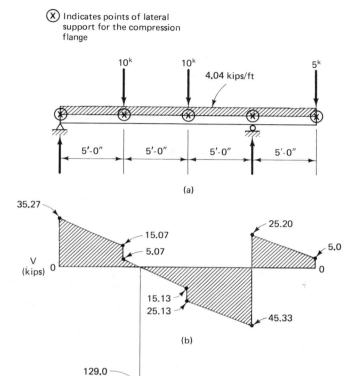

(X) Indicates points of lateral
support for the compression
flange

(a)

(b)

(c)

Figure 2-15 Load, shear, and moment diagrams.

The rapid solutions of Examples 2-7 and 2-8 have depended on the correct assumption of F_b. Since F_b depends, in part, on the section to be selected, it cannot always be predetermined. Various design aids have been developed to speed the design process. In Example 2-5, F_b was determined for a W21 × 50 beam of A36 steel which had inadequate lateral support. For ℓ of 144 in. (L_b = 12 ft), F_b = 17.1 ksi. For each value of L_b, a value for F_b could be determined. A plot of F_b versus L_b is shown in Fig. 2-16. The shape of this curve is typical for a compact cross section. A plot of M_R versus L_b, instead of F_b versus L_b, would have the same form as Fig. 2-16 since $M_R = F_b S_x$ and S_x is constant for the cross section.

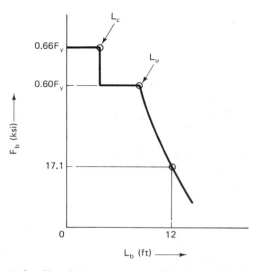

Figure 2-16 F_b versus L_b for a W21 × 50 (A36).

A family of these curves will be found in the AISCM, Part 2. These curves make up a very valuable design aid for beams and should be used whenever possible. The lightest beam section may be selected directly using only the applied moment M (ft-kips) and the unbraced length of compression flange L_b (ft). Note that the vertical axis for these curves is *total allowable moment*. This is the same as resisting moment, M_R.

Two other terms are illustrated in Fig. 2-16:

L_c = Maximum unbraced length (ft) of compression flange at which the allowable bending stress may be taken at $0.66F_y$ (for compact shapes), or as determined by AISCS formula 1.5-5a or 1.5-5b (for partially compact shapes).

L_u = Maximum unbraced length (ft) of the compression flange at which the allowable bending stress may be taken at $0.60F_y$.

For most shapes the value of L_u (in feet) is given as $20,000/[12(d/A_f)F_y]$ from AISCS formula 1.5-7. *For a few shapes,* L_u is given as $\sqrt{102,000/F_y} \times (r_T/12)$ from AISCS formula 1.5-6a, where this is more liberal. L_c and L_u are unique for each shape and also vary with F_y. They are tabulated properties and may be found in the AISCM, Part 2, Allowable Stress Design Selection Table, and in the Beam Tables of Part 2, for F_y of 36 ksi and 50 ksi. Their use greatly facilitates the determination of whether or not lateral support is adequate. However, the precise values of L_c and L_u are of no consequence when using the beam curves.

When using the curves, any shape represented by a curve that is found *above or to the right* of a particular M_R and L_b combination is a shape that is satisfactory. The lightest adequate shape is represented by a solid line. The dashed lines represent shapes that are also adequate but are heavier. The AISCM contains beam curves for

$F_y = 36$ ksi and $F_y = 50$ ksi. These curves may be used for the range of L_b from 0 up to the maximum value indicated in the curves.

Example 2-9

Select the lightest W shape for the beam shown in Fig. 2-17 for the following conditions:

(a) A36 steel, $L_b = 4$ ft.
(b) A36 steel, $L_b = 16$ ft.
(c) A441 steel, $L_b = 20$ ft.

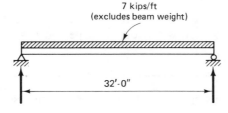

7 kips/ft
(excludes beam weight)

32'-0"

Figure 2-17   Beam load diagram.

Solution

(a) $F_y = 36$ ksi. Determine the applied moment for a simply supported, uniformly loaded, single-span beam. Neglect the beam weight temporarily.

$$M = \frac{wL^2}{8} = \frac{7.0(32)^2}{8} = 896 \text{ ft-kips}$$

From the beam curves, AISCM, Part 2, with $M = 896$ ft-kips and unbraced length $(L_b) = 4$ ft, select a W36 × 150. Note that all lines are horizontal at the left vertical axis. For a moment of 896 ft-kips, the W36 × 150 would be satisfactory for any L_b from 0 to 12.5 ft. Add the moment due to the beam weight:

$$\text{additional } M = \frac{wL^2}{8} = \frac{0.150(32)^2}{8} = 19 \text{ ft-kips}$$

$$\text{total } M = 896 + 19 = 915 \text{ ft-kips}$$

$$M_R \text{ for the W36} \times 150 \text{ with } L_b = 4 \text{ ft is } 1010 \text{ ft-kips}$$

$$1010 \text{ ft-kips} > 915 \text{ ft-kips} \qquad \textbf{O.K.}$$

Use W36 × 150

(b) $F_y = 36$ ksi. $M = 896$ ft-kips from part (a). $L_b = 16$ ft. From the beam curves, select a W36 × 160. For L_b of 16 ft, $M_R = 960$ ft-kips for this shape. Add the moment due to the beam weight:

$$\text{additional } M = \frac{wL^2}{8} = \frac{0.16(32)^2}{8} = 20.5 \text{ ft-kips}$$

$$\text{total } M = 896 + 20.5 = 917 \text{ ft-kips} < 960 \text{ ft-kips} \quad \textbf{O.K.}$$

Use W36 × 160.

(c) *Assume* that $F_y = 50$ ksi. $M = 896$ ft-kips from part (a). $L_b = 20$ ft. From the beam curves, select a W36 × 150 with $M_R = 960$ ft-kips. Add the beam weight moment:

$$\text{additional } M = \frac{0.15(32)^2}{8} = 19 \text{ ft-kips}$$

$$\text{total } M = 896 + 19 = 915 \text{ ft-kips} < 960 \text{ ft-kips} \quad \textbf{O.K.}$$

Check F_y: from the AISCM, Part 1, Tables 1 and 2, the
W36 × 150 is a Group 2 shape and $F_y = 50$ ksi   **O.K.**

Use W36 × 150.

When the curves are not applicable (e.g., when F_y equals some value other than 36 ksi or 50 ksi), F_b must be assumed and subsequently verified.

Example 2-10

Rework Example 2-9. Select the lightest W shape for the beam shown in Fig. 2-17. The framing system used indicates the use of an $L_b = 9.0$ ft. The steel is to be A572 Grade 60 ($F_y = 60$ ksi).

Solution   *Assume* that $F_b = 0.66F_y = 0.66(60) = 40$ ksi. Using the moment calculated in Example 2-9(a), which included a 150-lb/ft beam weight:

$$\text{required } S_x = \frac{M}{F_b} = \frac{915(12)}{40} = 275 \text{ in.}^3$$

From the AISCM, Part 2 (Allowable Stress Design Selection Table), select a W30 × 108 ($S_x = 299$ in.^{3}).

Now verify F_b. The section is compact since $F_y' = (—)$ (AISCM, Part 1), which indicates that it is in excess of 65 ksi. Therefore $F_y' > F_y$. Since L_c is not tabulated for a steel with $F_y = 60$ ksi, it must be computed and compared with $L_b = 9$ ft. With reference to the AISCS, Section 1.5.1.4.1, L_c is the smaller of

$$\frac{76b_f}{\sqrt{F_y}} = \frac{76(10.475)}{\sqrt{60}} = 102.8 \text{ in.}$$

and

$$\frac{20,000}{(d/A_f)F_y} = \frac{20,000}{3.75(60)} = 88.9 \text{ in.}$$

Therefore,

$$L_c = \frac{88.9}{12} = 7.41 \text{ ft}$$

Since $L_b > L_c$, F_b must be reduced to at least $0.60F_y$ and possibly lower. This will affect the required S_x with a subsequent change in the section selected. Assume the new F_b to be $0.60(60) = 36$ ksi:

$$\text{required } S_x = \frac{915(12)}{36} = 305 \text{ in.}^3$$

From the AISCM, Part 2, select a W30 $\times$ 116 ($S_x = 329$ in.^{3}). For this shape, a calculation for L_c (similar to that preceding) will show that $L_c = 8.3$ ft. Therefore, L_b is still greater than L_c. Now determine L_u as the larger of

$$\frac{20,000}{12(d/A_f)F_y} = \frac{20,000}{12(3.36)60} = 8.3 \text{ ft}$$

and

$$\sqrt{\frac{102,000}{F_y}} \left(\frac{r_T}{12}\right) = \sqrt{\frac{102,000}{60}} \left(\frac{2.64}{12}\right) = 9.1 \text{ ft}$$

Therefore, $L_u = 9.1$ ft and $F_b = 0.60F_y$ (as assumed), since

$$L_c < L_b < L_u$$

The beam weight included in the design moment is on the conservative side since the assumed 150 lb/ft is greater than the actual beam weight of 116 lb/ft. However, no modifications of the calculations are necessary. **Use a W30 $\times$ 116.**

Other beam design aids are available, such as the tables of *Uniform Load Constants,* also found in the AISCM, Part 2. Examples of their use will be found in the *Manual.*

The curves found in the AISCM, Part 2, may also be used for analysis-type problems as well as for design. Knowing the beam size and unbraced length (L_b), the total allowable moment may be obtained from the curves.

Example 2-11

Compute the allowable superimposed uniformly distributed load that may be placed on a W36 $\times$ 150 spanning 24 ft–0 in. The beam is laterally supported at its supports only ($L_b = 24$ ft–0 in.) and is A36 steel.

Solution   Entering the curves for a W36 $\times$ 150 with an $L_b = 24$ ft, the total allowable moment = 680 ft-kips. The moment due to beam's own weight is

$$M = \frac{0.150(24)^2}{8} = 10.8 \text{ ft-kips}$$

The allowable moment for superimposed load is

$$680 - 10.8 = 669.2 \text{ ft-kips}$$

Since the applied moment can equal the allowable moment, as a limit,

$$M = \frac{wL^2}{8} \qquad \text{or} \qquad w = \frac{8M}{L^2} = \frac{8(669.2)}{24^2} = 9.3 \text{ kips/ft}$$

2-8 SUMMARY OF PROCEDURE: BEAM DESIGN FOR MOMENT

Based on the foregoing examples, a general procedure may be established for the design of beams for moment.

1. Establish the conditions of load, span, and lateral support. This is best done with a sketch. Establish the steel type.
2. Determine the design moment. If necessary, complete shear and moment diagrams should be drawn. An estimated beam weight may be included in the applied load.
3. The beam curves should be used to select an appropriate section when possible. As an alternative, F_b must be estimated and the required section modulus determined:

$$\text{required } S_x = \frac{M}{F_b}$$

The section is then selected using the S_x table.
4. After the section has been selected, recompute the design moment, including the effect of the weight of the section. Check to ensure that the section selected is still adequate.
5. Check any assumptions that may have been made concerning F_y or F_b.
6. Be sure that the solution to the design problem is plainly stated.

2-9 SHEAR IN BEAMS

Except under very special loading conditions, all beams are subjected to shear as well as moment. In the normal process of design, beams are selected on the basis of the moment to be resisted and then checked for shear. Shear rarely controls a design unless loads are very heavy (and, possibly, close to the supports) and/or spans are very short. From strength of materials, the shear stress that exists within a beam may be determined from the general shear formula

$$f_v = \frac{VQ}{Ib}$$

where f_v = shear stress on a horizontal plane located with reference to the neutral axis (ksi)

V = vertical shear force at that particular section (kips)

Q = statical moment of area between the plane under consideration and the outside of the section, about the neutral axis (in.^{3})

I = moment of inertia of the section about the neutral axis (in.^{4})

b = thickness of the section at the plane being considered (in.).

This formula furnishes us with a *horizontal* shear stress at a point, which, as shown in any strength of materials text, is equal in intensity to a *vertical* shear stress at the same point in a beam.

Example 2-12

A W16 × 100 is subjected to a vertical shear of 80 kips. Determine the maximum shear stress and plot the distribution of shear stress for the entire cross section.

Solution   Design dimensions for the W16 × 100 are shown in Fig. 2-18(a).

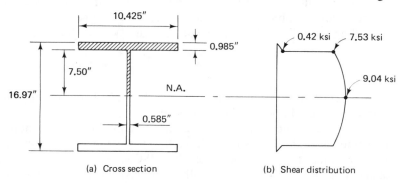

(a) Cross section   (b) Shear distribution

Figure 2-18   Shear stress.

Other properties are

$$A = 29.4 \text{ in.}^2$$

$$I_x = 1490 \text{ in.}^4$$

The maximum shear stress will be at the neutral axis (where Q is maximum). Taking Q for the crosshatched area about the neutral axis, we have

$$Q = 10.425(0.985)\left(7.50 + \frac{0.985}{2}\right) + 7.50(0.585)\left(\frac{7.50}{2}\right)$$

$$= 98.5 \text{ in.}^3$$

$$\text{maximum } f_v = \frac{VQ}{Ib} = \frac{80(98.5)}{1490(0.585)} = 9.04 \text{ ksi}$$

If shear stress values are calculated at other levels in the cross section and the results plotted, the shear distribution will appear as in Fig. 2-18(b).

Note that the flanges resist very low shear stresses. Even though the areas of the flanges are large, it is the web that predominantly resists the shear in wide-flange beams. For this reason the AISCS allows the use of an **average web shear** approach for the shear stress determination:

$$f_v = \frac{V}{dt_w}$$

where f_v = computed maximum shear stress (ksi)
V = vertical shear at the section considered (kips)
d = depth of the beam (in.)
t_w = web thickness of the beam (in.)

This method is approximate compared to the theoretically correct general shear formula, and assumes that the shear is resisted by the rectangular area of the web extending the full depth of the beam. For the W16 × 100 of Example 2-12,

$$f_v = \frac{V}{dt_w} = \frac{80}{16.97(0.585)} = 8.06 \text{ ksi}$$

This is less than the shear stress of 9.04 ksi calculated by the general shear stress formula and could be considered unsafe.

Allowable shear stresses are set intentionally low to account for the fact that the computed average shear stress will be lower than the actual shear stress. In beams, at locations other than at connections, the AISCS, Section 1.5.1.2.1, sets the allowable shear stress $F_v = 0.40F_y$. The localized effects of shear at connections are discussed in Chapter 7.

Example 2-13

The W18 × 50 beam shown in Fig. 2-19(a) has been designed for moment. The uniform load includes the beam weight. Check the beam for shear. Assume A36 steel and full lateral support.

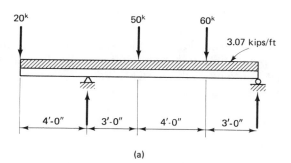

(a)

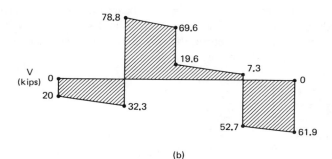

(b)   **Figure 2-19**   Load and shear diagrams.

Solution   The shear diagram is drawn as shown in Fig. 2-19(b). The maximum shear to which the beam is subjected is 78.8 kips. The W18 × 50 properties are

$$d = 17.99 \text{ in.}$$

$$t_w = 0.355 \text{ in.}$$

Calculating actual and allowable shear stresses, we have

$$f_v = \frac{V}{d t_w} = \frac{78.8}{17.99(0.355)} = 12.34 \text{ ksi}$$

$$F_v = 0.40 F_y = 0.40(36) = 14.4 \text{ ksi}$$
(which is rounded to 14.5 ksi in the AISCS, Appendix A)

$$f_v < F_v$$

W18 × 50 O.K. for shear.

The concept of **shear capacity** is also useful. The shear capacity of a beam may be determined by multiplying its web area ($d \times t_w$) by the allowable shear stress F_v. The AISCM calls this the **maximum permissible web shear** and (unfortunately) designates it V with no differentiation from the applied shear V. For the W18 × 50 of Example 2-13,

$$\text{shear capacity} = F_v d t_w = 14.5(17.99)(0.355) = 92.6 \text{ kips}$$

This may be readily compared with the maximum applied shear to verify that the beam is satisfactory. The maximum permissible web shear (V) is a tabulated quantity. Refer to the tables of Uniform Load Constants in the AISCM, Part 2, where V for each section is tabulated for $F_y = 36$ ksi and 50 ksi.

2-10 DEFLECTIONS

When a beam is subjected to a load that creates bending, the beam must sag or deflect as shown in Fig. 2-20. Although a beam is safe for moment and shear, it may be unsatisfactory because it is too flexible. Therefore, the consideration of the deflection of beams is another part of the beam design process.

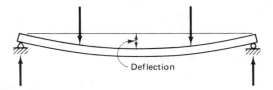

Deflection

Figure 2-20 Beam deflection.

Excessive deflections are to be avoided for many reasons. Among these are the effects on attached nonstructural elements such as windows and partitions, undesirable vibrations, and the proper functioning of roof drainage systems. Naturally, a visibly sagging beam tends to lessen one's confidence in both the strength of the structure and the skill of the designer.

To counteract the sag in a beam, an upward bend or **camber** may be given to the beam. This is commonly done for longer beams to cancel out the dead-load deflection and, sometimes, part of the live-load deflection. One production method involves cold bending of the beam by applying a point load with a hydraulic press or ram. For shorter beams, which are not intentionally cambered, the fabricator will process the beam so that any natural sweep within accepted tolerances will be placed so as to counteract expected deflections.

Normally, deflection criteria are based on some maximum limit to which the deflection of the beam must be held. This is generally in terms of some fraction of the span length. For the designer this involves a calculation of the expected deflection for the beam in question, a determination of the appropriate limit of deflection, and a comparison of the two.

The calculation of deflections is based on principles treated in most strength-of-materials texts. Various methods are available. For common beams and loadings the AISCM, Part 2, Beam Diagrams and Formulas, contains deflection formulas. The use of some of these will be illustrated in subsequent example problems.

The deflection limitations of specifications and codes are usually in the form of suggested guidelines because the strength adequacy of the beam is not at stake.

Traditionally, beams that have supported plastered ceilings have been limited to maximum *live-load* deflections of span/360. This is a requirement of the AISCS, Section 1.13.1. The span/360 deflection limitation is often used for live-load deflections in other situations. It is common practice, and in accordance with some codes, to limit maximum total deflection (due to live load *and* dead load) to span/240 for roofs and floors that support other than plastered ceilings.

The AISCS Commentary, Section 1.13.1, contains guidelines of another nature. It suggests:

1. The *depth* of fully stressed beams and girders in floors should, if practicable, be not less than $F_y/800$ times the span.
2. The *depth* of fully stressed roof purlins should, if practicable, be not less than $F_y/1000$ times the span, except in the case of flat roofs.

Further, it recommends that where human comfort is the criterion for limiting motion, as in the case of vibrations, the depth of a steel beam supporting large open floor areas free of partitions or other sources of damping should be not less than $\frac{1}{20}$ of the span.

Since the moment of inertia increases with the square of the depth, the guidelines for minimum beam depth limit deflections in a general way. The AISCS Commentary, Section 1.13.3, also contains a method for checking the flexibility of roof systems when **ponding,** the retention of water on flat roofs, is a consideration.

Example 2-14

Select the lightest W shape for the beam shown in Fig. 2-21. Assume full lateral support and A441 steel. Consider moment, shear, and deflection. Maximum allowable deflection for total load is to be span/360.

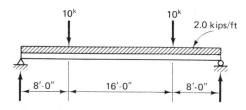

10^k 10^k 2.0 kips/ft

|← 8'-0" →|←   16'-0"   →|← 8'-0" →|

Figure 2-21   Beam load diagram.

Solution   The usual procedure is to design for moment and check for the other effects. Select the section for moment:

$$M = \frac{wL^2}{8} + Pa = \frac{2.00(32)^2}{8} + 10(8) = 336 \text{ ft-kips}$$

From the beam curves, assuming that $F_y = 50$ ksi, select a W21 × 62 that has an M_R of 349 ft-kips. From the AISCM, Part 1, Tables 1 and 2, $F_y = 50$ ksi. Check the additional moment due to the beam's own weight:

$$M = \frac{0.062(32)^2}{8} = 7.9 \text{ ft-kips}$$

$$\text{total } M = 336 + 7.9 = 343.9 \text{ ft-kips}$$

$$343.9 < 349 \qquad\qquad \textbf{O.K.}$$

Check the shear. For this beam, the maximum shear occurs at, and is equal to, the reaction. Therefore,

$$\text{maximum shear} = \frac{2(10) + 32(2.062)}{2} = 43.0 \text{ kips}$$

From the beam tables (Uniform Load Constants), Part 2, maximum permissible web shear for the W21 $\times$ 62 is 168 kips. Therefore, the shear strength is O.K.

Check deflection (Δ). From the AISCM properties tables, I for the W21 $\times$ 62 is 1330 in.[4]:

$$\text{maximum allowable } \Delta = \frac{\text{span}}{360} = \frac{32(12)}{360} = 1.07 \text{ in.}$$

From formulas in the AISCM, Part 2, Beam Diagrams and Formulas, the actual expected deflection may be calculated. Note that the units are kips and inches.

$$\Delta = \frac{5wL^4}{384EI} + \frac{Pa(3L^2 - 4a^2)}{24EI}$$

$$= \frac{5(2.062)(32)^4(12)^3}{384(29,000)(1330)} + \frac{10(8)(12)^3}{24(29,000)(1330)}[3(32)^2 - 4(8)^2]$$

$$= 1.26 + 0.42 = 1.68 \text{ in.} > 1.07 \text{ in.} \qquad \text{(no good)}$$

The moment of inertia (I) must be increased. Select a larger beam with an I value of

$$\frac{1.68}{1.07}(1330) = 2088 \text{ in.}^4$$

From the AISCM, Part 2, Moment of Inertia Selection Tables, select W24 $\times$ 76 ($I = 2100$ in.[4]). This shape has a higher S_x and greater shear capacity than the W21 $\times$ 62. Therefore, moment and shear are satisfactory. **Use W24 $\times$ 76.**

It is worthwhile noting, in the foregoing example, that the M_R of an A36 steel W24 $\times$ 76 (full lateral support) is 352 ft-kips. Therefore, the use of the higher strength A441 steel is not justified. This is sometimes the case where the design is governed by deflection (I) rather than by strength (F_y).

2-11 *HOLES IN BEAMS*

Beams are normally found as elements of a total structural system rather than as individual, isolated entities. They are penetrated by mechanical and electrical systems, are enveloped by nonstructural elements, and must be connected to other structural members. They must sometimes be cut to provide clear areas. The problem of holes in beams is a common one.

Among the more evident effects of holes in beams (or, generally, any decrease in cross-sectional area) is the capacity reduction. Two such reductions may be readily identified. Holes in beam webs reduce the shear capacity. Holes in beam flanges reduce the moment capacity.

The AISCS is not specific concerning a recommended design procedure for beams with web holes. The common procedure of reducing shear capacity in direct proportion to web area reduction is an oversimplification for other than small holes. One design procedure that has been proposed[2] involves the use of a reduced allowable shear stress, which, in turn, is based on various geometric parameters of the beam and hole. Should the hole be too large (f_v > reduced F_v), the beam may be reinforced locally by welding stiffening material around the perimeter of the hole.[3] Some recommendations have recently been made concerning design methods for unreinforced web holes.[4]

Wide flange beam with large web hole. Note how stiffeners have been welded around the perimeter of the hole for reinforcement.

Some general rules may be stated with regard to web holes. They should be located away from areas of high shear. For uniformly loaded beams, web holes near the center of the span will not be critical. Holes should be centered on the neutral axis to avoid high bending stresses. The holes should be round or have rounded corners (for rectangular holes) to avoid stress concentrations. The cutting of web holes in the field should not be allowed without the approval of the designer.

With respect to holes in the flanges of beams, it is the moment capacity that is affected. The cross-sectional property that governs moment capacity is moment of inertia I. The *web* of a wide flange beam contributes very little to the moment of inertia and the effect of web holes on moment capacity may be neglected. The effect of flange holes, however, is to reduce the moment of inertia. The calculation of the reduction is accomplished by subtracting from the gross moment of inertia the quantity Ad^2 for each hole, where

A = cross-sectional area of the hole (diameter × flange thickness) (in.^{2})

d = distance from the neutral axis to the centroid of the hole (in.)

The neutral axis shifts very little where holes exist in only one flange and may be assumed to remain at the centroid of the gross cross section. For our discussion, no distinction will be made as to whether the flange is tension or compression, even though beams are usually controlled by the strength of the compression flange. Finally, it is the conservative practice of some designers to consider both flanges to have holes (in a symmetrical pattern) even though only one does.

It is generally agreed that flange holes for bolts do not reduce the moment capacity of beams to the extent indicated by the reduced moment of inertia as described in the preceding paragraph. The AISCS, therefore, in Section 1.10.1, states that no reduction in moment of inertia shall be made for bolt holes in either flange unless the area of the holes in the flange exceeds 15% of the gross flange area. When the area of the holes does exceed this limit, only the excess area is used to reduce the moment of inertia.

Holes for bolts are normally punched with a diameter $\frac{1}{16}$ in. larger than the bolt diameter. Additionally, the AISCS, Section 1.14.4, directs that for net area computation, the hole diameter should be taken $\frac{1}{16}$ in. greater than the actual nominal diameter of the hole. Therefore, for purposes of analysis and design, hole diameters are taken as the fastener diameter plus $\frac{1}{8}$ in.

Example 2-15

Using the AISCS, determine the resisting moment M_R for a W18 × 71 which has two holes punched in each flange for 1-in.-diameter bolts. $F_b = 24$ ksi. Assume A36 steel.

Solution   The cross section is shown in Fig. 2-22. For design purposes, the holes are $1\frac{1}{8}$ in. in diameter. First, check whether the moment of inertia must be reduced based on the AISCS, Section 1.10.1:

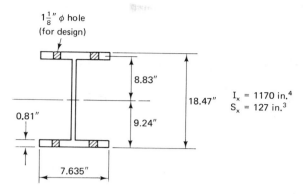

Figure 2-22 W18 × 71.

$$\text{hole area per flange} = 2(1.125)(0.81) = 1.82 \text{ in.}^2$$
$$15\% \times \text{flange area} = 0.15(7.635)(0.81) = \underline{0.93 \text{ in.}^2}$$
$$\text{excess hole area per flange} = 0.89 \text{ in.}^2$$

$$\text{net } I_x = 1170 - 2(0.89)(8.83)^2 = 1031 \text{ in.}^4$$

$$\text{net } S_x = \frac{\text{net } I_x}{c} = \frac{1031}{9.24} = 112 \text{ in.}^3$$

$$\text{reduced } M_R = F_b S_x = \frac{24(112)}{12} = 224 \text{ ft-kips}$$

Calculating the percent reduction in the resisting moment, noting that M_R for the gross section (from the AISCM) is the 254 ft-kips, yields

$$\frac{254 - 224}{254}(100) = 12\%$$

The resisting moment has been reduced by about 12%.

Neglecting the AISCS, a more conservative approach is frequently used. This allows the total hole area in the flanges to be considered in reducing the gross moment of inertia. Again, as discussed previously, it is general practice to consider both flanges to have holes (in a symmetrical pattern) even though only one does. This approach is often termed the **rational approach.**

Example 2-16

Recalculate the resisting moment M_R for the beam of Example 2-15 using the rational approach.

Solution

$$\text{hole area per flange} = 2(1.125)(0.81) = 1.82 \text{ in.}^2$$

$$\text{net } I_x = 1170 - 2(1.82)(8.83)^2 = 886.2 \text{ in.}^4$$

$$\text{net } S_x = \frac{\text{net } I_x}{c} = \frac{886.2}{9.24} = 95.9 \text{ in.}^3$$

$$\text{reduced } M_R = F_b S_x = \frac{24(95.9)}{12} = 191.8 \text{ ft-kips}$$

$$M_R \text{ for gross cross section} = 254 \text{ ft-kips} \qquad \text{(AISCM)}$$

The percent reduction in the resisting moment is:

$$\frac{254 - 191.8}{254}(100) = 24.5\%$$

2-12 WEB CRIPPLING

Web crippling is the localized failure of a beam web due to the introduction of an excessive load over a small length of the beam. Figure 2-23 illustrates the type of deformation failure expected. Practical and commonly used bearing lengths N are

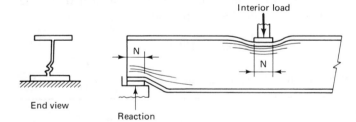

Interior load

End view

Reaction

Figure 2-23   Web crippling.

usually large enough to prevent web crippling from occurring. The AISCS, Section 1.10.10, requires that the compressive stress at the toe of the fillet (see Fig. 2-24) not exceed $0.75F_y$. An assumption is made that the load "spreads out" along 45° lines so that the critical area for stress, which occurs at the toe of the fillet, has a length of $(N + k)$ or $(N + 2k)$ for end reactions and interior loads, respectively, and a width of t_w. The dimension k, which locates the toe of the fillet, is tabulated for various shapes in the AISCM, Part 1. The controlling equations for web crippling may therefore be stated:

For interior loads: $\quad \dfrac{R}{t_w(N + 2k)} \leq 0.75F_y \qquad$ AISCS formula 1.10-8

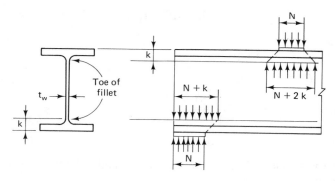

Figure 2-24 Web crippling.

For end reactions: $\dfrac{R}{t_w(N + k)} \le 0.75F_y$ AISCS formula 1.10-9

where R is the applied concentrated load or end reaction and N is the length of bearing. Should the web crippling stress be excessive, the problem may be corrected by increasing the bearing length or by designing *bearing stiffeners* (discussed in Chapter 10) or by selecting a beam with a thicker web. Web crippling should be checked under all concentrated loads and at supports where the beam is supported by walls or pedestals or at columns when the connection is a seated type.

Example 2-17

A W24 $\times$ 55 beam of A36 steel has an end reaction of 70 kips and is supported on a plate such that $N = 6$ in. (see Fig. 2-23). Check the beam for web crippling.

Solution   For this shape, $t_w = 0.395$ in. and $k = 1\frac{5}{16}$ in. = 1.31 in. Check the stress on the critical area:

$$\frac{R}{t_w(N + k)} = \frac{70}{0.395(6 + 1.31)} = 24.2 \text{ ksi}$$

The allowable web crippling stress is

$$0.75F_y = 0.75(36) = 27.0 \text{ ksi}$$

$$24.2 < 27.0 \qquad\qquad \textbf{O.K.}$$

There is some additional data on web crippling in the tables of Uniform Load Constants of the AISCM, Part 2. The tabulated quantity N_e is the length of bearing required to develop the maximum permissible web shear V. The tabulated value R (kips) is the maximum end reaction for $3\frac{1}{2}$ in. of bearing length and the tabulated R_i (kips) is the increase allowed in R for each additional inch of bearing. Example 2-17 could be reworked as follows using data from the AISCM, table of Uniform Load Constants:

$$R = 51.3 \text{ kips}$$

$$R_i = 10.7 \text{ kips}$$

$$\text{Required } N = 3.5 + \frac{70 - 51.3}{10.7} = 5.25 \text{ in.}$$

$$5.25 \text{ in.} < 6.0 \text{ in.} \qquad \textbf{O.K.}$$

The web crippling expressions (AISCS formulas 1.10-8 and 1.10-9) may be expressed in different forms:

To determine **allowable load** (based on allowable web crippling stress), for end reactions:

$$R = 0.75F_y(t_w)(N + k)$$

for interior loads:

$$R = 0.75F_y(t_w)(N + 2k)$$

To determine **minimum length of bearing** required, for end reactions:

$$\text{minimum } N = \frac{R}{0.75F_y(t_w)} - k$$

for interior loads:

$$\text{minimum } N = \frac{R}{0.75F_y(t_w)} - 2k$$

Example 2-18

Rework Example 2-17 by determining the minimum length of bearing required and comparing with the actual bearing length.

Solution

$$\text{minimum } N = \frac{R}{0.75F_y(t_w)} - k$$

$$= \frac{70}{27.0(0.395)} - 1.31$$

$$= 6.56 - 1.31 = 5.25 \text{ in.}$$

$$5.25 \text{ in.} < 6.0 \text{ in.} \qquad \textbf{O.K.}$$

The beam is satisfactory with respect to web crippling.

2-13 BEAM BEARING PLATES

Beams may be supported by connections to other structural members or they may rest on concrete or masonry supports such as walls or pilasters. When the support is of some material that is weaker than steel (such as concrete), it is usually necessary to spread the load over a larger area so as not to exceed the **allowable bearing stress** F_p. This is achieved through the use of a **bearing plate.** The plate must be large enough so that the actual bearing pressure f_p under the plate is less than F_p. Also, the plate must be thick enough so that the bending stress in the plate at the assumed critical section (see Fig. 2-25) is less than the allowable bending stress F_b. An assumption is made that the pressure developed under the plate is uniformly distributed.

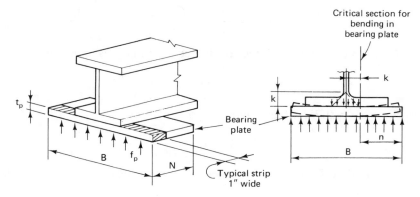

Figure 2-25   Beam bearing plate.

F_b, from the AISCS, Section 1.5.1.4.3, is $0.75F_y$. The allowable bearing pressure, F_p, for masonry or concrete may be obtained from the AISCS, Section 1.5.5, as follows:

For a plate covering the full area of concrete support:

$$F_p = 0.35f'_c$$

For a plate covering less than the full area of concrete support:

$$F_p = 0.35f'_c\sqrt{\frac{A_2}{A_1}} \leq 0.7f'_c$$

where f'_c = specified compressive strength of concrete (ksi)
A_1 = bearing area (loaded area)(in.^{2})
A_2 = *full* cross-sectional area of concrete support (in.^{2}) which is concentric with and geometrically similar to A_1

For a further explanation of bearing on concrete, see Reference 5.

The bending stress in the plate, at the critical section, may be determined with reference to Fig. 2-25. The moment at the critical section for a 1-in.-wide strip of depth t_p (in.) which acts like a cantilever beam is

$$M = \text{(actual bearing pressure)} \times \text{(area)} \times \text{(moment arm)}$$

$$= f_p \times (n \times 1) \times \frac{n}{2}$$

$$= \frac{f_p n^2}{2}$$

The bending stress is determined from the flexure formula:

$$f_b = \frac{Mc}{I} = \frac{(f_p n^2/2)(t_p/2)}{1(t_p^3)/12} = \frac{3f_p n^2}{t_p^2}$$

As a limit, $f_b = F_b$. Solving for the required thickness, we have

$$\text{required } t_p = \sqrt{\frac{3f_p n^2}{F_b}}$$

Since $F_b = 0.75F_y$, this may be rewritten

$$\text{required } t_p = \sqrt{\frac{3f_p n^2}{0.75F_y}} = 2n\sqrt{\frac{f_p}{F_y}}$$

A procedure for the design of beam bearing plates is given in the AISCM, Part 2.

Example 2-19

A W16 × 40 is to be supported on a concrete wall as shown in Fig. 2-26. $f'_c = 3000$ psi. The beam reaction is 55 kips. Design a bearing plate for the beam. Assume a 2-in. edge distance from the edge of the plate to the edge of the wall (maximum $N = 6$ in.). All steel is A36.

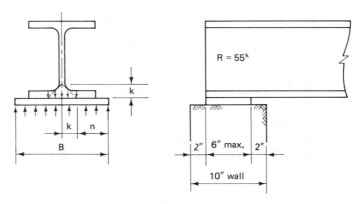

Figure 2-26 Beam bearing plate design.

Solution   Use the procedure given in the AISCM, Part 2:

$$F_y = 36 \text{ ksi}$$

$$t_w = 0.305 \text{ in.}$$

$$k = 1\tfrac{3}{16} = 1.19 \text{ in.}$$

1. From web crippling, the minimum required N is

$$N = \frac{R}{0.75F_y t_w} - k = \frac{55}{0.75(36)(0.305)} - 1.19 = 5.49 \text{ in.}$$

Use $N = 6$ in.

2. $F_b = 0.75F_y = 0.75(36) = 27 \text{ ksi}$

Since the area of the support and the bearing area (A_2 and A_1) are unknown, conservatively assume that $F_p = 0.35f'_c = 0.35(3) = 1.05$ ksi.

3. The required support area is

$$A = \frac{R}{F_p} = \frac{55}{1.05} = 52.4 \text{ in.}^2$$

4. The required $B = A/N = 52.4/6 = 8.73$ in. Use $B = 9.0$ in.

5. $f_p = \dfrac{R}{B(N)} = \dfrac{55}{6(9)} = 1.02$ ksi

6. $n = \dfrac{B}{2} - k = \dfrac{9.0}{2} - 1.19 = 3.31$ in.

$$\text{required } t_p = 2n\sqrt{\frac{f_p}{F_y}} = 2(3.31)\sqrt{\frac{1.02}{36}} = 1.11 \text{ in.}$$

See the AISCM, Part 1, Bars and Plates—Product Availability, for information on plate availability.

7. Use a bearing plate $1\tfrac{1}{8} \times 6 \times 0' - 9$.

It may be possible, if reactions are small, to support a beam, in a bearing situation, without the use of a bearing plate. Bearing pressure, web crippling, and flange bending are the considerations. The critical section for flange bending is again assumed to be at a distance k from the center of the section.

Example 2-20

A W24 × 76 is to be supported on a 12-in.-wide concrete wall such that there is bearing 8 in. wide. $f'_c = 3000$ psi. The beam reaction is 25 kips. Determine if a bearing plate is required. Assume A36 steel.

Solution   A diagram of the beam is shown in Fig. 2-27. Beam properties and dimensions are

$$F_y = 36 \text{ ksi}$$

$$b_f = 8.99 \text{ in.}$$

$$t_f = 0.68 \text{ in.}$$

$$k = 1\frac{7}{16} = 1.44 \text{ in.}$$

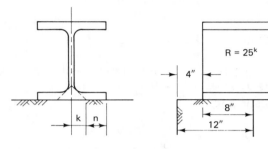

Figure 2-27   Beam without bearing plate.

Check the bearing pressure:

$$f_p = \frac{25}{8.99(8)} = 0.35 \text{ ksi}$$

$$F_p = 0.35f'_c = 0.35(3.0) = 1.05 \text{ ksi} > 0.35 \text{ ksi} \qquad \textbf{O.K.}$$

Web crippling is easily checked in the AISCM, Part 2, Table of Uniform Load Constants, recalling that the tabulated R, 58.7 kips, is the maximum end reaction for a $3\frac{1}{2}$ in. bearing length.

$$58.7 \text{ kips} > 25 \text{ kips} \qquad \textbf{O.K.}$$

The bending stress in the flange may be determined using the formula developed previously for f_b in the bearing plate (the flange acts exactly as does the plate).

$$n = \frac{b_f}{2} - k = \frac{8.99}{2} - 1.44 = 3.06 \text{ in.}$$

$$f_b = \frac{3f_p n^2}{t_f^2} = \frac{3(0.35)(3.06)^2}{(0.68)^2} = 21.26 \text{ ksi}$$

$$F_b = 0.75F_y = 0.75(36) = 27 \text{ ksi} > 21.26 \text{ ksi} \qquad \textbf{O.K.}$$

Therefore, this beam may be used on a bearing length of 8 in. without a bearing plate.

REFERENCES

1. S. H. Marcus, *Basics of Structural Steel Design* (Reston, VA: Reston Publishing Co., Inc., 1977), pp. 241–249.

2. J. E. Bower, "Recommended Design Procedures for Beams with Web-Openings," *AISC Engineering Journal,* Vol. 8, No. 4, October 1971.

3. R. G. Redwood, "Simplified Plastic Analysis for Reinforced Web Holes," *AISC Engineering Journal,* Vol. 8, No. 4, October 1971.

4. *Rectangular Concentric and Eccentric Un-reinforced Web Penetrations in Steel Beams,* United States Steel Corporation, P.O. Box 86, 600 Grant Street, Pittsburgh, PA 15230.

5. L. Spiegel, and G. F. Limbrunner, *Reinforced Concrete Design* (Englewood Cliffs, NJ: Prentice-Hall, Inc., 1980), p. 294.

PROBLEMS

Note: In the following problems and sketches, the given loads are *superimposed* loads. That is, they do not include the weights of the beams (unless noted otherwise).

2-1.   A floor framing plan is shown. Draw load diagrams for beams B1 and B2 and girder G1. The structural members will support a 6-in.-thick reinforced concrete floor slab. The live load is 300 psf. Assume reinforced concrete to weigh 150 pcf. Assume an additional load of 10 psf to account for the weight of the beams.

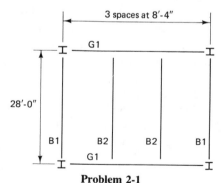

Problem 2-1

2-2.   The floor framing plan for a brewery process tank platform is shown. The tank legs impose loads of 40 kips at the locations shown. The floor is a 6-in. reinforced concrete slab and the design live load is to be 200 psf. Draw load diagrams for beams B1 through B8. Assume reinforced concrete to weigh 150 pcf. Also, assume an additional load of 10 psf to account for the weight of the beams.

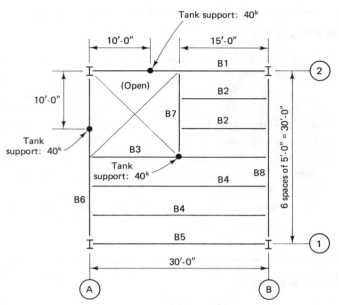

Problem 2-2

2-3. Calculate the maximum bending stress in a W18 × 40 beam that spans 36 ft and supports two equal concentrated loads of 12 kips each placed at the third points. Be sure to include the beam weight.

2-4. A W30 × 108 simply supported beam spans 32 ft and supports a superimposed uniformly distributed load of 4 kips/ft. Assume A36 steel. Determine the maximum bending stress. Be sure to include the beam weight.

2-5. Using A588 steel, list two compact shapes, two noncompact shapes, and two partially compact shapes. Consider any W, M, S, or HP shape.

2-6. Determine F_b for a W14 × 90 shape. Assume full lateral support. Use
(a) A36 steel.
(b) A572-Grade 42 steel.
(c) A572-Grade 60 steel.

2-7. A W21 × 68 supports the loads shown. F_b = 22 ksi. Is the beam satisfactory?

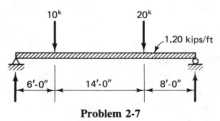

Problem 2-7

2-8. A W16 × 40 supports the loads shown. Find the maximum allowable load P (kips). F_b = 24 ksi.

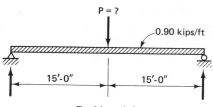

Problem 2-8

2-9. A W21 × 44 supports the loads shown. Determine if the beam is satisfactory. $F_b = 24$ ksi.

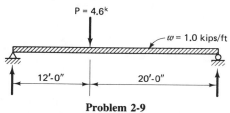

Problem 2-9

2-10. Compute F_b for the welded plate girder shown. Assume adequate lateral support for the compression flange. Use
   (a) A36 steel.
   (b) A572-Grade 60 steel.

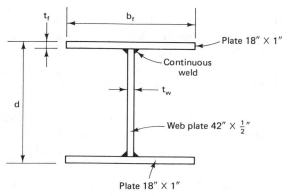

Problem 2-10

2-11. A W21 × 44 beam supports the loads shown. Assume A36 steel. Find the maximum uniformly distributed load w (kips/ft) that the beam can support. The beam's compression flange has lateral supports spaced 4 ft apart.

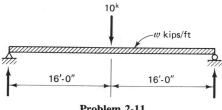

Problem 2-11

2-12. The beam shown is of A36 steel and has full lateral support. The beam is composed of two W10 × 22 shapes bolted together at the flanges. Compute the maximum bending stress. Is the beam adequate?

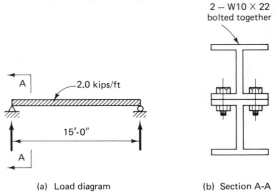

(a) Load diagram   (b) Section A-A

Problem 2-12

2-13. A W21 × 57 simple span A36 steel beam is laterally supported at the end supports only. Compute F_b if the beam is used for the following span lengths:
 (a) 6 ft–0 in
 (b) 12 ft–0 in.
 (c) 25 ft–0 in. Use $C_b = 1.0$.

2-14. Find the allowable uniformly distributed superimposed load that may be placed on a simply supported W30 × 99 beam that spans 30 ft. Lateral support exists at the ends only. The steel is A242. Use $C_b = 1.0$ and consider moment only.

2-15. A W12 × 30 of A36 steel is on a simple span of 22 ft. Determine L_c and L_u. Find F_b for $L_b = 14$ ft and 22 ft. Plot a curve of F_b as a function of L_b. Assume that $C_b = 1.0$.

2-16. Select the lightest W shape to support a uniformly distributed load of 4 kips/ft on a simple span of 25 ft. Assume A36 steel and full lateral support.

2-17. Rework Problem 2-16, given a load of 10 kips/ft, a span of 32 ft, and A242 steel.

2-18. Select the lightest W shape for the beam shown. Assume A36 steel and full lateral support.

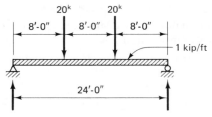

Problem 2-18

2-19. Select the lightest W shape for a beam that is to span 24 ft between simple supports and will overhang one support by 8 ft. The beam supports a uniform load of 2.3 kips/ft and a concentrated load of 4 kips at the end of the overhang. Assume A242 steel and full lateral support.

2-20. Select the lightest W shape for the beam shown. Assume A36 steel and full lateral support.

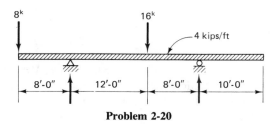

Problem 2-20

2-21. A 6-in. reinforced concrete floor is supported on beams and girders with a layout for an *interior* bay, as shown. The floor is designed for a live load of 150 psf. Assume A36 steel and full lateral support for all members. Select the lightest W shapes for the beams and girders.

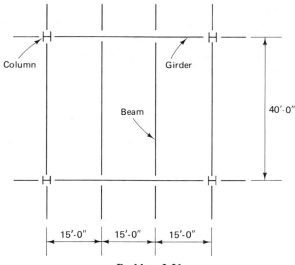

Problem 2-21

2-22. Design the beams for the tank support platform of Problem 2-2. The floor slab provides lateral support for all beams except B1 and B6 in the area adjacent to the opening under the tank. Assume A36 steel. Consider moment only.

2-23. A simply supported beam is to support a uniform load of 2.1 kips/ft on a span of 24 ft. Lateral support exists at the reactions and at the third points. Assume A36 steel. Select the lightest W shape.

2-24. Select the lightest W shape for the beam shown. Lateral support exists at the reactions and at the concentrated loads. Assume A36 steel.

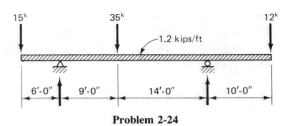

Problem 2-24

2-25. A W16 × 57 beam is subjected to a shear V of 78 kips. Plot the horizontal shear stress distribution for this condition. Compare with the shear stress as determined using the average web shear approach.

2-26. A W14 × 34 beam of A36 steel is on a simple span of 6 ft–6 in. Assume full lateral support. Calculate the maximum superimposed uniformly distributed load that may be placed on the beam. Consider moment and shear.

2-27. Select the lightest W shape beam to support a uniformly distributed load of 1 kip/ft on a simple span of 30 ft. Lateral support exists at the ends and at midspan. Assume A36 steel. Consider moment and shear.

2-28. Select the lightest W shape beam to support a uniformly distributed load of 5 kips/ft on a simple span of 20 ft. Lateral support exists at the ends only. Assume A36 steel. Consider moment and shear.

2-29. Select the lightest W shape for the beam shown. Assume A36 steel and full lateral support. Consider moment and shear.

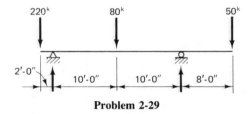

Problem 2-29

2-30. Select the lightest W shape for the beam shown. Lateral support exists at points A and B only. Assume A36 steel. Consider moment and shear.

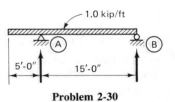

Problem 2-30

2-31. Select the lightest W shape for the beam shown. Assume A36 steel and full lateral support. Consider moment and shear.

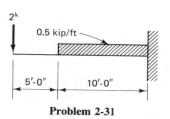

Problem 2-31

2-32. Select the lightest W shape for the beam shown. Assume A36 steel. Consider moment and shear. Assume
  (a) Full lateral support.
  (b) Lateral support at the ends only.

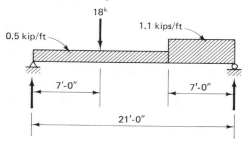

Problem 2-32

2-33. A W14 × 74 of A36 steel supports a uniformly distributed load of 3 kips/ft on a simple span of 24 ft. Assume full lateral support. The maximum total allowable deflection is span/360. Determine if the beam is adequate. Consider moment, shear, and deflection.

2-34. A W16 × 100 of A36 steel is on a simple span of 18 ft. Lateral support exists at the ends only. Determine the maximum superimposed uniformly distributed load that may be placed on the beam. The maximum allowable total deflection is span/360. Consider moment, shear, and deflection.

2-35. An S12 × 50 of A36 steel is on a simple span of 30 ft. Assume full lateral support. The beam supports a uniformly distributed load of 0.25 kip/ft. Determine the maximum concentrated load (point load) that can be placed at midspan. The maximum allowable total deflection is span/360. Consider moment, shear, and deflection.

2-36. Select the lightest W shape for a beam that supports a uniformly distributed load of 400 lb/ft on a simple span of 22 ft. Assume A36 steel and full lateral support. The maximum allowable total deflection is span/360. Consider moment, shear, and deflection.

2-37. Check deflection for the beam selected in Problem 2-18 and redesign if necessary. The maximum allowable deflection for total load is span/800.

2-38. Rework Problem 2-36 but assume that the load is 4 kips/ft, the span is 24 ft, the steel is A441, and the maximum allowable total deflection is span/1200.

2-39. Select the lightest W shape for a beam that supports a 20-kip concentrated load (point load) at the center of a 20-ft simple span. Assume A36 steel and that lateral support exists at the ends only. The maximum allowable total deflection is span/1000. Consider moment, shear, and deflection.

2-40. Select the lightest W shape for the beam shown. Assume A36 steel and full lateral support. The maximum allowable total deflection is span/240. Consider moment, shear, and deflection.

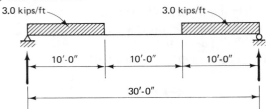

Problem 2-40

2-41. Select the lightest W shape for B1 and G1, which support the floor for a typical interior bay in a small industrial building as shown in the plan view. Use A36 steel. The floor will be a 6-in. reinforced concrete slab. Design loads are 150 psf live load and 20 psf dead load (partitions and ceilings). Assume full lateral support for the beams. Assume lateral support for the girders only where the beams are connected to them. The maximum allowable deflection for total load is span/360. Consider moment, shear, and deflection.

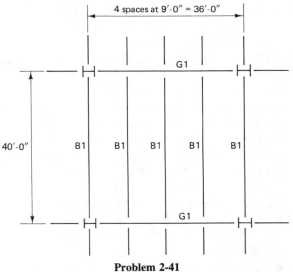

Problem 2-41

2-42. A W21 × 93 of A588 steel has two holes in each flange for $\frac{7}{8}$-in.-diameter bolts. Find the resisting moment M_R for the beam. Assume full lateral support.

2-43. Select the lightest W shape for a beam that will support a uniformly distributed load of 2.5 kips/ft on an 18-ft simple span. Use A36 steel and assume that the beam has full lateral support. The beam is a spandrel beam supporting an exterior wall. Therefore, it requires a bottom flange detail which uses two $\frac{3}{4}$-in.-diameter bolts in any given plane through the bottom flange.

(a) Use the AISC Specifications.

(b) Use the rational approach.

2-44. Select the lightest W shape for the beam shown. Assume A36 steel and full lateral support. There will be two holes in each flange for $\frac{3}{4}$-in.-diameter bolts located 9 ft–0 in. from the left support.

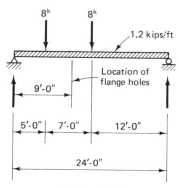

Problem 2-44

2-45. The steel for a building is being erected. You are the engineer's representative on the site. A question has arisen and a quick answer is necessary. A simple span beam W16 × 45, A36 steel, is subjected to a maximum moment of 121.5 ft-kips. The beam is laterally supported ($L_b = 0$). Would you permit the drilling of two holes in the tension flange (for $\frac{7}{8}$-in.-diameter bolts)? Show all computations for your answer. Assume that the given maximum moment includes the beam's own weight.

2-46. A simply supported beam is to span 15 ft. It will support a uniformly distributed load of 2 kips/ft over the full span and a concentrated load (point load) of 60 kips at midspan. Assume A36 steel and full lateral support.

(a) Select the lightest W shape. Consider moment and shear.

(b) Determine the minimum bearing length required at the supports and under the concentrated load.

2-47. The beam in Problem 2-29 is to be supported on bearing plates such that the bearing length N is 12 in. Check web crippling. If the proposed bearing length is inadequate, specify the revised required bearing length.

2-48. Design a beam bearing plate for a W18 × 50 beam supported by a masonry wall. $F_p = 0.25$ ksi. The beam reaction is 49 kips and the length of bearing N is limited to 10 in. Use A36 steel.

2-49. Design a beam bearing plate for a W30 × 116 beam supported by a concrete wall. $F_p = 0.75$ ksi. The beam reaction is 105 kips and the length of bearing N is limited to 6 in. Use A36 steel.

2-50. A W12 × 65 beam is to be supported on a brick (and mortar) pier. The beam reaction is 4 kips. F_p = 300 psi. Determine if a bearing plate is required. The beam is A36 steel and has a bearing length of $3\frac{1}{2}$ in.

2-51. A W18 × 106 beam is supported on a concrete wall. The reaction is 40 kips. The maximum bearing length N is 8 in. Use F_p = 750 psi and A36 steel. Design a beam bearing plate, if required.

2-52. A beam is supported on a concrete wall as shown. Find the maximum allowable reaction. Assume A36 steel. Concrete f'_c = 3000 psi. Use the AISCS for F_p.

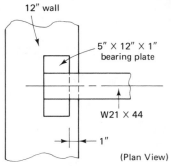

Problem 2-52

3 Special Beams

3-1 LINTELS

In the construction of walls, it is necessary to provide beams over openings, such as doors or windows. These beams, commonly called **lintels,** are required in order to support the weight of the wall and any other loads above the opening. Lintels are usually found only in building construction. Their loadings and design considerations deserve special attention. Structural steel lintels that support masonry walls will be treated here. However, lintels are not limited to this type of material or loading.

Lintels may be composed of various structural steel shapes, as shown in Fig. 3-1. The type used will depend on the wall to be supported and the span between supports. The simple plate lintel shown in Fig. 3-1(a) may be satisfactory for only the very shortest of spans while the wide-flange lintels shown in Fig. 3-1(e) and (f) will be appropriate for the longer spans.

The load that a lintel supports should be carefully considered. If the wall is sufficiently continuous (both horizontally and vertically), it will probably support

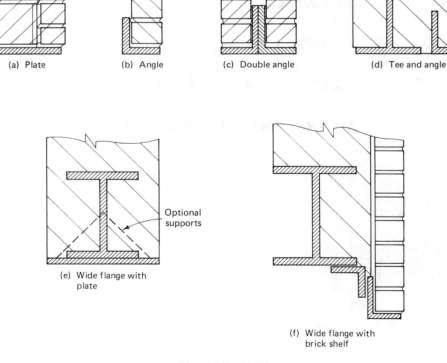

(a) Plate   (b) Angle   (c) Double angle   (d) Tee and angle

Optional supports

(e) Wide flange with plate

(f) Wide flange with brick shelf

Figure 3-1   Lintels.

itself by an arching action that develops above the lintel, leaving only an approximately triangular section of the wall to be supported as shown in Fig. 3-2(a). One rule of thumb suggests that the wall should be continuous above the opening a distance at least equal to the span of the lintel. Substantial piers and/or supports must also be furnished *adjacent* to the opening and must be capable of supporting the horizontal and vertical forces transmitted by the arch action. This action is usually assumed to develop only after the mortar has set and attained adequate strength. If the assumption is made that the load area is bounded by two sloping 45° lines, as shown, the triangular wall area will have a base of L and a height of $L/2$, where L is the span length of the lintel. Should there be an interruption in the wall that prevents the arch action from developing [such as shown in Fig. 3-2(b)], the full weight of the wall above the lintel should be included.

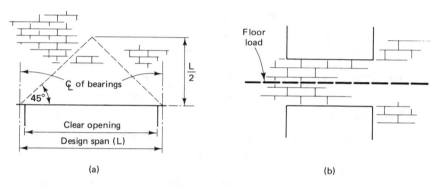

Figure 3-2 Lintel loadings.

Lintel beam design is based on both practical and theoretical considerations. Lintels are assumed to act as simply supported beams with a span length equal to the distance center to center of bearings. The ends of the lintel, beyond the opening, are generally made to bear on supporting walls a distance of at least 4 in. and even up to 8 in. The length of the bearing will depend on the span length and load to be supported as well as on the bearing capacity of the supporting material.

The queston as to whether the lintel's compression flange has adequate lateral support is controversial since the top compression flange is embedded in masonry and lateral support is uncertain. It is generally recommended that some form of temporary support or shoring be used for the lintel until the mortar has set and the arching action has developed above the opening. Where lateral buckling is of concern due to long spans and heavy loads, a special investigation should be made and a reduced F_b considered.

It is usually desirable to have lintels fully or partially buried in walls and hidden from view from at least one side. If two shapes are used, as in Fig. 3-1(c) and (d), they need not be mechanically fastened together. Individual loose lintel members are easier and more convenient to place than heavier, built-up sections. The selection of the two members should be based on approximately equal deflections of the individual parts so that unsightly cracking does not develop in the walls.

When selecting angles or tees to support brick walls, the preferred length of the horizontal outstanding leg or half-flange is $3\frac{1}{2}$ in., so that the steel edge does not project beyond the edge of the bricks while still providing sufficient width to support the bricks. In addition, if possible, the horizontal leg thickness should not exceed $\frac{3}{8}$ in., so that it can be buried in a mortar joint which is usually less than $\frac{1}{2}$ in. in thickness.

Shear is generally neglected in the design of steel lintels. However, it may be checked in the same way that normal beams are checked by using the average web shear approach, as discussed in Chapter 2 of this text.

$$f_v = \frac{V}{dt_w}$$

where d = full depth of vertical leg
  t_w = thickness of vertical leg
  f_v and V are as previously defined

The allowable shear stress will still be $0.40F_y$.

Example 3-1

Design a lintel to span a clear opening of 8 ft–0 in. The masonry wall to be supported is 8 in. thick and weighs 130 lb/ft³ [see Fig. 3-2(a) for reference]. Use an inverted structural tee. Assume A36 steel, a 6-in. bearing on each side of the opening, and an allowable deflection of span/240. Assume that arching action can develop.

Solution   The lintel is loaded as shown in Fig. 3-3. The span length may be taken as the distance center to center of supports: 8 ft–6 in. Assume the weight of lintel to be 20 lb/ft. The maximum moment due to the triangular load (see the

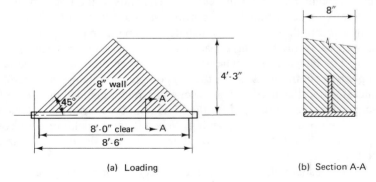

(a) Loading   (b) Section A-A

Figure 3-3   Lintel design.

AISCM, Part 2, Beam Diagrams and Formulas, Case 3) and the moment due to the lintel weight are added:

$$\text{maximum } M = \frac{WL}{6} + \frac{wL^2}{8}$$

1. The total weight of the triangular load is

$$W = \frac{8}{12}(8.5)\left(\frac{4.25}{2}\right)(130) = 1565 \text{ lb}$$

2. The maximum bending moment is

$$M = \frac{1565(8.5)}{6} + \frac{20(8.5)^2}{8} = 2400 \text{ ft-lb}$$

3. Determine the required section modulus. An inverted tee used as a beam places the stem (or web) in compression. The rather thin stem may be subject to localized buckling. Tees used in this way are governed by the provisions of the AISCS, Section 1.9.1 and Appendix C, which provide that

$$F_b \leq 0.60F_yQ_s$$

where Q_s is a reduction factor based on the width–thickness ratio of the stem. Q_s is tabulated as a property for the structural tees.

Assume that $F_b = 0.60F_y = 22$ ksi. This must be verified later. Calculate the required section modulus:

$$\text{required } S_x = \frac{M}{F_b} = \frac{2400(12)}{22,000} = 1.31 \text{ in.}^3$$

4. Select an appropriate WT. The tee selected should have a nominal flange width no larger than 8 in. Therefore, try a WT5 × 16.5. There is no value tabulated for Q_s, which means that the tee complies with the AISCS, Section 1.9.1.2. Therefore, the assumed F_b is satisfactory. The assumed lintel weight is slightly conservative.

5. Check the deflection. The deflection due to lintel weight is very small and is therefore neglected.

$$\text{allowable } \Delta = \frac{\text{span}}{240} = \frac{8.5(12)}{240} = 0.43 \text{ in.}$$

$$\text{actual } \Delta = \frac{WL^3}{60EI} = \frac{1565(8.5)^3(1728)}{60(29,000,000)(7.71)} = 0.12 \text{ in.}$$

$$0.12 \text{ in.} < 0.43 \text{ in.} \qquad \textbf{O.K.}$$

Use WT5 × 16.5.

6. Check the shear. The maximum shear exists at the reaction:

$$\text{maximum } V = \frac{W}{2} + \frac{wL}{2}$$

$$= \frac{1565}{2} + \frac{16.5(8.5)}{2} = 853 \text{ lb}$$

For the WT5 × 16.5, $d = 4.865$ in. and $t_w = 0.290$ in.

$$f_v = \frac{V}{dt_w} = \frac{853}{4.865(0.290)} = 605 \text{ psi} = 0.605 \text{ ksi}$$

$$F_v = 0.40F_y = 0.40(36) = 14.4 \text{ ksi}$$

Since 14.4 ksi > 0.605 ksi, the WT is satisfactory for shear.

Example 3-2

Design a lintel to carry a 12-in.-thick brick wall over a clear opening of 12 ft–6 in. Use A36 steel. The maximum allowable deflection is span/240. Use a structural tee (WT) and an angle (L), as shown in Fig. 3-4(b). The weight of the wall is 120 lb/ft³. The bearing length at the supports is 8 in.

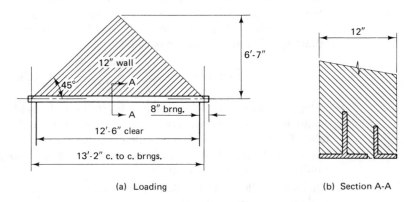

(a) Loading   (b) Section A-A

Figure 3-4 Lintel design.

Solution

1. Assume a triangular load due to wall arching action and refer to Fig. 3-4 for load diagram. Assume 40 lb/ft for lintel weight. The total triangular load on the lintel is W.

$$W = \frac{1}{2}(13.17)(6.58)\left(\frac{12}{12}\right)(120) = 5200 \text{ lb}$$

2. The maximum bending moment is

$$M = \frac{WL}{6} + \frac{wL^2}{8} = \frac{5200(13.17)}{6} + \frac{40(13.17)^2}{8} = 12{,}280 \text{ ft-lb}$$

3. Determine the required section modulus. Assume that, $F_b = 0.60F_y = 22$ ksi or 22,000 psi:

$$\text{required } S_x = \frac{M}{F_b} = \frac{12{,}280(12)}{22{,}000} = 6.70 \text{ in.}^3$$

4. Figure 3-4(b) shows that the tee will support two-thirds of the wall. Therefore, the tee should supply two-thirds of the S_x. However, F_b for the tee may be affected by a Q_s factor. Q_s may be thought of as effectively reducing the S_x of a shape by the same ratio that it reduces the F_b:

$$\text{reduced } S_x = (\text{original } S_x)Q_s$$

In the selection process, the structural tee should have a flange width of 8 in., or slightly less, and the horizontal leg of the angle should not exceed 4 in. This will make the sum of the two dimensions slightly less than the wall thickness of 12 in.

For the tee:

$$\text{required } S_x = \frac{2}{3}(6.70) = 4.47 \text{ in.}^3$$

Select a WT7 × 24 ($S_x = 4.48$ in.^{3}, $b_f = 8.03$ in., and there is no value tabulated for Q_s).

For the angle:

$$\text{required } S_x = \frac{1}{3}(6.70) = 2.23 \text{ in.}^3$$

An L5 × 3½ × ⅜ ($S_x = 2.29$ in.^{3}) would be appropriate, but Q_s must be considered as with the WT. From AISCM, Part 1, Properties of Double Angles, the tabulated Q_s for *angles separated* is 0.982. (This is conservative for bending members.) Therefore

$$\text{reduced } S_x = 0.982(2.29) = 2.25 \text{ in.}^3 > 2.23 \text{ in.}^3 \qquad \textbf{O.K.}$$

The total lintel weight is

$$24 + 10.4 = 34.4 \text{ lb/ft}$$

Therefore, the assumed weight of 40 lb/ft is slightly conservative.

5. Check the deflection for the selected steel shapes. The shapes should be selected so that the deflections are approximately equal. If one of the shapes is overly stiff, relative to the other, it will support a dis-

proportionate amount of the load and will be overstressed. Since the WT supports two-thirds of the wall load, for equal deflections its moment of inertia I should be twice the I of the angle. For the WT, $I_{WT} = 24.9$ in.^{4}. Therefore, I for the angle should be approximately $24.9/2 = 12.5$ in.^{4}. I for the selected angle is 7.78 in.^{4}. Therefore, select another angle with I of approximately 12.5 in.^{4}, a minimum S_x of 2.23 in.^{3}, and a horizontal leg of $3\frac{1}{2}$ in.

An L6 $\times$ $3\frac{1}{2}$ $\times$ $\frac{3}{8}$ will be selected. The weight of the lintel will therefore be

$$24 + 11.7 = 35.7 \text{ lb/ft}$$

For the angle, $I_x = 12.9$ in.^{4}, $S_x = 3.24$ in.^{3}, and $Q_s = 0.911$.

$$\text{Reduced } S_x = 0.911(3.24) = 2.95 \text{ in.}^3 > 2.23 \text{ in.}^3 \qquad \textbf{O.K.}$$

The total I_x is

$$24.9 + 12.9 = 37.8 \text{ in.}^4$$

Neglecting deflection due to the weight of the lintel,

$$\Delta_{\text{actual}} = \frac{WL^3}{60EI} = \frac{5200(13.17)^3(1728)}{60(29,000,000)(37.8)} = 0.31 \text{ in.}$$

$$\Delta_{\text{allowable}} = \frac{\text{span}}{240} = \frac{13.17(12)}{240} = 0.66 \text{ in.}$$

$$0.31 \text{ in.} < 0.66 \text{ in.} \qquad \textbf{O.K.}$$

Use a WT7 $\times$ 24 and an L6 $\times$ $3\frac{1}{2}$ $\times$ $\frac{3}{8}$. Shapes are to be loose and the 6-in. leg is to be vertical.

6. Check the shear. The area stressed in shear is assumed to be the web of the WT and the vertical leg of the angle. The maximum shear occurs at the reaction.

$$\text{maximum } V = \frac{W}{2} + \frac{wL}{2}$$

$$= \frac{5200}{2} + \frac{35.7(13.17)}{2} = 2835 \text{ lb}$$

$$f_v = \frac{V}{\text{area stressed in shear}}$$

$$= \frac{2835}{6.895(0.340) + 6(\frac{3}{8})} = 617 \text{ psi} = 0.617 \text{ ksi}$$

$$F_v = 0.40F_y = 0.40(36) = 14.4 \text{ ksi}$$

Since 14.4 ksi > 0.617 ksi, the lintel is satisfactory for shear.

The lintel of Example 3-2 could also be designed to be made up from three individual angles. In Fig. 3-4(b), a double angle would replace the WT shape. [Refer also to Fig. 3-1(c) and (d).] The selection of the angles would proceed as follows:

$$\text{required } S_x = 6.70 \text{ in.}^3 \text{ (no change)}$$

Try three angles L5 $\times$ 3$\frac{1}{2}$ $\times$ $\frac{3}{8}$ (with the 5-in. leg vertical).

$$S_x = 3(2.29) = 6.87 \text{ in.}^3$$

The foregoing assumes that $Q_s = 1.0$, which would be the case for angles in contact back to back. However, from the AISCM, Part 1, Properties of Double Angles, $Q_s = 0.982$ for angles separated. Therefore, assuming that this Q_s applies to all three angles:

$$S_x = 6.87(0.982) = 6.75 \text{ in.}^3$$

$$6.75 \text{ in.}^3 > 6.70 \text{ in.}^3 \qquad \textbf{O.K.}$$

Check the deflection:

$$\Delta_{\text{allowable}} = 0.66 \text{ in. (no change)}$$

$$\Delta_{\text{actual}} = \frac{WL^3}{60EI} = \frac{5200(13.17)^3(1728)}{60(29,000,000)(3 \times 7.78)} = 0.51 \text{ in.}$$

$$0.51 \text{ in.} < 0.66 \text{ in.} \qquad \textbf{O.K.}$$

The three angles are identical, support equal loads, and therefore, deflect equally. The weight for the three-angle lintel will be

$$3(10.4) = 31.2 \text{ lb/ft}$$

It is slightly lighter than the WT–angle combination (35.7 lb/ft) of Example 3-2. Additionally, the three-angle lintel would be easier to handle and place.

3-2 FLITCH BEAMS

Beams that are composed of more than one type of material are called **composite beams.** One type of composite beam, called a **flitch beam,** which has been used for many years, is made up of wood sections reinforced with structural steel shapes or plates. One resulting cross section is shown in Fig. 3-5(a). Flitch beams are sometimes found in wood-framed structures where long clear openings are desired (such as over a single-door two-car garage in residential construction). They have also been found to be useful in rehabilitation projects where existing wood members must be strengthened. This is accomplished by bolting steel plates or channels on the outside of the existing wood beams.

The design of a flitch beam with a cross section such as shown in Fig. 3-5 involves the selection of both wood and steel components. The planks and the steel plate are bolted together so that they *deflect* together. Since the deflection and the

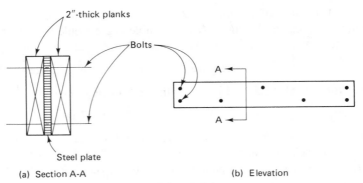

(a) Section A-A   (b) Elevation

Figure 3-5 Flitch beam.

curvature of the planks and the plate are the same, the pattern of the *bending strains* will be identical over the cross section. The strains will be maximum at the top and bottom of the flitch beam and will vary linearly to zero at the neutral axis. From strength of materials, for elastic composite members that are equally strained, it can be shown that the induced stresses are proportional to the modulus of elasticity values.

$$\frac{f_s}{f_w} = \frac{E_s}{E_w} = n$$

from which

$$f_s = nf_w$$

where f_s = bending stress in the steel
f_w = bending stress in the wood
E_s = modulus of elasticity for steel (29,000,000 psi)
E_w = modulus of elasticity for wood
n = modular ratio

The relationship between the stress in the steel and the stress in the wood is determined once n is calculated. The value for E_s is taken to be 29,000,000 psi. The value for E_w varies with the species of wood. Table 3-1 lists some design values for commonly used structural lumber. For other species, see Reference 1. Both wood and steel have allowable bending stresses which should not be exceeded. We will denote the allowable bending stress in the wood as F_{bw}. Since the ratio between the induced maximum bending stresses is fixed, the bending stress in one material may be at the allowable while the bending stress in the other material will usually be at less than the allowable. Once the bending stresses in the materials (whether at allowable or less) are known, analysis or design may proceed. In general, A36 steel would be used with a maximum allowable bending stress of $0.6F_y$.

The ratio of *loads* supported by each of the two materials is the same as ratio of the respective resisting moments. Therefore, the ratio of the induced *shears* in each material is the same as the ratio of the respective resisting moments. The *shear*

TABLE 3.1 RECOMMENDED DESIGN VALUES FOR VISUALLY GRADED STRUCTURAL LUMBER (2 TO 4 IN. THICK, 5 IN. AND WIDER)

Species or group	Grade	Allowable stress		E_w (psi)	n
		Bending F_{bw} (psi)	Horizontal shear, F_{vw} (psi)		
Douglas fir– larch	Select Structural	1800	95	1,800,000	16.1
	No. 1 and Appearance	1500	95	1,800,000	16.1
	No. 2	1250	95	1,700,000	17.1
	No. 3	725	95	1,500,000	19.3
Hem– Fir	Select Structural	1400	75	1,500,000	19.3
	No. 1 and Appearance	1200	75	1,500,000	19.3
	No. 2	1000	75	1,400,000	20.7
	No. 3	575	75	1,200,000	24.2
Eastern hemlock	Select Structural	1550	85	1,200,000	24.2
	No. 1 and Appearance	1300	85	1,200,000	24.2
	No. 2	1050	85	1,100,000	26.4
	No. 3	625	85	1,000,000	29.0
Eastern spruce	Select Structural	1200	70	1,500,000	19.3
	No. 1 and Appearance	1000	70	1,500,000	19.3
	No. 2	825	70	1,400,000	20.7
	No. 3	475	70	1,200,000	24.2

capacity of each material must be greater than the respective induced shear. Recalling that for *rectangular sections*

$$\text{maximum } f_v = \frac{3V}{2A}$$

and that shear capacity may be expressed as

$$V_{\text{capacity}} = \frac{2}{3}F_v A = \frac{F_v A}{1.5}$$

the foregoing may be expressed as governing equations for shear:

$$V_{\text{capacity(wood)}} = \frac{F_{vw} A_w}{1.5} \geq \text{shear in wood}$$

$$V_{\text{capacity(steel)}} = \frac{F_{vs} A_s}{1.5} \geq \text{shear in steel}$$

where f_v = shear stress
 V = applied vertical shear force
 A = rectangular area of wood (A_w) or steel (A_s)
 F_v = allowable shear stress for wood (F_{vw}) or steel (F_{vs})

Table 3-2 contains properties of dressed lumber (S4S) which will be needed for analysis and design problems. For properties of other size sections, see Reference 1, standard strength of materials text books, or determine the properties by calculation.

TABLE 3-2 PROPERTIES OF DRESSED LUMBER (S4S)

Nominal size (in.)	Dressed size (S4S) (in.)	S_x[a] (in.^{3})	Weight per foot[b] (lb)
2 × 6	$1\frac{1}{2} \times 5\frac{1}{2}$	7.56	2.0
2 × 8	$1\frac{1}{2} \times 7\frac{1}{4}$	13.14	2.6
2 × 10	$1\frac{1}{2} \times 9\frac{1}{4}$	21.39	3.4
2 × 12	$1\frac{1}{2} \times 11\frac{1}{4}$	31.64	4.1

[a]Centroidal axis x-x is parallel to the short dimension.
[b]Based on 35 lb/ft^3.

Example 3-3

Determine the adequacy (with respect to moment and shear) of a flitch beam composed of two S4S 2 × 12 planks reinforced with a $\frac{1}{2}$-in. steel plate as shown in Fig. 3-6. Assume that continuous lateral support is furnished by a floor system supported by the beam. Use A36 steel with F_b = 22 ksi. Planks are No. 2 Eastern hemlock. The beam is on a simple span of 16 ft and supports a uniformly distributed load of 0.7 kip/ft which includes the weight of the beam.

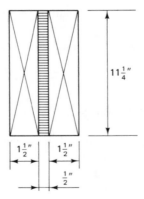

Figure 3-6   Flitch beam analysis.

Solution   From Table 3-1, E_w = 1,100,000 psi.

$$n = \frac{E_s}{E_w} = \frac{29,000,000}{1,100,000} = 26.4$$

Assume that the wood is stressed to its allowable bending stress (see Table 3-1):

$$f_w = F_{bw} = 1050 \text{ psi}$$
$$f_s = nf_w = 26.4(1050) = 27,720 \text{ psi}$$

But the allowable bending stress for the steel, F_b, is 22 ksi or 22,000 psi. Therefore, the maximum bending stress in the steel must be limited to 22,000 psi and the maximum bending stress in the wood may be calculated as

$$f_w = \frac{f_s}{n} = \frac{22,000}{26.4} = 833 \text{ psi}$$

The two bending stresses that exist without either stress being excessive have now been established. The resisting moment for the wood M_{Rw} (at a stress less than the allowable stress) is

$$M_{Rw} = f_w S_x \qquad \text{(use } S_x \text{ from Table 3-2)}$$
$$= \frac{0.833(2)(31.64)}{12} = 4.39 \text{ ft-kips}$$

For the steel plate:

$$S_x = \frac{I}{c} = \frac{bh^2}{6} = \frac{0.5(11.25)^2}{6} = 10.55 \text{ in.}^3$$

$$M_{Rs} = F_b S_x = \frac{22(10.55)}{12} = 19.34 \text{ ft-kips}$$

For the flitch beam:

$$M_R = M_{Rw} + M_{Rs} = 4.39 + 19.34 = 23.73 \text{ ft-kips}$$

Therefore, the total resisting moment is comprised as follows:

$$M_{Rw}: \quad \frac{4.39}{23.73}(100) = 18.5\%$$

$$M_{Rs}: \quad \frac{19.34}{23.73}(100) = 81.5\%$$

The flitch beam supports a uniformly distributed load of 0.7 kip/ft (which includes the weight of the beam) on a span of 16 ft. The maximum applied moment may be calculated as

$$M = \frac{wL^2}{8} = \frac{0.70(16)^2}{8} = 22.4 \text{ ft-kips}$$

Since 22.4 ft-kips < 23.7 ft-kips, the beam is satisfactory with respect to moment.

The maximum applied shear may be calculated as

$$V = \frac{wL}{2} = \frac{0.70(16)}{2} = 5.6 \text{ kips}$$

This shear is divided between the two materials in the same ratio as is moment; therefore, for wood

$$V_w = 0.185(5.6) = 1.04 \text{ kips}$$

and for steel

$$V_s = 0.815(5.6) = 4.56 \text{ kips}$$

The allowable shear stress for steel is $F_{vs} = 14{,}500$ psi and for wood is $F_{vw} = 85$ psi. The shear capacities may be calculated as

$$V_{\text{capacity(wood)}} = \frac{F_{vw}A_w}{1.5} = \frac{85(2)(1.5)(11.25)}{1.5} = 1913 \text{ lb} = 1.91 \text{ kips}$$

$$V_{\text{capacity(steel)}} = \frac{F_{vs}A_s}{1.5} = \frac{14{,}500(0.5)(11.25)}{1.5} = 54{,}400 \text{ lb} = 54.4 \text{ kips}$$

Since 1.04 kips < 1.91 kips and 4.56 kips < 54.4 kips, both materials are satisfactory for shear.

Example 3-4

A flitch beam is to support a uniform load of 600 lb/ft (this includes an estimated flitch beam weight of 30 lb/ft) on a simple span of 14 ft. Assume that 2 × 12 S4S planks of No. 1 Hem–Fir will be used. Use A36 steel with F_b of 22 ksi. Design the flitch beam. Assume continuous lateral support.

Solution   Determine the applied moment:

$$M = \frac{wL^2}{8} = \frac{600(14)^2}{8} = 14{,}700 \text{ ft-lb}$$

Determine if a flitch beam is required. From Table 3-1, $F_{bw} = 1200$ psi.

$$\text{required } S_x = \frac{M}{F_{bw}} = \frac{14{,}700(12)}{1200} = 147 \text{ in.}^3$$

This would require five 2 × 12 planks. Therefore, use a flitch beam composed of two 2 × 12s and a steel plate. The plate will be $11\frac{1}{4}$ in. deep to match the depth of the planks. The thickness of the plate must be determined. From Table 3-1, $n = 19.3$. If the steel is at the allowable bending stress ($F_b = 22$ ksi),

$$f_w = \frac{f_s}{n} = \frac{F_b}{n} = \frac{22}{19.3} = 1.140 \text{ ksi} = 1140 \text{ psi}$$

For this wood, from Table 3-1, $F_{bw} = 1200$ psi. Therefore,

$$f_s = F_b = 22 \text{ ksi}$$

$$f_w = 1140 \text{ psi} < F_{bw} \qquad\qquad\qquad \textbf{O.K.}$$

The resisting moment of the wood is

$$M_{Rw} = f_w S_x = \frac{1.140(2)(31.64)}{12} = 6.01 \text{ ft-kips}$$

The moment to be resisted by the steel plate is

$$\text{required } M_{Rs} = 14.70 - 6.01 = 8.69 \text{ ft-kips}$$

The actual resisting moment of the steel plate is

$$M_{Rs} = F_b S_x = F_b\left(\frac{th^2}{6}\right)$$

Equating the required and the actual M_{Rs} values and solving for the required thickness t, we have

$$\text{required } t = \frac{6M_{Rs}}{F_b h^2} = \frac{6(8.69)(12 \text{ in./ft})}{22(11.25)^2} = 0.22 \text{ in.}$$

Try a plate $\frac{1}{4}$ in. $\times$ 11-$\frac{1}{4}$ in. and check the adequacy of the flitch beam for shear. The actual resisting moment of the steel plate is

$$M_{Rs} = F_b S_x = \frac{22(0.25)(11.25)^2}{6(12 \text{ in./ft})} = 9.67 \text{ ft-kips}$$

Therefore, the total resisting moment is comprised as

$$M_{Rw}: \quad \frac{6.01}{6.01 + 9.67}(100) = 38.3\%$$

$$M_{Rs}: \quad \frac{9.67}{6.01 + 9.67}(100) = 61.7\%$$

The maximum applied shear is

$$V = \frac{wL}{2} = \frac{600(14)}{2} = 4200 \text{ lb}$$

For the steel, the shear stress is calculated as

$$f_{vs} = \frac{3V_s}{2A_s} = \frac{3(0.617)(4200)}{2(0.25)(11.25)} = 1382 \text{ psi}$$

and for the wood

$$f_{vw} = \frac{3V_w}{2A_w} = \frac{3(0.383)(4200)}{2(2)(1.5)(11.25)} = 71.5 \text{ psi}$$

The allowable shear stresses for the steel and the wood are $F_{vs} = 14,500$ psi and $F_{vw} = 75$ psi. Therefore, both are satisfactory for shear. Use two 2×12 S4S planks of No. 1 Hem–Fir and a steel plate $\frac{1}{4}$ in. $\times 11\frac{1}{4}$ in.

The flitch beam should be bolted together to ensure that the individual parts act as a unit. Bolts $\frac{3}{4}$ in. in diameter staggered on 2 ft–0 in. centers will be sufficient for applications of the type discussed. Flitch beams composed of heavier plates and larger timbers will require further investigation.

3-3 COVER-PLATED BEAMS

Sometimes, available rolled shapes will have inadequate bending strength and will not satisfy the requirements for a given beam. Also, depth restrictions may require the use of shallower rolled shapes which do not possess the required bending strength. Although the W36 is the deepest shape rolled in the United States, rolled shapes of 40 in. nominal depth in standard sections are also available. Additionally, modern technology in the steel mill has developed to the point where tailor-made shapes for beams and columns are available. These shapes can range up to approximately 44 in. deep with linear weights up to 848 lb/ft (see Reference 7, Chapter 1). These large rolled wide-flange beams may offer considerable cost savings when compared to welded plate girders or cover-plated beams. They are appropriate for use in some situations. If the available hot-rolled shapes (standard or tailor-made) are inadequate, the other traditional methods may be used to solve the problem.

First, it may be possible to double-up two shapes, that is, place them side by side to develop the required strength. In some structures that contain repetitive beams, it may be possible to decrease the lateral spacing and thereby decrease the load carried by each beam. Second, a **plate girder** may be devised by welding together plates of the required sizes and often in the shape typical of wide-flange sections. The design of plate girders is discussed in Chapter 10 of this text. Third, an appropriate W shape can be strengthened by adding *cover plates* to its flanges as shown in Fig. 3-7. In the past, the attachment of the cover plates has been by riveting and many of these cover-plated beams may still be observed. However, cover-plated beams (and plate girders) are now fabricated predominantly by welding.

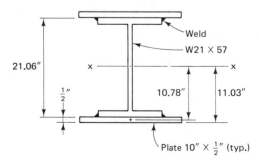

Figure 3-7 Cover-plated beam analysis.

The analysis of a cover-plated beam involves the determination of the moment capacity, $M_R = F_b I/c = F_b S$.* The moment of inertia (and section modulus) are readily determined using familiar methods from strength of materials. The allowable bending stress for members in which the cover plate is wider than the flange must be determined in accordance with the requirements of the AISCS, Section 1.5.1.4. In addition, cover plates possibly may not extend for the full length of the beam. Therefore, the strength of the section without cover plates must be checked against the applied moment which it must resist.

Example 3-5

Find the resisting moment of the A36 steel cross section shown in Fig. 3-7. The cover plates are attached with continuous welds as shown. Assume full lateral support for the compression flange.

Solution   Properties of the W21 × 57:

$$d = 21.06 \text{ in.}$$

$$b_f = 6.56 \text{ in.}$$

$$I_W = 1170 \text{ in.}^4$$

$$S_W = 111 \text{ in.}^3$$

Note that a W subscript is used to distinguish I and S (with respect to the x-x axis) for the W shape alone. Find I and S for the cover-plated cross section using the familiar transfer formula for moment of inertia from strength of materials. In its general form, the transfer formula is written

$$I = \sum I_c + \sum Ad^2$$

Rewriting this expression for the symmetrical cover-plated cross section, we have

$$I = I_W + 2(A)\left(\frac{d}{2} + \frac{t}{2}\right)^2$$

where I = moment of inertia of the cover-plated cross section with respect to the
x-x axis
A = area of one cover plate
t = thickness of one cover plate
I_w and d are as described previously

---

*In this section, for our discussion of cover-plated beams, all bending calculations will be with respect to the strong $x - x$ axis of the cross sections. The x subscript for I and S will be omitted.

Therefore,

$$I = 1170 + 2(10)(0.5)\left(\frac{21.06}{2} + \frac{0.5}{2}\right)^2 = 2332 \text{ in.}^4$$

$$S = \frac{I}{c} = \frac{2332}{11.03} = 211.4 \text{ in.}^3$$

Note that the moments of inertia of the plates about their own centroidal axes parallel to the x-x axis are very small and have been neglected. We will next determine F_b. The AISCS, Section 1.5.1.4.1, covers compactness for built-up members. Since the W21 × 57 is itself compact, only the width–thickness ratio of the compression flange must be checked. Note that according to the AISCS, Section 1.5.1.4.1 (item 2) that the width–thickness ratio may not exceed $65/\sqrt{F_y}$ and that according to the AISCS, Section 1.9.1.1, the width of the projecting element is taken from the free edge to the weld. Therefore,

$$\frac{\text{width}}{\text{thickness}} = \frac{(10 - 6.56)/2}{0.5} = 3.44$$

$$\frac{65}{\sqrt{F_y}} = \frac{65}{\sqrt{36}} = 10.8$$

$$3.44 < 10.8 \qquad\qquad\qquad\qquad \textbf{O.K.}$$

Also in the AISCS Section 1.5.1.4.1 (Item 3), the width-thickness ratio of stiffened elements may not exceed $190/\sqrt{F_y}$. The portion of the plate between the welds is considered a stiffened element and the distance between the welds is the width:

$$\frac{\text{width}}{\text{thickness}} = \frac{6.56}{0.5} = 13.12$$

$$190/\sqrt{F_y} = 190/\sqrt{36} = 31.7$$

$$13.12 < 31.7 \qquad\qquad\qquad\qquad \textbf{O.K.}$$

Therefore, $F_b = 0.66F_y = 24.0$ ksi. Calculate the resisting moment for the built-up section:

$$M_R = F_b S = 24.0(211.4) = 5074 \text{ in.-kips}$$

$$= \frac{5074}{12} = 422.8 \text{ ft-kips}$$

The *design* of symmetrical cover-plated beams involves the proportioning of the required plates, which, when added to the chosen rolled shape, will result in moment capacity sufficient to enable the beam to resist the applied moment. In Fig. 3-8, the

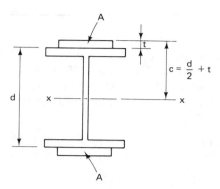

Figure 3-8 Cover-plated beam cross-section.

rolled shape has been strengthened with two plates each having area A and thickness t. The areas A are to be chosen so that $M_R \geq M$, where M is the applied moment.

$$M_R = F_b S$$

$$\text{required } S = \frac{M}{F_b}$$

As before, let I_w and S_w be properties of the W shape (about the x-x axis) and S and I be properties of the cover-plated section (also about the x-x axis). The quantities d, t, and c are as shown in Fig. 3-8.

$$I = I_w + 2A\left(\frac{d}{2} + \frac{t}{2}\right)^2$$

$$S = \frac{I}{c} = \frac{I_w + 2A[(d/2) + (t/2)]^2}{(d/2) + t}$$

Since t is usually unknown, an approximate expression for S may be written by assuming t to be small compared to d. Neglecting t, we have

$$\text{approximate furnished } S = \frac{I_w + 2A(d/2)^2}{d/2} = \frac{I_w}{d/2} + Ad = S_w + Ad$$

If the required S is equated to the foregoing, we have

$$\text{required } S = S_w + Ad$$

and (*approximately*)

$$\text{required } A = \frac{\text{required } S - S_w}{d}$$

The area determined by the foregoing expression will be on the low side. An analysis check must be made following selection of the plates.

Example 3-6

A simply supported beam is to support a uniform load of 2.4 kips/ft and a 10-kip concentrated load at midspan. The span is 40 ft, A36 steel, with full lateral support for the compression flange. Maximum overall beam depth is not to exceed 20 in. Design a symmetrical beam cross section for maximum applied moment.

Solution   Compute the maximum applied moment M and required S assuming that $F_b = 0.66F_y = 24$ ksi. Include an estimated beam weight of 150 lb/ft.

$$M = \frac{wL^2}{8} + \frac{PL}{4} = \frac{2.55(40)^2}{8} + \frac{10(40)}{4} = 610 \text{ ft-kips}$$

$$\text{required } S = \frac{M}{F_b} = \frac{610(12)}{24} = 305 \text{ in.}^3$$

With reference to the Allowable Stress Design Selection Table, AISCM, Part 2, a W21 will be too deep for the stated conditions and the largest W18 (W18 × 119) has insufficient moment capacity ($M_R = 462$ ft-kips, $S_w = 231$ in.^{3}). Therefore, design a cover-plated beam. Use the W18 × 119 and select appropriate cover plates.
    Properties of the W18 × 119:

$$d = 18.97 \text{ in.}$$

$$I_w = 2190 \text{ in.}^4$$

$$S_w = 231 \text{ in.}^3$$

$$b_f = 11.265 \text{ in.}$$

The *approximate* required area A for each cover plate is

$$A = \frac{\text{required } S - S_w}{d} = \frac{305 - 231}{18.97} = 3.90 \text{ in.}^2$$

As noted previously, because of simplifying assumptions, this calculated required area will be on the low side. Therefore, try two cover plates which are 9 in. × $\frac{1}{2}$ in. The cover-plated section is shown in Fig. 3-9. Compute the section modulus S which is furnished by the cover-plated section:

$$I = I_w + 2(A)\left(\frac{d}{2} + \frac{t}{2}\right)^2$$

$$= 2190 + 2(9)(0.50)(9.74)^2 = 3044 \text{ in.}^4$$

$$\text{furnished } S = \frac{I}{c} = \frac{3044}{9.99} = 304.7 \text{ in.}^3$$

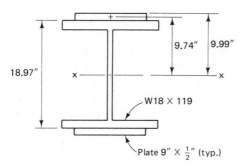

Figure 3-9 Cover-plated beam analysis.

The required section modulus was 305 in.^{3}. This is close enough to be considered satisfactory. For a check of beam weight (150 lb/ft was assumed) use a unit weight of steel of 490 lb/ft^3 or see Table for Weight of Rectangular Sections, AISCM, Part 1:

$$\text{beam weight} = 119 + \frac{2(9)(0.5)(12)}{1728}(490) = 150 \text{ lb/ft}$$

Check the width/thickness ratio of the plate as a stiffened element of the compression flange:

$$\frac{\text{width}}{\text{thickness}} = \frac{9.0}{0.5} = 18.0$$

$$190/\sqrt{F_y} = 190/\sqrt{36} = 31.7$$

$$18.0 < 31.7 \qquad\qquad\qquad\qquad \textbf{O.K.}$$

The cover-plated section qualifies for $F_b = 0.66F_y$ as assumed. Therefore, use a W18 × 119 with top and bottom cover plates 9 × 1/2.

In the foregoing example, equal-area plates were chosen. In some situations this will not be the case. For example, where a concrete slab acts in conjunction with the compression flange of a composite plate girder, the compression flange of the plate girder will normally be smaller than the plate used on the tension side. The design of such an unsymmetrical section is necessarily a trial-and-error procedure. The previous approach for symmetrical sections does not apply. For a discussion of the required welded connection between the cover plate and the flange, the reader is referred to Chapter 10 of this text.

Cover plates may extend the full length of the beam. However, this is not necessary and they may be discontinued in areas where the applied moment is low enough so that the resisting moment of the wide-flange section is sufficient. The point where the resisting moment of the wide-flange section is equal to the applied moment is called the **theoretical cutoff point.** The AISCS, Section 1.10.4, requires that partial-length cover plates be extended a definite length a' beyond the theoretical cutoff point. As a maximum, a' is to be taken as two times the plate width. This is the length required when there is no weld across the end of the plate, but continuous welds along both edges of the cover plate in the length a'. For the determination of a', reference should be made to the AISCS, Section 1.10.4.

The determination of the theoretical cutoff point for the cover plates involves the superposition of the applied moment M diagram and the moment capacity M_R of the wide-flange section. The solution may be either graphical or mathematical. If the moment diagram is easily defined mathematically (such as a straight line or a parabola), the mathematical solution will be simpler. The following example illustrates a graphical solution. For a mathematical solution of the similar problem of bar cutoffs in reinforced concrete beams, see Reference 2.

Example 3-7

Determine the theoretical cutoff point for the cover plates for the beam designed in Example 3-6.

Solution   Figure 3-10 shows the applied moment M diagram drawn to scale. Moments were computed at the eighth points for purposes of drawing the diagram. M_R for the W18 × 119, 462 ft-kips, is superimposed. The point at which the M diagram crosses the M_R line is the theoretical cutoff point. This is seen to be 11 ft from the left support. Symmetry exists. The required total length of cover plate would be 18 ft plus any required extensions at the ends of the plates.

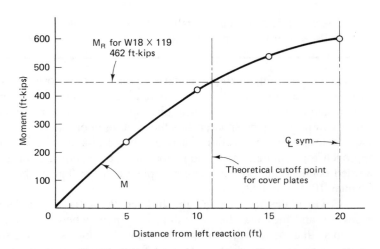

Figure 3-10   Determination of theoretical cutoff point for cover plates.

3-4 UNSYMMETRICAL BENDING

Thus far, the beams that we have considered have been loaded so that bending occurs about the strong axis (the *x-x* axis). That is, the loads have been applied in the *plane* of the weak axis. This is the normal situation for gravity loads on wide-flange beams which are oriented with their webs in the vertical plane. The *x-x* and *y-y* axes of beam cross sections are also called the **principal axes.** Occasionally, beams are subjected to loads that are not in the plane of the weak axis. Or a beam may have to support two or more systems of loads which are applied simultaneously but in different directions. When this occurs, the beam is said to be subjected to **unsymmetrical bending,** or bending about two axes. Unsymmetrical bending may further be defined as bending about any other than one of the principal axes. Some examples are shown in Fig. 3-11.

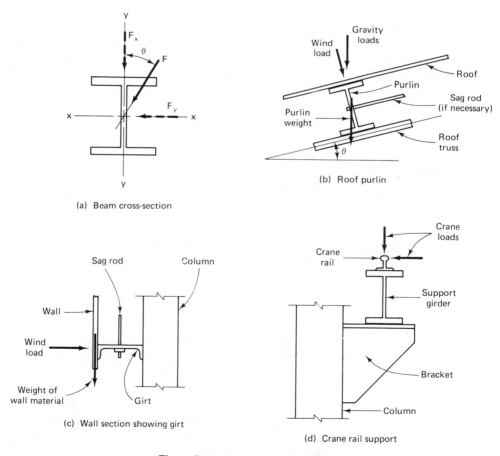

(a) Beam cross-section

(b) Roof purlin

(c) Wall section showing girt

(d) Crane rail support

Figure 3-11   Unsymmetrical bending.

The beam of Fig. 3-11(a) supports a load that passes through the centroid of the cross section. The roof purlin of Fig. 3-11(b), supported on the sloping top chord of a roof truss, must support roof loads, wind load, and its own weight, applied as shown. Note that the wind load is assumed to be applied perpendicular to the roof surface. The girt of Fig. 3-11(c) must support vertical and horizontal loads, as shown, in addition to its own weight. The support girder for the crane rail, shown in Fig. 3-11(d), must support vertical loads as well as the lateral thrust due to the moving crane.

In the case where the load passes through the centroid of the cross section, [Fig. 3-11(a)], it may be broken into its components which are parallel and perpendicular to the principal (x-x and y-y) axes. Thus (where θ is the angle between the applied force and the y-y axis)

$$F_x = F[\cos \theta]$$

$$F_y = F[\sin \theta]$$

and moments about the principal axes M_x and M_y may be found. Note that the subscripts are such that, for example, F_x is the force that creates bending (M_x) about the x-x axis. The stresses may be calculated separately for bending about each axis and added algebraically. Thus

$$f_b = \frac{M_x}{S_x} \pm \frac{M_y}{S_y}$$

The $\pm$ sign indicates that the stresses may be additive (i.e., both tension or both compression) or they may be of opposite sign and be subtractive.

In most cases, the applied load will not be positioned so that its line of action passes through the centroid of the cross section. It is more common for the load to be applied at the top flange. When this occurs, the top flange must resist most of the lateral force component, as shown in Fig. 3-11(b) and (d). Actually, this situation results in twisting of the beam. However, it is commonly assumed that the top flange acts alone in resisting the lateral force component. The formula for bending stress in the top flange then becomes

$$f_b = \frac{M_x}{S_x} \pm \frac{M_y}{S_{y(\text{top flange})}}$$

For typical wide-flange shapes, the section modulus of the top flange about the y-y axis ($S_{y(\text{top flange})}$) is approximately equal to $S_y/2$. Hence the formula is commonly written

$$f_b = \frac{M_x}{S_x} \pm \frac{M_y}{S_y/2}$$

Since the *allowable* bending stresses with respect to the x-x and y-y axes are different, the AISCS, Section 1.6, utilizes an interaction formula of stress ratios for the purpose of establishing a design criterion. For low axial stress ($f_a/F_a \le 0.15$), the applicable formula is

$$\frac{f_a}{F_a} + \frac{f_{bx}}{F_{bx}} + \frac{f_{by}}{F_{by}} \leq 1.0 \qquad \text{AISCS formula 1.6-2}$$

where f_a is the computed axial stress, F_a is the allowable axial stress permitted in the absence of bending, f_b and F_b are as previously defined, and the x and y subscripts refer to the axis about which bending takes place. For the beams that will be considered here, axial stress will be zero; therefore, the preceding formula reduces to

$$\frac{f_{bx}}{F_{bx}} + \frac{f_{by}}{F_{by}} \leq 1.0$$

To utilize the interaction formula for the design of an unsymmetrically loaded beam, a trial-and-error process must be used. The beam size must be estimated, then checked by the interaction formula and revised if necessary. Some design aids can be used and computer programs are available to help in the selection process.[3]

Example 3-8

Check the adequacy of a W10 × 22 which carries a uniform (gravity) load of 0.50 kip/ft on a simple span of 15 ft. The beam is placed on a slope of 4 : 12 as shown in Fig. 3-12. Use A36 steel and assume that the load passes through the centroid of the section. Further, assume allowable bending stresses of $F_{bx} = 24$ ksi and $F_{by} = 27$ ksi.

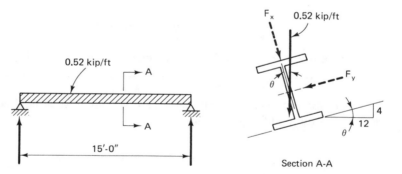

Figure 3-12   Purlin analysis.

Solution   The beam weight has been included in the uniform load of 0.52 kip/ft shown in Fig. 3-12.

Properties of the W10 × 22:

$$S_x = 23.2 \text{ in.}^3$$

$$S_y = 3.97 \text{ in.}^3$$

Resolve the load into its components F_x and F_y:

$$\theta = \tan^{-1}\left(\frac{4}{12}\right) = 18.4°$$

$$F_x = 0.52(\cos \theta) = 0.49 \text{ kip/ft}$$

$$F_y = 0.52(\sin \theta) = 0.16 \text{ kip/ft}$$

Calculate the component moments:

$$M_x = \frac{w_x L^2}{8} = \frac{0.49(15)^2}{8} = 13.8 \text{ ft-kips}$$

$$M_y = \frac{w_y L^2}{8} = \frac{0.16(15)^2}{8} = 4.5 \text{ ft-kips}$$

Calculate the actual bending stresses:

$$f_{bx} = \frac{M_x}{S_x} = \frac{13.8(12)}{23.2} = 7.1 \text{ ksi}$$

$$f_{by} = \frac{M_y}{S_y} = \frac{4.5(12)}{3.97} = 13.6 \text{ ksi}$$

Check AISC formula 1.6-2 (with $f_a = 0$):

$$\frac{f_{bx}}{F_{bx}} + \frac{f_{by}}{F_{by}} \leq 1.0$$

$$\frac{7.1}{24} + \frac{13.6}{27} = 0.8$$

Therefore, the W10 × 22 is satisfactory because 0.8 < 1.0.

For *design* purposes, a reasonable approximation for a required beam size may be obtained by modifying the relation

$$f_b = \frac{M_x}{S_x} \pm \frac{M_y}{S_y/2}$$

Assuming that it is desired to select a section on the basis of S_x, we set $f_b = F_b$ (proper subscripts to be included shortly) and solve for the required S_x.

$$F_b = \frac{M_x}{S_x} \pm \frac{2M_y}{S_y} = \frac{1}{S_x}\left[M_x \pm 2M_y\left(\frac{S_x}{S_y}\right)\right]$$

$$\text{required } S_x = \frac{M_x}{F_b} \pm \frac{2M_y}{F_b}\left(\frac{S_x}{S_y}\right)$$

Ignoring the minus sign and introducing the proper subscripts, we have

$$\text{required } S_x = \frac{M_x}{F_{bx}} + \frac{2M_y}{F_{by}}\left(\frac{S_x}{S_y}\right)$$

Knowing M_x, M_y, F_{bx}, and F_{by} and estimating a ratio of S_x/S_y (with the aid of Table 3-3), one can compute an approximate required section modulus (S_x). A member may then be selected and a check made using the AISCS interaction equation.

TABLE 3-3 S_x/S_y RATIOS FOR SHAPES

Shape	Nominal depth, d (in.)	Approximate range of S_x/S_y [a]
W	4–5	3
	6	3–5
	8	3–7
	10–14	Over 50 lb/ft: 2.5–5 Under 50 lb/ft: 3.5–11
	16–18	5–11
	21–24	6–13
	27–36	7–11
S	6–8	d (depth)
	10–18	0.75d
	20–24	0.6d

[a]Ratio decreases as weight increases for same nominal depth.

As a guide in selecting which of the numerical values of S_x/S_y to use, the smaller numbers in each W group apply to shapes with relatively wide flanges and square profiles.

The following design example utilizes an overly simplified crane loading. Longitudinal forces are neglected, as is the fact that there would be two (or more) wheel loadings on the rail. See the AISCS, Section 1.3.4, for brief discussion on minimum crane runway loads. Also, Reference 4 contains an excellent discussion of some of the aspects of crane runway design.

Example 3-9

Select a wide flange shape to be used as a bridge crane runway girder. The girder is on a simple span of 20 ft. Assume that the crane wheel imparts a vertical load of 16 kips and a lateral load of 1.6 kips, applied at the top flange of the girder. A standard 85-lb/yd rail will be used. Refer to Fig. 3-13, and use the AISC specification and A36 steel. (Neglect shear and deflection.)

Solution   Use the previously developed formula for approximate required S_x to select a member. Then check by AISC formula 1.6-2. Assume a girder weight

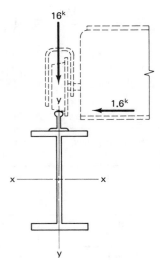

Figure 3-13 Bridge crane runway girder.

of 60 lb/ft. Note that rail weights are given in pounds per yard. The weight per foot is therefore

$$\frac{85}{3} = 28.3 \text{ lb/ft}$$

from which the total uniform load (girder and rail) is

$$60 + 28.3 = 88.3 \text{ lb/ft}$$

The applied moments with respect to the *x-x* and *y-y* axes are calculated as

$$M_x = \frac{wL^2}{8} + \frac{PL}{4} = \frac{0.0883(20)^2}{8} + \frac{16(20)}{4} = 84.4 \text{ ft-kips}$$

$$M_y = \frac{PL}{4} = \frac{1.6(20)}{4} = 8.0 \text{ ft-kips}$$

F_{by}, from the AISCS, Section 1.5.1.4.3, is

$$0.75F_y = 0.75(36) = 27.0 \text{ ksi}$$

Assuming that the top flange is not laterally braced between end supports, we note that F_{bx} will probably be reduced below $0.66F_y$ or $0.60F_y$. Having no other guideline, and subject to later change, assume that $F_{bx} = 0.60F_y = 22.0$ ksi. For a span of 20 ft, a beam depth in the range 10 to 14 in. would be a reasonable minimum. Therefore, from Table 3-3, pick an estimated S_x/S_y ratio of 4.0.

$$\text{required } S_x = \frac{M_x}{F_{bx}} + \frac{2M_y}{F_{by}}\left(\frac{S_x}{S_y}\right)$$

$$= \frac{84.4(12)}{22} + \frac{2(8)(12)}{27}(4.0) = 74.5 \text{ in.}^3$$

Try a W14 × 53:

$$S_x = 77.8 \text{ in.}^3$$

$$S_y = 14.3 \text{ in.}^3$$

$$F_{by} = 27.0 \text{ ksi}$$

Utilizing the beam curves of the AISCM, Part 2, with $L_b = 20$ ft, the allowable moment M_R may be obtained from the W14 × 53 curve and since $M_R = F_{bx}S_x$,

$$F_{bx} = \frac{M_R}{S_x} = \frac{124(12)}{77.8} = 19.1 \text{ ksi}$$

Also, the actual bending stress

$$f_{bx} = \frac{M_x}{S_x} = \frac{84.4(12)}{77.8} = 13.0 \text{ ksi}$$

Assuming that the top flange resists the lateral load,

$$f_{by} = \frac{M_y}{S_y/2} = \frac{8.0(12)}{14.3/2} = 13.4 \text{ ksi}$$

Check AISC formula 1.6-2 (with $f_a = 0$);

$$\frac{f_{bx}}{F_{bx}} + \frac{f_{by}}{F_{by}} = \frac{13.0}{19.1} + \frac{13.4}{27} = 1.18 > 1.0 \qquad \textbf{N.G.}$$

A different section must be selected. Try a W12 × 58.

$$S_x = 78.0 \text{ in.}^3$$

$$S_y = 21.4 \text{ in.}^3$$

$$F_{bx} = \frac{143(12)}{78.0} = 22.0 \text{ ksi}$$

$$f_{bx} = \frac{84.4(12)}{78.0} = 12.98 \text{ ksi}$$

$$F_{by} = 27.0 \text{ ksi}$$

$$f_{by} = \frac{8.0(12)}{21.4/2} = 8.97 \text{ ksi}$$

AISCS formula 1.6-2:

$$\frac{f_{bx}}{F_{bx}} + \frac{f_{by}}{F_{by}} = \frac{12.98}{22.0} + \frac{8.97}{27} = 0.92 < 1.0 \qquad \textbf{O.K.}$$

Use a W12 × 58.

As shown in Fig. 3-11(b) and as stated previously, a common example of unsymmetrical bending (biaxial bending) is the purlin supported on the sloping top chord of roof trusses. The force parallel to the roof surface must be carried in transverse bending of the purlin (bending with respect to the y-y axis). Since the section modulus in this direction is small for most sections, it is usually a more economical design to brace the purlins with sag rods that serve as intermediate supports for the loading parallel to the roof surface. Sag rods are usually placed at the third points or the middle of the purlin span. Hence the purlins act as three-span or two-span continuous beams with respect to bending about the y-y axis and as simple beams, spanning from truss to truss, with respect to bending about the x-x axis. This in effect reduces the bending moment with respect to the y-y axis and would lend itself to the use of a lighter purlin section.

REFERENCES

1. National Forest Products Association, *National Design Specification for Wood Construction,* 1982.
2. L. Spiegel, and G. F. Limbrunner, *Reinforced Concrete Design* (Englewood Cliffs, NJ: Prentice-Hall, Inc., 1980), pp. 148–150.
3. J. L. Cranmer, Jr., "Micro-computer Program: Roof Purlin Design," *Civil Engineering,* Vol. 53, No. 2, Feb. 1982, pp. 60–61.
4. D. T. Ricker, "Tips for Avoiding Crane Runway Problems," *Engineering Journal, American Institute of Steel Construction,* Vol. 19, No. 4, 1982, pp. 181–205.

PROBLEMS

3-1. Design the lightest double-angle lintel to support an 8-in. wall. The weight of wall is 120 lb/ft^3. Use A 36 steel. The angles are in contact. The maximum allowable deflection = span/240. Assume that arching action can develop in the wall.
   (a) Clear opening is 6 ft–8 in. with a 4-in. bearing at each end of the lintel.
   (b) Clear opening is 8 ft–0 in. with a 6-in. bearing at each end of the lintel.

3-2. Design the lightest single-angle lintel to support a 4-in. masonry partition wall over a clear opening of 5 ft–0 in. The weight of wall is 120 lb/ft^3. Use A36 steel. Assume a 4-in. bearing at each end of the lintel. The maximum allowable deflection = span/240. Assume that arching action can develop in the wall.

3-3. Design the lightest lintel to support a 12-in. brick wall over a clear opening of 12 ft–0 in. The weight of wall is 120 lb/ft³. Use A36 steel. Assume an 8-in. bearing at each end of the lintel. The maximum allowable deflection = span/240. Select a lintel composed of a WT and an L. Assume that arching action can develop in the wall.

3-4. Design the lightest structural tee (WT) lintel to support an 8-in. wall over a clear opening of 12 ft–4 in. The weight of wall is 125 lb/ft³. Use A36 steel. Assume an 8-in. bearing at each end of the lintel. The maximum allowable deflection = span/240. Assume that arching action can develop in the wall.

3-5. A flitch beam is composed of two 2 × 12 (S4S) planks of No. 1 Eastern spruce and an A36 steel plate $11\frac{1}{4} \times \frac{3}{8}$ in. Use $F_b = 22,000$ psi.
(a) Find the resisting moment M_R (ft-kips) for the flitch beam.
(b) Find the allowable uniformly distributed load (lb/ft) for the flitch beam on a design span length of 14 ft–0 in.
(c) Check the shear.

3-6. A 6 × 10 in. beam of Select Structural Douglas fir is reinforced by the addition of two A36 C10 × 15.3 channels bolted to the sides of the beam, as shown. Assume that $F_b = 22,000$ psi for the steel. Use nominal dimensions for the wood beam. $F_{bw} = 1600$ psi for this thickness and $E_w = 1,600,000$ psi.
(a) Find the resisting moment for the wood beam alone.
(b) Find the resisting moment for the reinforced beam.
(c) Find the percent increase in resisting moment due to the addition of the two channels to the 6 × 10 in. beam.

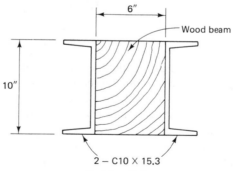

Problem 3-6

3-7. A flitch beam is made up of two wood beams and a steel plate, as shown. The steel plate is A36 steel and the wood is Select Structural Eastern spruce. If W equals the total uniformly distributed load supported by the beam, what portion of the load will be supported by the steel and what portion by the wood? Use nominal dimensions for the wood. The materials are so connected that they will act together as a unit. $F_{bw} = 1050$ psi and $E_w = 1,400,000$ psi.

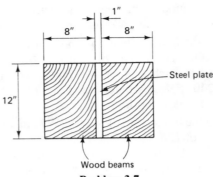

Problem 3-7

3-8. Design a flitch beam having the cross section shown for a maximum bending
moment of 96 ft-kips. The beam is limited to a maximum depth of 12 in. The
steel is to be A36 with an allowable bending stress of $F_b = 22,000$ psi. Available steel plates are 1 in. thick. The wood is to be Select Structural Hem–Fir
with an allowable bending stress $F_{bw} = 1300$ psi. $E_w = 1,300,000$ psi. Use
nominal dimensions for the wood.

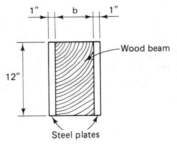

Problem 3-8

3-9. A flitch beam over a double garage door is to support roof trusses that super-
impose a load (assumed to be uniformly distributed) of 900 lb/ft on the beam.
The design span length is 15 ft–8 in. Design the flitch beam using two S4S wood
sections of No. 1 Hem–Fir and an A36 steel plate.
   (a) Use 2 × 12s.
   (b) Use 2 × 10s.

3-10. An S12 × 31.8 has a section modulus with respect to its x-x axis of 36.4 in.^{3}.
What thickness of plate 6 in. wide must be welded to each flange to double the
value of the section modulus?

3-11. An A36 W12 × 35 simply supported beam must support a uniformly distributed load of 4.0 kips/ft over a span length of 20 ft. Determine the size of cover
plate that should be welded to each flange if the compression flange has full
lateral support.

3-12. Compute the resisting moment with respect to the x-x axis for the A36 steel member shown. The cover plates are attached with continuous fillet welds and the compression flange has full lateral support.

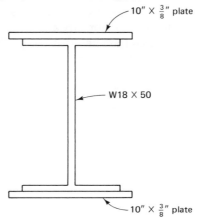

Problem 3-12

3-13. An A36 W24 × 76 is used as a 30-ft simple span beam. The beam compression flange has full lateral support. Compute the superimposed uniformly distributed load that the beam can support if

    (a) A 6 × $\frac{3}{4}$ plate is added to the top flange and a 12 × $\frac{3}{8}$ plate is added on the bottom flange.

    (b) A 12 × $\frac{3}{8}$ plate is added on the bottom flange only.

3-14. A beam has been designed to span 40 ft on simple supports. The beam is an A36 W30 × 116. Due to unforeseen circumstances, the design load must be changed to 3.8 kips/ft uniform load. It is desired to use the same beam strengthened with cover plates on each flange. Determine the cover plate size and required length. Assume full lateral support.

3-15. Check the adequacy of a W8 × 35 which supports a uniform load of 0.325 kip/ft on a simple span of 18 ft, as shown. Use allowable bending stresses of F_{bx} = 24 ksi and F_{by} = 27 ksi.

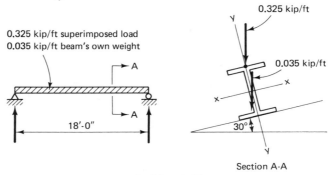

Problem 3-15

3-16. Select a W10 section to serve as a purlin which is simply supported on a 20-ft span between roof trusses. The roof is assumed to support a superimposed dead load of 15 psf and a snow load of 35 psf. The slope of the roof truss top chord is 1 vertically to 2 horizontally and the purlins are spaced 10 ft–0 in. on centers. Use A36 steel and assume that the purlin compression flange has full lateral support.

4 Tension Members

4-1 *INTRODUCTION*

The proportioning of tension members is among the simpler of the problems that face the structural designer. However, although easy to proportion, tension members, and structures in which the main load-carrying members are in tension, require great care in the design and *detailing* of their connections. Some catastrophic structural failures have been directly attributed to poor tension member connection details. Tension members do not have the inherent stability problems of beams and columns. A tensile load applied along the longitudinal axis of the member tends to hold the member in alignment, thereby making stability of minor concern. Tension members therefore do not generally require the *bracing* usually associated with beams and columns. The resulting tension member structures are less *redundant,* and the potential for sudden failure exists if there is any inadequacy present, such as a weakness in a connection.

Of most concern in the *selection* of tension members, is the choice of the configuration of the cross section so that the connections will be simple and efficient. Also, the connection should transmit the load to the member with as little eccentricity as possible.

Examples of tension members may be found in many structures. They include hangers for catwalks and storage bins, truss web and chord members, cables for direct

This frame is braced in both the vertical and horizontal planes with structural steel angles. The angles are bolted to gusset plates which are welded to the frame. (Courtesy of the American Hot Dip Galvanizers Association)

support of roofs, sag rods, tie rods, and various types of braces. Most of the common hot-rolled structural steel shapes may be used as tension members. Small bracing members may be circular threaded steel rods or flexible members such as wire ropes or cables. Single angles, double angles, tees, and channels may also be used for bracing purposes, as well as light truss tension members. Large truss members may consist of W shapes as well as an almost infinite variety of built-up members. Suspension bridges at one time were supported on chains composed of long plates of rectangular cross section having enlarged ends through which large-diameter steel pins were inserted as connectors. These are the *pin-connected members* referred to in the AISCS, Section 1.5.1.1. The "eye-bar chains" have been superseded by more efficient multiwire steel cables which are either spun in place or prefabricated.

4-2 TENSION MEMBER ANALYSIS

The direct stress formula is the basis for tension member analysis (and design). It may be written for *stress,*

$$f_t = \frac{P}{A}$$

or for *tensile capacity,*

$$P_t = F_t A$$

where f_t = computed tensile stress
P = applied axial load
P_t = axial tensile load capacity (or maximum allowable axial tensile load)
F_t = allowable axial tensile stress
A = cross-sectional area of axially loaded tension member (either net area A_n or gross area A_g)

A_n, net area, is illustrated in Fig. 4-1 and is logically the cross-sectional area actually available to be stressed in tension. Net area may be visualized by imagining that the tension member (a plate in this case) fractures along the line shown in Fig. 4-1(a). The net area, shown crosshatched in Fig. 4-1(b), is then calculated:

$$A_n = (\text{gross area}) - (\text{area of holes})$$

$$= bt - 2(d_h t)$$

where b and t are plate width and thickness and d_h is hole diameter for analysis purposes. Recall from Section 2-11 of this text that a hole diameter for analysis purposes is $\frac{1}{16}$ in. greater than the actual hole diameter and normally $\frac{1}{8}$ in. greater than the fastener diameter (AISCS, Section 1.14.4).

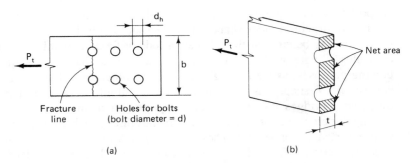

Figure 4-1 Net area.

The direct stress formula applies directly to homogeneous axially loaded tension members. Its use is based on the assumption that the tensile stress is uniformly distributed over the net section of the tension member, despite the fact that high stress concentrations are known to exist (at working loads) around the holes in a tension member. The commonly used structural steels are sufficiently ductile so that they undergo yielding and stress redistribution. This will result in a uniform stress distribution at ultimate load.

The allowable tensile stress F_t takes into consideration two types of failure. First, the member may rupture on the least net area as shown in Fig. 4-1. This is the classical and historical approach to tension member analysis. For this type of failure, the AISCS, Section 1.5.1.1, states that $F_t = 0.50F_u$ on the net area, where F_u is the specified minimum tensile strength of the steel. Second, the tension member may undergo uncontrolled yielding of its gross area *without rupture*. Excessive elongation of a tension member is undesirable in that it normally results in deformation of the structure and can lead to failure in other parts of the structural system. For this type of failure, the AISCS, Section 1.5.1.1 establishes an allowable tensile stress $F_t = 0.60F_y$ on the gross cross-sectional area of the member. These allowable stresses do not apply to pin-connected members (such as eye bars or plates connected with relatively large pins, as discussed in the AISCS, Section 1.14.5), threaded steel rods, or flexible tension members such as cables and wire rope. The axial tensile load capacity of the member is based on the lesser of the two allowable load values.

Example 4-1

Find the axial tensile load capacity P_t of the lapped, bolted tension member shown in Fig. 4-2. Bolts are $\frac{3}{4}$-in. diameter and the plate material is A36 steel ($F_u = 58 - 80$ ksi from the AISCM, Part 1, Table 1). Assume that the fasteners are adequate and do not control the tensile capacity.

Solution

$$P_t = F_t A_n \quad \text{or} \quad F_t A_g$$

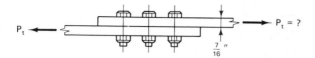

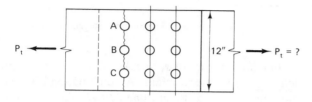

Figure 4-2   Tension member analysis.

Based on gross area:

$$P_t = 0.60F_y A_g$$

$$= 22\left(\frac{7}{16}\right)(12) = 115.5 \text{ kips}$$

Based on net area, visualizing a transverse fracture through holes A, B, and C in Fig. 4-2:

$$A_n = A_g - A_{\text{holes}}$$

$$= \left(\frac{7}{16}\right)(12) - 3\left(\frac{3}{4} + \frac{1}{8}\right)\left(\frac{7}{16}\right) = 4.10 \text{ in.}^2$$

$$P_t = 0.50F_u A_n$$

Using the lower limit of the F_u range (conservative):

$$P_t = 0.50(58)(4.10) = 118.9 \text{ kips}$$

Therefore, the capacity of this tension member is 115.5 kips as controlled by general yielding of the gross area.

In Example 4-1, the critical net area on which fracture could logically be expected was easy to visualize. In some cases, the fasteners will be arranged so that the controlling fracture line will be something other than transverse, as shown in Fig. 4-3. This situation can occur when fasteners are staggered in order to accommodate a desired size or shape of connection. Note in Fig. 4-3 that there are two possible failure lines across the width of the plate. These may be defined as lines *ABCD* and *ABE*. The distance between the holes perpendicular to the longitudinal axis is defined as the **gage distance** g and the distance between the holes parallel to the longitudinal axis is the **pitch,** or **spacing,** s. For large values of s, line *ABE* will be the more critical

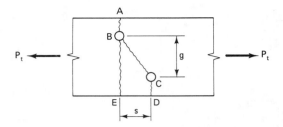

Figure 4-3   Staggered holes.

failure line (smaller net area). For small values of s, line $ABCD$ will be more critical. Actually, both gage and pitch will affect the problem. A combination of shear and tensile stresses acts on the sloping line BC of failure line $ABCD$. The interaction of these stresses presents a rather complicated theoretical problem. The AISCS uses a simplified method of analysis in this situation which is based on the studies and observations of V. H. Cochrane and T. A. Smith in the early part of the twentieth century.[1] The AISCS, Section 1.14.2.1, stipulates that where a fracture line contains within it a diagonal line, the net width of the part should be obtained by deducting from the gross width the diameters of all the holes along the fracture line and adding, for each diagonal line, the quantity

$$\frac{s^2}{4g}$$

where s and g are as previously defined. An expression for **net width** w_n may be written

$$w_n = w_g - \sum d_h + \sum \frac{s^2}{4g}$$

where w_g represents gross width. The foregoing formula for w_n is convenient to use with members of uniform thickness. If the formula is multiplied by thickness t, it becomes

$$w_n t = w_g t - \sum d_h t + \sum \frac{s^2 t}{4g}$$

Or, since $w_n t = A_n$ and $w_g t = A_g$,

$$A_n = A_g - \sum d_h t + \sum \frac{s^2 t}{4g}$$

The latter formula for A_n is more useful since it provides net area directly and is also applicable with members that do not have uniform thickness (i.e., channels). In a determination of critical net area where multiple possible failure lines exist, the critical net area is, of course, the least net area.

Example 4-2

Determine the critical net width w_n for the plate shown in Fig. 4-4. Fasteners will be 1-in.-diameter bolts.

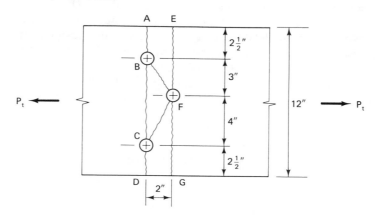

Figure 4-4   Net width calculation.

Solution   Use the formula for net width:

$$w_n = w_g - \sum d_h + \sum \frac{s^2}{4g}$$

For illustrative purposes, all possible failure lines will be checked. Practically, some may be seen by inspection to be noncritical. Note how failure lines have been designated with letters. For *analysis purposes,* the diameter of the holes $d_h = 1\frac{1}{8}$ in. = 1.13 in.

Line	Net width
ABCD	$12 - 2(1.13) + 0 = 9.74$ in.
EFG	$12 - 1(1.13) + 0 = 10.87$ in.
ABFG	$12 - 2(1.13) + \dfrac{2^2}{4(3)} = 10.07$ in.
EFCD	$12 - 2(1.13) + \dfrac{2^2}{4(4)} = 9.99$ in.
ABFCD	$12 - 3(1.13) + \dfrac{2^2}{4(3)} + \dfrac{2^2}{4(4)} = 9.19$ in. (controls)

Therefore, the critical net width for this plate is 9.19 in. The critical net area would be found by multiplying the critical net width by the plate thickness.

To shorten the calculations, as a general rule, the transverse section having the greatest number of holes should be checked first. Follow this by checking every zigzag line (section) which has more holes than the initially checked transverse section. With reference to Example 4-2, it is seen that this procedure would have eliminated the checking of noncritical lines *EFG, ABFG,* and *EFCD.*

Finally, the failure line should be considered in a logical way with regard to the transmission of the load through it. In Fig. 4-5, the tension member carries a load P. Each of the three bolts may be assumed to transmit $P/3$ into the supporting member. Failure lines *ABC, ABFG, DEBC,* and *DEBFG* would each carry the full load P. Failure line *DEFG,* however, would carry only $\frac{2}{3}P$ since the bolt at B transmits the other $P/3$.

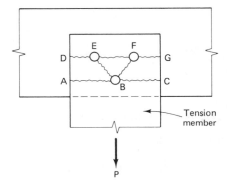

Figure 4-5   Three-bolt tension member connection.

4-3 EFFECTIVE NET AREA

For some tension members, such as rolled shapes, which do not have all elements of the cross section connected to the supporting members, the failure load is less than would be predicted by the product $A_n F_u$. The phenomenon to which this situation is generally attributed is called **shear lag** and is illustrated in Fig. 4-6. Note that the angle is connected along only one leg. This leads to a concentration of stress along that leg and has left part of the unconnected leg unstressed or stressed very little. Studies have shown that the shear lag effect diminishes as the length of the connection is increased.

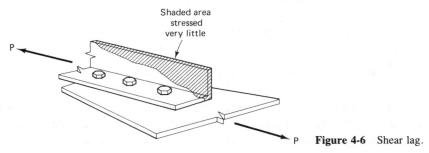

P   **Figure 4-6**   Shear lag.

The AISCS, Section 1.14.2.2, accounts for the effect of shear lag through the use of an **effective net area** A_e. This applies to axially loaded tension members where the load is transmitted by bolts through some, but not all, of the cross-sectional elements of the member. The effective net area is computed from

$$A_e = C_t A_n$$

where A_n = net area of the member
$\quad\quad C_t$ = reduction coefficient (see Table 4-1)

TABLE 4-1 VALUES FOR REDUCTION
COEFFICIENT, C_t

Case 1	W, M, S shapes or their tees. Connection is to the flanges. Minimum of three bolts per line in the direction of stress.	$C_t = 0.90$
Case 2	All shapes and built-up cross sections not meeting the requirements of case 1. Minimum of three bolts per line in the direction of stress.	$C_t = 0.85$
Case 3	All members whose connections have only two bolts per line in the direction of stress.	$C_t = 0.75$

Additionally, for relatively short connection fittings such as splice plates, gusset plates, and beam-to-column fittings subjected to tensile force, the effective net area shall be taken as the actual net area except that it shall not be taken as greater than 85% of the gross area. Therefore, for these short plates and fittings subjected to tension, C_t does not apply and

$$A_e = A_n \text{ (not to exceed } 0.85A_g)$$

For other than short tension members, if all elements are connected to the supporting members, $A_e = A_n$.

Example 4-3

A tension member in a truss is to be composed of a W8 × 24 and will be connected with two lines of $\frac{3}{4}$-in.-diameter bolts in each flange as shown in Fig. 4-7. Assume three bolts per line and A36 steel. Find the tensile load capacity P_t.

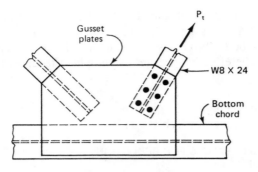

Figure 4-7 Truss connection.

Solution   Properties of the W8 × 24:

$$A_g = 7.08 \text{ in.}^2$$

$$d = 7.93 \text{ in.}$$

$$b_f = 6.495 \text{ in.}$$

$$t_f = 0.40 \text{ in.}$$

1. Based on gross area:

$$P_t = F_t A_g$$

$$P_t = 0.60 F_y A_g$$

$$= 22(7.08) = 156 \text{ kips}$$

2. Based on effective net area:

$$P_t = F_t A_e$$

$$P_t = 0.50 F_u A_e$$

$$A_e = C_t A_n$$

$$A_n = A_g - 4\left(\frac{7}{8}\right)(0.40) = 7.08 - 1.40 = 5.68 \text{ in.}^2$$

For C_t evaluation, the member is covered by case 1 (Table 4-1) if $b_f \geq \frac{2}{3}d$:

$$\frac{2}{3}d = 0.67(7.93) = 5.31 \text{ in.}$$

$$b_f = 6.495 \text{ in.} > 5.31 \text{ in.} \qquad\qquad \textbf{O.K.}$$

Therefore, $C_t = 0.90$ and

$$P_t = 0.50 F_u C_t A_n$$

$$= 0.50(58)(0.90)(5.68) = 148 \text{ kips}$$

For this member, $P_t = 148$ kips as controlled by a rupture failure based on the least net area.

Example 4-4

Find the tensile load capacity P_t of the double-angle tension member composed of 2L5 × 5 × $\frac{3}{4}$, shown in Fig. 4-8(a) and (b). Bolts are $\frac{3}{4}$-in. diameter and the steel is A36. Assume that the fasteners are adequate and do not control the tensile capacity.

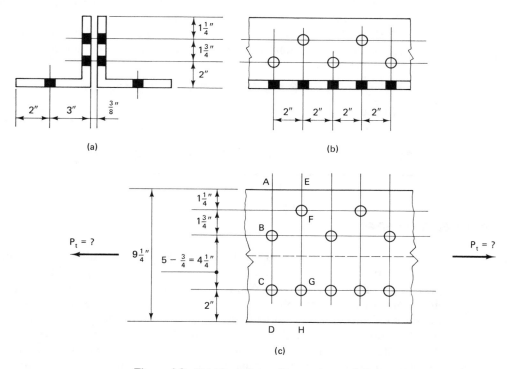

Figure 4-8   Double-angle tension member analysis.

Solution   Assuming each angle to be flattened out into a single plate, as shown in Fig. 4-8(c), the least net area may be computed. Note that the gross width is taken as the sum of the leg widths minus the angle thickness as per the AISCS, Section 1.14.3. The least net area may be determined by computing the least net width and multiplying by the angle thickness. Hole diameter $d_h = \frac{3}{4} + \frac{1}{8} = \frac{7}{8}$ in. = 0.875 in. Gages may be obtained from the AISCM, Part 4.

Line	*Net width*
ABCD	$9.25 - 2(0.875) + 0 = 7.5$ in.

$$EFBCD \qquad 9.25 - 3(0.875) + \frac{2^2}{4(1.75)} = 7.2 \text{ in.}$$

$$EFBGH \qquad 9.25 - 3(0.875) + \frac{2^2}{4(1.75)} + \frac{2^2}{4(4.25)} = 7.43 \text{ in.}$$

Line *EFBGH* could be seen by inspection to be less critical than line *EFBCD*. The critical (least) net width is 7.2 in. The critical (least) net area for one angle is $7.2(0.75) = 5.4$ in.^{2}. Since there are two angles, the critical net area is $5.4(2) = 10.8$ in.^{2}.

There is no reduction factor C_t since each leg of both angles is connected and transmitting tension. The tensile capacity may be computed by (based on gross area)

$$P_t = 0.60F_yA_g = 22(13.9) = 306 \text{ kips}$$

or (based on net area):

$$P_t = 0.50F_uA_n = 0.50(58)(10.8) = 313 \text{ kips}$$

Therefore, the capacity of the double-angle tension member is 306 kips.

4-4 LENGTH EFFECTS

As mentioned earlier in this chapter, tension members do not suffer from the problems of instability and buckling as do compression members and beams. Therefore, *length* plays a minor role. The AISCS, Section 1.8.4, suggests *upper limits* for the *slenderness ratios* of tension members. Recall that the **slenderness ratio** is the ratio of the member's unbraced length to its least radius of gyration. Staying within the recommended limits is not essential for the structural integrity of the members. However, it will afford some member resistance to undesirable vibrations as well as some member stiffness during shipping and erection. The recommended upper limits for the slenderness ratio ℓ/r are:

For main members	240
For lateral bracing and other secondary members	300

For purposes of definition, a **main member** may be considered to be a primary, gravity-load-carrying member, the failure of which would result in the collapse of the structure or of a substantial part of the structure. A **secondary member** is a less important member which generally does not carry gravity loads but would carry lateral loads such as wind loads. The failure of a single secondary member would not necessarily lead to structural collapse because secondary members are usually redundant.

The recommended upper limits on ℓ/r are *preferred,* but are not mandatory and apply to tension members other than steel rods and cables.

Example 4-5

In Example 4-3, assume that the W8 $\times$ 24 tension member is 20 ft long. Determine if the member's slenderness ratio is within the AISCS recommendations.

Solution   Assume a main member and maximum preferred $\ell/r = 240$. Calculate actual ℓ/r. Use the least radius of gyration.

$$\frac{\ell}{r} = \frac{20(12)}{1.61} = 149 < 240 \qquad \textbf{O.K.}$$

Note that the slenderness ratio is unitless. Since r is tabulated in units of inches, ℓ must be converted to inches.

4-5 DESIGN OF TENSION MEMBERS

The design, or selection, of adequate tension members involves the following:

1. Provision of adequate net area (A_n or A_e)
2. Provision of adequate gross area
3. Provision of adequate radius of gyration to meet the preferred ℓ/r limits.
4. Provision of a cross-sectional shape such that the connections can be simple, practical, and economical

The first three of the foregoing considerations involve strength and slenderness and are easily calculated using the principles already discussed in this chapter. The latter involves the way the member will fit into and be affected by the structure of which it is a part. It could involve a considerable amount of judgment on the part of the designer. The actual design of a tension member (particularly in trusses) must be based on *assumed* end conditions. After a member is selected and the end connection designed, *it may be necessary to revise the design.*

Example 4-6

A double-angle tension member *BC* will be used as a web member in a light truss, as shown in Fig. 4-9. The tensile load will be 44 kips. Use A36 steel. The

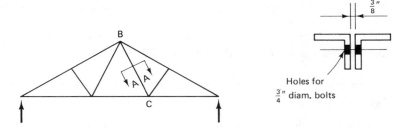

Section A-A

Figure 4-9 Tension member design.

length is 13.4 ft. Fasteners will be $\frac{3}{4}$-in.-diameter bolts and will connect the double-angle member to a $\frac{3}{8}$-in.-thick gusset plate.

Solution The required effective net area, based on fracture, is

$$\text{required } A_e = \frac{P}{F_t} = \frac{P}{(0.50)F_u} = \frac{44}{0.50(58)} = 1.52 \text{ in.}^2$$

Assuming three or more fasteners and one gage line as shown in section A-A of Fig. 4-9, $C_t = 0.85$. The actual net area required is

$$A_e = C_t A_n$$

$$\text{required } A_n = \frac{A_e}{C_t} = \frac{1.52}{0.85} = 1.79 \text{ in.}^2$$

The relationship between gross area and net area will depend on the thickness of the angle, which is unknown at this time.

The minimum required gross area, based on general yielding, is

$$\text{required } A_g = \frac{P}{F_t} = \frac{P}{0.60F_y} = \frac{44}{22} = 2.00 \text{ in.}^2$$

Assuming a main member, the minimum required radius of gyration may be calculated:

$$\text{maximum } \frac{\ell}{r} = 240$$

$$\text{minimum required } r = \frac{\ell}{240} = \frac{13.4(12)}{240} = 0.67 \text{ in.}$$

From the AISCM tables for properties of double angles, try $2L2\frac{1}{2} \times 2 \times \frac{1}{4}$, long legs back to back (LLBB).

$$A_g = 2.13 \text{ in.}^2 > 2.00 \text{ in.}^2 \qquad\qquad \textbf{O.K.}$$

$$r_x = 0.784 \text{ in.} > 0.67 \text{ in.} \qquad\qquad \textbf{O.K.}$$

Check the actual net area (refer to Fig. 4-9):

$$A_n = 2.13 - 2\left(\frac{1}{4}\right)\left(\frac{7}{8}\right) = 1.69 \text{ in.}^2 < 1.79 \text{ in.}^2 \qquad \textbf{N.G.}$$

Try 2L3 × 2 × $\frac{1}{4}$ (LLBB):

$$A_g = 2.38 \text{ in.}^2 > 2.00 \text{ in.}^2 \qquad \textbf{O.K.}$$

$$r_x = 0.957 \text{ in.} \qquad r_y = 0.891 \text{ in.} > 0.67 \text{ in.} \qquad \textbf{O.K.}$$

$$A_n = 2.38 - 2\left(\frac{1}{4}\right)\left(\frac{7}{8}\right) = 1.94 \text{ in.}^2 > 1.79 \text{ in.}^2 \qquad \textbf{O.K.}$$

Use 2L3 × 2 × $\frac{1}{4}$ (LLBB).

Note that the angles were selected with long legs back to back in Example 4-6. For unequal leg angles, this configuration will result in a better balance between the radius of gyration values for the x-x and y-y axes. If the unbraced lengths are different for the two axes, it may be more economical to arrange the angles with the short legs back to back.

4-6 THREADED RODS IN TENSION

Rods of circular cross section are commonly used for tension members. These are sometimes called tie rods or sag rods, depending on application. The connection to other structural members is made by threading the end of the rod and installing a nut. The thread reduces the cross-sectional area available to carry tension. Historically, design of threaded rods has been based on the cross-sectional area (called the **root area**) at the base of the threads. In Fig. 4-10, the root area has a diameter K. More recently, threaded rod design was based on an area greater than the root area but less than the cross-sectional area of the unthreaded body of the rod. This area was called the **tensile stress area.** In each case, an allowable tensile stress was determined as a fraction of the yield stress (i.e., $0.60F_y$ applied to the tensile stress area).

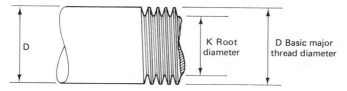

Figure 4-10   Threaded rod—partial view.

Rods with enlarged ends, called **upset rods** (shown in Fig. 4-11), have been used to circumvent the problem of reduced cross-sectional area under the threads. The

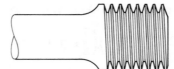

Figure 4-11 Upset rod.

enlarged end is proportioned so that the strength at the threads is greater than the strength of the unthreaded body of the rod. The forging process used to create the enlarged end is expensive. Therefore, upset rods are less common in construction applications than they are in machine applications.

Under the current AISC Specification, the allowable tensile stress for threaded rods, other than upset rods, is taken as $0.33F_u$ and this is applied to the *unthreaded nominal body area* of the rod (area of diameter D in Fig. 4-10). Refer to the AISCS, Table 1.5.2.1. For upset rods, the threaded upset end must be proportioned so that its capacity is *greater than* the capacity of the unthreaded body of the rod. The capacity of the threaded upset end is calculated from

$$P_t = F_t A_D = 0.33 F_u A_D$$

where A_D is the gross area of the upset end. The capacity of the unthreaded body of the rod is calculated from

$$P_t = F_t A_D = 0.60 F_y A_D$$

where A_D is the gross area of the body of the rod. Therefore, the capacity of a properly proportioned upset end rod may be calculated from the latter formula.

Circular rods are available in most steels commonly used in construction. For useful data on threads, the reader is referred to the threaded fastener data contained in the AISCM.

In the selection of threaded rods, it should be noted that there are no slenderness ratio recommendations by the AISCS. A common rule of thumb is to use a rod diameter not less than $1/500$ of the rod length. The minimum size of rod should be limited to $\frac{5}{8}$ in. in diameter since smaller rods are easily damaged during construction. Also, the minimum design load for a threaded fastener (and, therefore, for a threaded rod) is 6 kips, as per the AISCS, Section 1.15.1.

Example 4-7

A hanging storage bin weighing 30 tons will be supported by three circular threaded rods. Use A36 steel. Determine the required rod diameter and specify the required threads.

Solution   Each rod will carry 10 tons (or 20 kips). The required unthreaded nominal body area (gross A_D) is

$$\text{required } A = \frac{P}{0.33F_u} = \frac{20}{0.33(58)} = 1.04 \text{ in.}^2$$

From the AISCM, Part 4, threaded fastener data, basing the selection on gross area A_D, use a $1\frac{1}{4}$-in.-diameter rod. Using the standard thread designation, the required thread will be **$1\frac{1}{4}$–7 UNC 2A.**

Example 4-8

Rework Example 4-7 using upset rods. Determine the required rod size and the required diameter of the upset end. Specify the required thread.

Solution   The required nominal body area (gross area A_D) of the rod before upsetting is

$$\text{required } A = \frac{P}{0.60F_y} = \frac{20}{22.0} = 0.91 \text{ in.}^2$$

Use a $1\frac{1}{8}$-in.-diameter rod ($A_D = 0.994$ in.²). The selection of the upset end will be based on the capacity of the nominal body area, which is

$$0.60(F_y)(0.994) = 22.0(0.994) = 21.9 \text{ kips}$$

The required area of the upset end (gross area A_D) is

$$\text{required } A = \frac{P}{0.33F_u} = \frac{21.9}{0.33(58)} = 1.14 \text{ in.}^2$$

Use a $1\frac{1}{4}$-in.-diameter upset end ($A_D = 1.227$ in.²). The required thread is, as before, **$1\frac{1}{4}$–7 UNC 2A.**

REFERENCE

1. William McGuire, *Steel Structures* (Englewood Cliffs, NJ: Prentice-Hall, Inc., 1968), pp. 308–318.

PROBLEMS

4-1.   Compute the tensile capacity for the double-angle member shown. The bolts are $\frac{7}{8}$-in. diameter and the steel is A36.

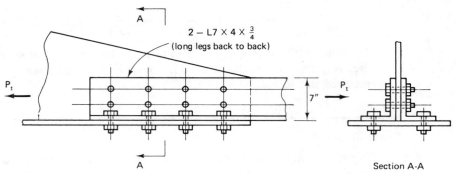

Problem 4-1

4-2. Compute the tensile capacity for the connection shown. (Neglect bolt shear and bearing on the plates.) The bolts are $\frac{7}{8}$-in. diameter and the steel is A36.

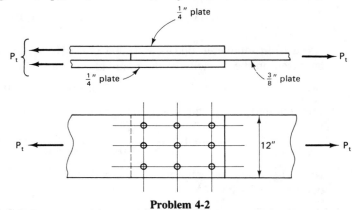

Problem 4-2

4-3. Compute the net area for the plates shown. Assume $\frac{3}{4}$-in. bolts.

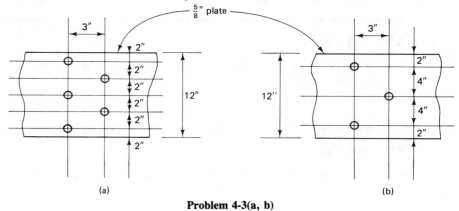

Problem 4-3(a, b)

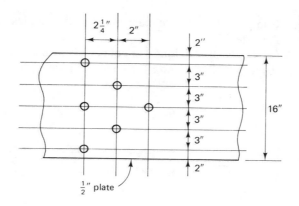

(c)

Problem 4-3(c)

4-4. Compute the tensile capacity for the A36 steel plate shown. Bolts are $\frac{7}{8}$-in. diameter.

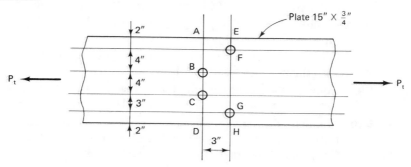

Problem 4-4

4-5. Compute the tensile capacity for the angles and conditions shown. Use A36 steel.

(a) L4 × 4 × $\frac{1}{2}$, connected with $\frac{3}{4}$-in.-diameter high-strength bolts.

(b) L6 × 4 × $\frac{5}{8}$, short leg connected with $\frac{7}{8}$-in.-diameter high-strength bolts.

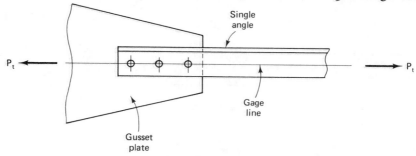

Problem 4-5

4-6. Compute the tensile capacity for the angles and conditions shown. Use A36 steel.
(a) L5 × 3 × $\frac{3}{8}$, long leg connected with $\frac{7}{8}$-in.-diameter high-strength bolts.
(b) L6 × 4 × $\frac{5}{8}$, long leg connected with $\frac{3}{4}$-in.-diameter high-strength bolts.

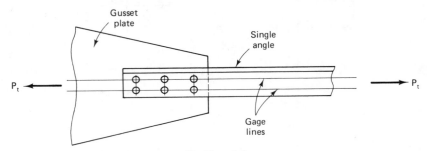

Problem 4-6

4-7. A truss tension member is to be composed of a W6 × 15 and will be connected with two gage lines of $\frac{3}{4}$-in.-diameter bolts in each flange, as shown in Fig. 4-7. Assume three bolts per gage line and A36 steel. Compute the tensile capacity P_t.

4-8. Compute the tensile capacity for the double-angle member shown. The bolts are $\frac{7}{8}$-in. diameter and the steel is A36.

4-9. Compute the tensile capacity for the double-angle member shown. The bolts are $\frac{3}{4}$-in. diameter and the steel is A441. Use standard gages as per the AISCM.

4-10. Compute the tensile capacity for the main member double channel (two C10 × 20) shown. The channels are $\frac{1}{2}$ in. back to back and are 16 ft–0 in. long. Check the slenderness ratio. The steel is A36.

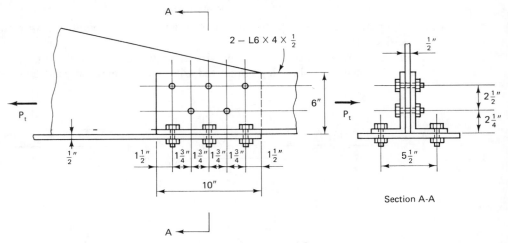

Problem 4-8

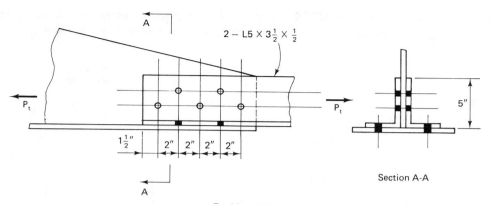

Problem 4-9

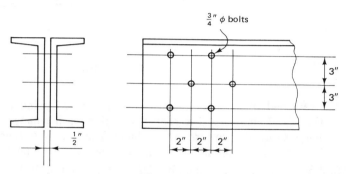

Problem 4-10

4-11. Using the AISCS, determine the maximum recommended unbraced length for an A36 single-angle main tension member $L6 \times 4 \times \frac{1}{2}$.

4-12. Two A36 $L5 \times 3\frac{1}{2} \times \frac{1}{2}$ are connected to a gusset plate with $\frac{3}{4}$-in.-diameter bolts, as shown. Compute the tensile capacity P_t.

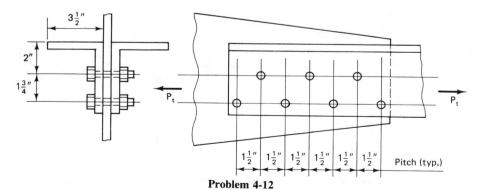

Problem 4-12

4-13. Using the AISCS, compute the maximum recommended unbraced length for the A36 main tension members having cross sections as shown.

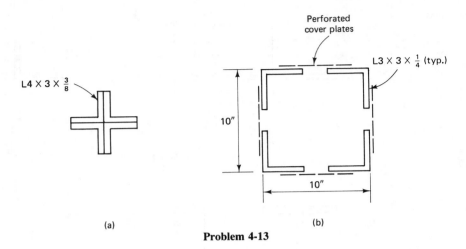

(a)   (b)

Problem 4-13

4-14. A welded main tension member is fabricated using one S5 × 10 and two C5 × 6.7 channels. The shapes are joined at the flange tips with the channel flanges turned in to form a box shape. The steel is A36 and the member length is 32 ft–0 in. Compute the tensile load capacity and check the slenderness ratio.

4-15. For the truss shown, member *DE* is composed of 2L4 × 3 × $\frac{1}{4}$. All joints are assumed to be pin-connected with one gage line of $\frac{3}{4}$-in.-diameter high-strength bolts. The steel is A36 and at each end of *DE* assume a minimum of three bolts per gage line. Is *DE* adequate?

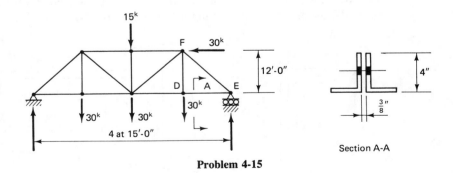

Section A-A

Problem 4-15

4-16. Determine if the A36 double-angle tension member shown can resist a tensile load of 64 kips. The bolts are $\frac{3}{4}$-in. diameter.

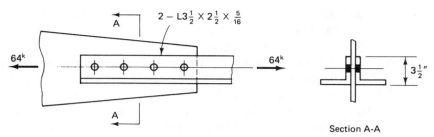

Problem 4-16

4-17. Select a W8 shape to resist a tensile load of 120 kips. The member is to be A36 steel and 27 ft–0 in. long. Assume a main member with end connections consisting of two gage lines of $\frac{3}{4}$-in.-diameter bolts per flange. Assume a minimum of three bolts per gage line.

4-18. Select a W shape to resist a tensile load of 250 kips. Use A36 steel. The member length is 25 ft–0 in. Assume a main member with end connections consisting of two gage lines of $\frac{3}{4}$-in.-diameter bolts per flange. Assume a minimum of three bolts per gage line.

4-19. Select the lightest A36 single-angle main tension member using $\frac{7}{8}$-in.-diameter high-strength bolts. Assume an end connection consisting of one gage line with a minimum of three bolts.
 (**a**) Tensile load = 55 kips, length = 10 ft–6 in.
 (**b**) Tensile load = 42 kips, length = 12 ft–0 in.

4-20. Rework Problem 4-19 but select a double-angle member, placing angles $\frac{3}{8}$ in. back to back. The cross section is similar to that shown for Problem 4-16.

4-21. Select a square tube to serve as a tension member. Use A501 steel with $F_y = 36$ ksi. The tensile load is 75 kips and the member length is 21 ft–0 in. Assume a main member with welded end connections.

4-22. Select a main double-angle tension member like that shown to resist a load $P = 180$ kips. Use A36 steel and $\frac{7}{8}$-in.-diameter high-strength bolts. The length of the member is 15 ft–0 in.

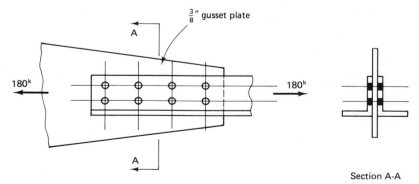

Problem 4-22

4-23. Select an A36 threaded tie rod to resist a tensile load of 15 kips.
   (a) Use standard threaded ends.
   (b) Use upset threaded ends.

4-24. A hanging storage bin weighing 52 tons is to be suspended by three circular threaded rods. Select suitable A36 steel rods and specify the rod diameter and recommended threads.

4-25. An overhead walkway is suspended by four circular threaded rods. The maximum load carried by each rod is 17.5 kips. The rods are 15 ft–0 in. long and are to be A36 steel. Select suitable rods if
   (a) Standard threaded ends are used.
   (b) Upset threaded ends are used.

4-26. Cross-bracing is to stabilize a steel frame, as shown. Select a suitable round threaded rod of A36 steel and specify recommended threads. (Assume pinned connections for beams and columns.)

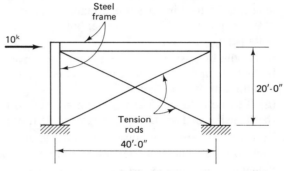

Problem 4-26

5 Axially Loaded Compression Members

5-1 INTRODUCTION

Structural members that carry compressive loads are sometimes given names which identify them as to their function. Compression members that serve as bracing are commonly called **struts.** Other compression members may be called **posts** or **pillars.** Trusses are composed of members that are in compression and members that are in tension. These may be either chord or web members. Of primary interest in this chapter will be the main vertical compression members in building frames which are called **columns.** Additionally, double-angle compression members will be discussed. Columns are compression members which have their length dimension considerably larger than their least cross-sectional dimension.

In this chapter we consider members which are subjected to **axial** (concentric) **loads**; that is, the loads are coincident with the longitudinal centroidal axis of the member. This is a special case and one that exists rarely, if at all. However, where *small* eccentricities exist, it may be assumed that an appropriate factor of safety will compensate for the eccentricity and the column may be designed as though it were axially loaded. Columns may support varying amounts of axial load and bending moment. If we consider the range of possible combinations of load and moment supported on columns, then at one end of the range is the axially loaded column. This column carries no moment. At the other end of the range is the member that carries

These wide-flange columns are being raised for a large two-story test facility building. Note that beams are already positioned in readiness to be lifted into place as the next step in the erection sequence. (Courtesy of the IBM Corporation.)

only moment with no (or very little) axial load. (As a moment-carrying member, it could be considered to be a **beam.**) When a column carries both axial load and moment, it is called a **beam-column.** Beam-columns are considered in Chapter 6.

Commonly used cross sections for steel compression members include most of the rolled shapes. For larger loads it is common to use a built-up cross section. In addition to providing increased cross-sectional area, the built-up sections allow a designer to tailor to specific needs the radius of gyration (r) values about the x-x and y-y axes. Typical compression member cross sections are shown in Fig. 5-1. The dashed lines shown on the cross sections of Fig. 5-1(f) and (g) represent tie plates, lacing bars, or perforated cover plates and do not contribute to the cross-sectional properties. Their function is to hold the main longitudinal components of the cross section in proper relative position and to make the built-up section act as a single unit.

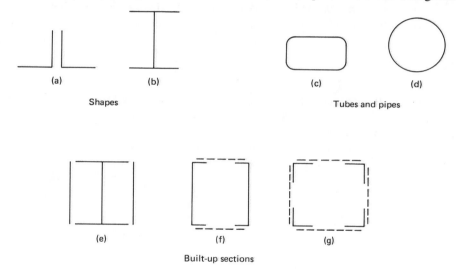

Figure 5-1   Compression member cross sections.

In dealing with compression members, the problem of **stability** is of great importance. Unlike tension members, where the load tends to hold the members in alignment, compression members are very sensitive to factors that may tend to cause lateral displacements or buckling. The situation is similar in ways to the lateral buckling of beams. The buckling problem is intensified and the load-carrying capacity is affected by such factors as eccentric load, imperfection of material, and initial crookedness of the member. Also, **residual stresses** play a role. These are the variable stresses that are "locked up" in the member as a result of the method of manufacture, which involves unequal cooling rates within the cross section. The parts that cool first (such as the flange tips) will have residual compression stresses, while parts that cool last (the junction of the flange and web) will have residual tension stresses. Residual stresses may also be induced by nonuniform plastic deformation caused by cold working, such as in the straightening process and the cambering process.

5-2 IDEAL COLUMNS

The development of the theory of elastic column behavior took place long ago. Leonard Euler (1707–1783), a Swiss mathematician, is credited with many significant contributions in the field of Newtonian mechanics, among them the derivation of the elastic column buckling formula. His theory was presented in 1759 and is still the basis for the analysis and design of slender columns. The buckling of a long or slender column may be demonstrated by loading an ordinary wooden yardstick in compression. The yardstick will buckle (constituting failure) as shown in Fig. 5-2(c), but will not fracture if the load is not increased beyond the buckling load. Upon release of the load, it will return to its original shape. Simply stated, Euler's formula gives the buckling load P_e for a pin-ended, homogeneous, initially straight, long column of an elastic material which is concentrically loaded. This is considered to be the ideal column. The Euler buckling load was expressed as

$$P_e = \frac{\pi^2 EI}{\ell^2}$$

where P_e = concentric load that will cause initial buckling
π = mathematical constant (3.1416)
E = material modulus of elasticity
I = least moment of inertia of the cross section
ℓ = length of the column from pin end to pin end

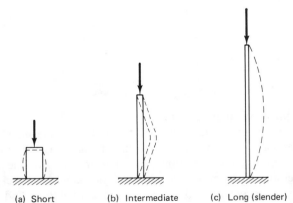

(a) Short   (b) Intermediate   (c) Long (slender)   **Figure 5-2** Column types and failure modes.

Tests have verified that the Euler formula accurately predicts buckling *load*, where the buckling *stress* is less than (approximately) the proportional limit of the material and adherance to the basic assumptions is maintained. Since the buckling stress must be compared with the proportional limit, Euler's formula is commonly written in terms of stress. This may easily be derived from the preceding buckling load formula, recognizing that $I = Ar^2$.

$$f_e = \frac{\pi^2 E}{(\ell/r)^2}$$

where f_e = uniform compressive stress at which initial buckling occurs.
   r = least radius of gyration of the cross section $\sqrt{I/A}$, where A is the gross cross-sectional area

It will be recalled that ℓ/r is termed the **slenderness ratio.**
   It is convenient to classify columns into three broad categories according to their modes of failure. These are shown in Fig. 5-2. Long or slender columns [Fig. 5-2(c)] have already been discussed. They fail by elastic buckling, where buckling occurs at compressive stresses within the elastic range. A very short and stocky column, as shown in Fig. 5-2(a), will obviously not fail by elastic buckling. It will crush or squash due to general yielding and compressive stresses will be in the inelastic range. If yielding is the failure criterion, the failure load may be determined as the product of F_y and cross-sectional area. This column is called a **short column.** A column that falls between these two extremes, as shown in Fig. 5-2(b), will fail by inelastic buckling when a localized yielding occurs. This will be initiated at some point of weakness or crookedness. This type of column is called an **intermediate column.** Its failure strength cannot be determined using either the elastic buckling criterion of the long column or the yielding criterion of the short column. It will be designed and analyzed using empirical formulas based on extensive test results.
   As a review of column behavior as treated in strength of materials, ideal columns will first be analyzed and designed using Euler's formula. This is not to be construed as the modern and practical approach to column analysis and design, but merely as a basis for the more modern methods. The AISC approach will follow later in this chapter.

Example 5-1

   Determine the Euler buckling load (P_e) for an axially loaded W14 × 22 shown in Fig. 5-3. The column has pinned ends. Assume A36 steel with a proportional limit of 34 ksi. The column length is 12 ft.

   Solution   Properties of the W14 × 22:

$$A = 6.49 \text{ in.}^2$$

$$r_y = 1.04 \text{ in.}$$

$$I_y = 7.00 \text{ in.}^4$$

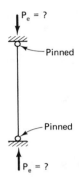

Figure 5-3 Euler column analysis.

Solve for the buckling stress:

$$f_e = \frac{\pi^2 E}{(\ell/r)^2} = \frac{\pi^2(29{,}000)}{[(12 \times 12)/1.04]^2} = 14.93 \text{ ksi}$$

14.94 ksi < 34 ksi (Euler's formula applies)

$$P_e = f_e A = 14.93(6.49) = 96.9 \text{ kips}$$

or

$$P_e = \frac{\pi^2 EI}{\ell^2} = \frac{\pi^2(29{,}000)(7.00)}{(12 \times 12)^2} = 96.6 \text{ kips}$$

5-3 EFFECTIVE LENGTHS

The Euler formula gives the buckling load for a column that has pinned ends. A practical column, in addition to being nonperfect in other respects, may have end conditions (end supports) which provide restraint of some magnitude and will not allow the column ends to rotate freely. Figure 5-4(a) and (b) show column supports which offer very little resistance to column end rotation. These are essentially pinned end supports. The supports shown in Fig. 5-4(c), (d), and (e) will provide significant resistance to column end rotation. The use of the Euler formula may be extended to columns having other than pinned ends through the use of an **effective length.** This concept is illustrated in Fig. 5-5, where a column having rigid (or fixed) ends is shown. The deflected shape of the buckled column is shown in Fig. 5-5(a). Inflection points will exist at the quarter points of the column length. These are points of zero moment and may be theoretically replaced with pins without affecting the equilibrium or the deflected shape of the column. If the central portion of the fixed-ended column is considered separately, as in Fig. 5-5(b), it is seen that it behaves as a pin-ended column of length $\ell/2$. The Euler critical load for the fixed-ended column is then seen to be the same as for a pin-ended column of length $\ell/2$. The length $\ell/2$ is said to be the effective length of the fixed-ended column and the **effective length factor** K is $\frac{1}{2}$

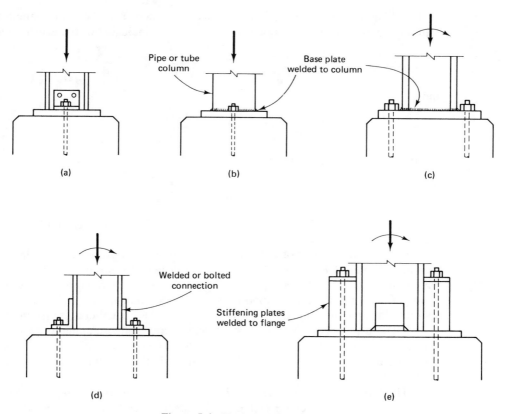

Figure 5-4 Typical column supports.

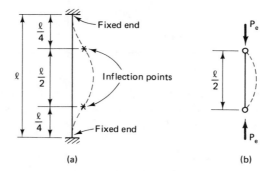

Figure 5-5 Column effective length.

or 0.50. The effective length is written as $K\ell$, where ℓ is the actual length of the column. Euler's formula may be rewritten with the inclusion of the effective length:

$$P_e = \frac{\pi^2 EI}{(K\ell)^2}$$

and

$$f_e = \frac{\pi^2 E}{(K\ell/r)^2}$$

For the fixed-ended column just discussed,

$$P_e = \frac{\pi^2 EI}{(0.5\ell)^2} = \frac{4\pi^2 EI}{\ell^2}$$

It is seen that the buckling load is increased by a factor of 4 when rigid end supports are furnished for a column.

Other combinations of column end conditions are covered in the AISCS Commentary. Table C1.8.1 provides theoretical K values for six idealized conditions in which joint rotation and translation are either fully realized or nonexistent. Since there is no perfectly rigid column support and no perfect pin support, the referenced table also provides recommended design values for K where ideal conditions are approximated. These values are slightly higher than the ideal values and therefore are conservative (the predicted P_e will be on the low side). The reader should carefully study the end condition criteria and the buckled shapes of the columns as shown in Table C1.8.1.

The Euler formula for buckling load may also be adapted to result in an expression for an allowable compressive load capacity. This will be termed, for convenience, P_a. A factor of safety (F.S.) is introduced (see Section 1-6 of this text for a discussion of factor of safety):

$$P_a = \frac{P_e}{\text{F.S.}}$$

$$P_a = \frac{\pi^2 EI}{(K\ell)^2(\text{F.S.})}$$

Example 5-2

A W10 × 49 column of A36 steel has end conditions that approximate the fixed-pinned condition (fixed at the bottom, pinned at the top, no sidesway). Assume a proportional limit of 34 ksi.

   (a) If the length of the column is 26 ft, find the allowable compressive load capacity, P_a, using the Euler formula and a factor of safety of 2.0.

(b) What is the minimum length of this column at which the Euler formula would still be valid?

Solution   From the AISCS, Table C1.8.1, K (for design) = 0.80. For the W10 × 49, A = 14.4 in.^{2}, I_y = 93.4 in.^{4}, and r_y = 2.54 in.

(a) Find f_e first and check the applicability of Euler's formula.

$$f_e = \frac{\pi^2 E}{(K\ell/r)^2} = \frac{\pi^2(29,000)}{[0.80(26 \times 12)/2.54]^2} = 29.6 \text{ ksi}$$

29.6 ksi < 34 ksi (Euler's formula applies)

$$P_a = \frac{P_e}{\text{F.S.}} = \frac{f_e A}{\text{F.S.}} = \frac{29.6(14.4)}{2.0} = 213 \text{ kips}$$

(b) Find the length at which f_e equals the proportional limit.

$$f_e = \frac{\pi^2 E}{(K\ell/r)^2}$$

$$\ell = \sqrt{\frac{\pi^2 E}{f_e (K/r)^2}} = \sqrt{\frac{\pi^2(29,000)}{34(0.8/2.54)^2}} = 291 \text{ in.} = 24.3 \text{ ft}$$

Example 5-3

Use the Euler formula to select a W-shape column to support an axial load of 50 kips. The length is 12 feet and the ends are pinned. Use A36 steel with a proportional limit assumed to be 34 ksi. Check the applicability of the Euler formula. Assume a factor of safety = 3.0 (*Note:* not AISC).

Solution   Select a column that has an allowable compressive load capacity P_a of at least 50 kips. Assume that the Euler formula applies.

$$P_a = \frac{\pi^2 E I}{(K\ell)^2 (\text{F.S.})}$$

$$\text{required } I = \frac{P_a (K\ell)^2 (\text{F.S.})}{\pi^2 E} = \frac{50(1.0 \times 12 \times 12)^2(3.0)}{\pi^2(29,000)} = 10.9 \text{ in.}^4$$

Try W6 × 20:

$$I_y = 13.3 \text{ in.}^4$$
$$A = 5.87 \text{ in.}^2$$
$$r_y = 1.50 \text{ in.}$$

Check the applicability of the Euler formula and the capacity of the W6 × 20.

$$f_e = \frac{\pi^2 E}{(K\ell/r)^2} = \frac{\pi^2(29,000)}{(144/1.50)^2} = 31.06 \text{ ksi}$$

31.06 ksi < 34 ksi (Euler's formula applies)

Calculating the buckling load, we have

$$P_e = 31.06(5.87) = 182 \text{ kips}$$

from which the allowable compressive load capacity is

$$P_a = \frac{P_e}{\text{F.S.}} = \frac{182}{3.0} = 60.8 \text{ kips} > 50 \text{ kips} \qquad \textbf{O.K.}$$

5-4 AISCS ALLOWABLE STRESSES FOR COMPRESSION MEMBERS

In the preceding sections, the Euler formula was used to analyze and design columns. In each case, the applicability of the approach was checked. In each case, the Euler formula did result in f_e less than the proportional limit. In effect, all the columns were slender columns. Practical columns, however, generally do not fall into this category. The practical analysis/design method must concern itself with the entire possible range of slenderness ratio $K\ell/r$. Theoretical formulas are not applicable for intermediate and short columns because of many material and geometric uncertainties. Their strengths cannot be predicted accurately theoretically; therefore, the results of extensive testing and experience must be utilized.

Figure 5-6 shows a plot of the failure stresses of columns versus their $K\ell/r$ ratios

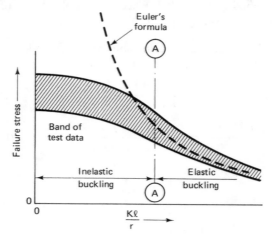

Figure 5-6 Column test data.

as determined by testing. Since no two practical columns are identical, the failure stresses are expected to fall within a range of values for a particular $K\ell/r$ value. Columns with $K\ell/r$ values to the right of line A-A have their failure stresses closely predicted by the Euler formula. They are subject to elastic buckling where the buckling occurs at a stress less than the proportional limit. Columns with $K\ell/r$ values to the left of line A-A fail by inelastic buckling (yielding occurs), and a departure of the test data from the curve that represents the Euler formula is noted.

The AISCS allowable stresses for compression members, as found in Section 1.5.1.3, may be shown diagrammatically as in Fig. 5-7. The maximum $K\ell/r$ is limited to 200 for compression members. The shape of the curve closely follows the shape of the curve of Fig. 5-6. It is essentially the same curve with a factor of safety applied. Allowable axial compressive stress on the gross section is denoted F_a. The value of $K\ell/r$ that separates elastic buckling from inelastic buckling has been arbitrarily established as that value at which the Euler buckling stress (f_e) is equal to $F_y/2$. This $K\ell/r$ value is denoted as C_c. Its value may be determined as follows:

$$f_e = \frac{\pi^2 E}{(K\ell/r)^2}$$

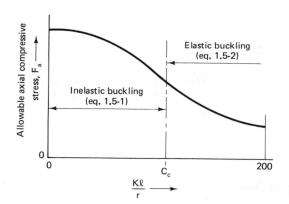

Figure 5-7   AISCS F_a versus $\dfrac{k\ell}{r}$.

Let $f_e = F_y/2$ and let $K\ell/r = C_c$; then

$$\frac{F_y}{2} = \frac{\pi^2 E}{(C_c)^2}$$

from which

$$C_c = \sqrt{\frac{2\pi^2 E}{F_y}}$$

Table 5 in Appendix A of the AISCS lists values of C_c for various values of F_y.

For column $K\ell/r$ values less than C_c, F_a is determined by

$$F_a = \frac{\left[1 - \dfrac{(K\ell/r)^2}{2C_c^2}\right]F_y}{\dfrac{5}{3} + \dfrac{3(K\ell/r)}{8C_c} - \dfrac{(K\ell/r)^3}{8C_c^3}} \qquad \text{AISCS formula 1.5-1}$$

For column $K\ell/r$ values greater than C_c,

$$F_a = \frac{12\pi^2 E}{23(K\ell/r)^2} \qquad \text{AISCS formula 1.5-2}$$

This is the familiar Euler formula for buckling stress with a factor of safety of $23/12$ or 1.92 incorporated. For axially loaded bracing and secondary members, when ℓ/r exceeds 120 (K is taken as 1.0),

$$F_{as} = \frac{F_a[\text{by formula } 1.5\text{-}1 \text{ } or \text{ } 1.5\text{-}2]}{1.6 - (\ell/200r)}$$

where F_{as} denotes the allowable axial compressive stress for secondary members.

Fortunately, the AISCM contains tables which are most useful in the determination of F_a. See Tables 3-36 and 3-50 in Appendix A.

5-5 ANALYSIS OF COLUMNS (AISC)

Several example problems will demonstrate the analysis method using the AISCS and available tables. It should be noted that, in accordance with the AISCS, all column analysis and design is based on the *gross* cross-sectional area of the column.

Example 5-4

Find the allowable compressive load capacity P_a for a W12 × 120 column which has a length of 16 ft. Use A36 steel. The ends are pinned.

Solution　For the W12 × 120:

$$A = 35.3 \text{ in.}^2$$

$$r_y = 3.13 \text{ in.}$$

$$\frac{K\ell}{r} = \frac{1.0(16)(12)}{3.13} = 61$$

The $K\ell/r$ value has been rounded to the nearest whole number for table use. Interpolation is not considered to be warranted. From the AISCS, Appendix A, Table 3-36, $F_a = 17.33$ ksi. Therefore,

$$P_a = F_a A = 17.33(35.3) = 612 \text{ kips}$$

Example 5-5

A W10 $\times$ 68 column of A36 steel is to carry an axial load of 300 kips. The length is 20 ft. Determine if the column is adequate if
(a) The ends are pinned.
(b) The ends are fixed.

Solution   For the W10 $\times$ 68:

$$A = 20.0 \text{ in.}^2$$

$$r_y = 2.59 \text{ in.}$$

(a) $K = 1.0$ from the AISCS, Table C1.8.1, and

$$\frac{K\ell}{r} = \frac{1.0(20)(12)}{2.59} = 93$$

from Table 3-36, $F_a = 13.84$ ksi:

$$P_a = F_a A = 13.84(20.0) = 277 \text{ kips}$$

$$277 \text{ kips} < 300 \text{ kips} \qquad\qquad \textbf{N.G.}$$

(b) $K = 0.65$ from the AISCS, Table C1.8.1, and

$$\frac{K\ell}{r} = \frac{0.65(20)(12)}{2.59} = 60$$

from Table 3-36, $F_a = 17.43$ ksi:

$$P_a = F_a A = 17.43(20.0) = 349 \text{ kips}$$

$$349 \text{ kips} > 300 \text{ kips} \qquad\qquad \textbf{O.K.}$$

Example 5-6

Find the compressive axial load capacity for a built-up column that has a cross section as shown in Fig. 5-8. The steel is A36, the length is 18 ft, and the ends are assumed to be fixed-pinned (totally fixed at bottom; rotation free, translation fixed at top). The column is a main member.

Solution   Properties of the W12 $\times$ 65:

$$A = 19.1 \text{ in.}^2$$

$$d = 12.12 \text{ in.}$$

$$b_f = 12.00 \text{ in.}$$

$$I_x = 533 \text{ in.}^4$$

$$I_y = 174 \text{ in.}^4$$

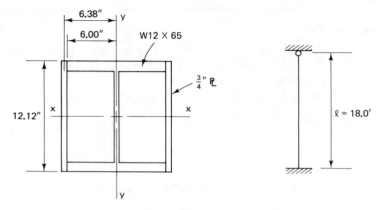

Figure 5-8 Built-up column.

Determine the least moment of inertia for the built-up cross section.

$$I = \sum I_c + \sum Ad^2$$

$$I_x = 533 + 2\left(\frac{1}{12}\right)(0.75)(12.12)^3 = 756 \text{ in.}^4$$

$$I_y = 174 + 2(0.75)(12.12)(6.38)^2 = 914 \text{ in.}^4$$

The x-x axis controls since its moment of inertia is smaller. Note in the I_x calculation that the Ad^2 terms are zero for both the W shape and the plates since the centroidal axes of these component parts coincide with the composite centroidal axis. In the I_y calculation, the I_c term for the plate has been neglected since it is very small. Calculating the radius of gyration, we have

$$\text{total } A = 19.1 + 2(12.12)(0.75) = 37.28 \text{ in.}^2$$

$$r_x = \sqrt{\frac{I_x}{A}} = \sqrt{\frac{756}{37.28}} = 4.50 \text{ in.}$$

With an effective length factor K of 0.80, the capacity may be calculated as usual:

$$\frac{K\ell}{r_x} = \frac{0.8(18)(12)}{4.50} = 38.4 \qquad \text{(use 38.0)}$$

Reference to Table 3-36 of the AISCM, Appendix A, yields $F_a = 19.35$ ksi:

$$P_a = AF_a = 37.28(19.35) = 721 \text{ kips}$$

Columns are sometimes braced differently about the major and minor axes as shown by column AB in Fig. 5-9. If all connections to the column are assumed to be

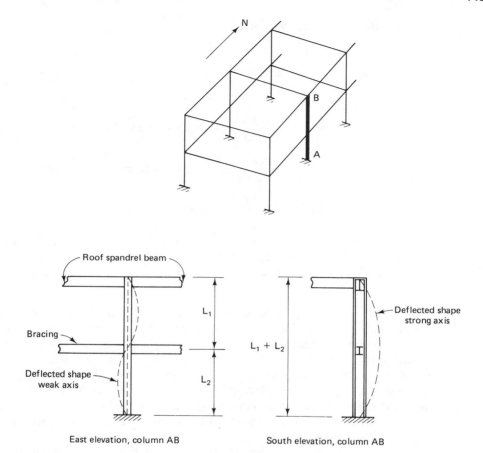

Figure 5-9 Column unbraced lengths.

simple (pinned) connections, the deflected shapes for buckling about the two axes will be as shown. Note that the column is braced so that the unbraced length for weak axis buckling is less than the unbraced length for buckling about the strong axis. In this situation, either axis may control depending on which has the associated larger $K\ell/r$ ratio. Naturally, if there is no reasonable certainty that the bracing will not be removed, the column should be designed with the bracing neglected.

Example 5-7

Find the allowable compressive axial load capacity for a W10 × 88 that has an unbraced length of 24 ft with respect to axis x-x and 12 ft with respect to axis y-y. Assume an A36 steel main member, pin-connected at the top and fixed at the bottom. (Assume that the column is pin-connected at midheight.)

Solution   For the W10 × 88:

$$A = 25.9 \text{ in.}^2$$

$$r_y = 2.63 \text{ in.}$$

$$r_x = 4.54 \text{ in.}$$

$$\frac{K\ell_y}{r_y} = \frac{1(12)(12)}{2.63} = 54.8 \text{ (top part of column)}$$

$$= \frac{0.8(12)(12)}{2.63} = 43.8 \text{ (bottom part of column)}$$

$$\frac{K\ell_x}{r_x} = \frac{0.8(24)(12)}{4.54} = 50.8$$

Since 54.8 is the controlling slenderness ratio, $F_a = 17.90$ ksi (using $K\ell/r = 55.0$). $P = AF_a = 25.9(17.9) = 463.6$ kips.

A common type of column used in one-story commercial construction is the unfilled circular steel pipe column. With equal stiffness in all directions, it is an efficient compression member. However, connections to the pipe column may require special considerations. Three categories of pipe for structural purposes are manufactured: standard, extra strong, and double-extra strong. Steel pipe may be manufactured to ASTM A501 or ASTM A53 (Type E and S, Grade B). For design purposes the yield stress for each steel may be taken as $F_y = 36$ ksi.

Square and rectangular structural tubing are also commonly used as building columns. The tubing is manufactured to $F_y = 46$ ksi under ASTM A500 Grade B. The tubular members are also relatively efficient and have an advantage in that end-connection details are simpler than with the pipe columns.

Example 5-8

Find the allowable compressive load capacity for a 10-in. standard steel pipe column which has an unbraced length of 15 ft. Ends are pin-connected ($K = 1$), and the steel is A501 ($F_y = 36$ ksi).

Solution   For the 10-in. standard steel pipe (AISCM, Part 1),

$$A = 11.9 \text{ in.}^2$$

$$r = 3.67 \text{ in.}$$

$$\frac{K\ell}{r} = \frac{1(15)(12)}{3.67} = 49$$

$$F_a = 18.44 \text{ ksi}$$
$$P_a = AF_a = 11.9(18.44) = 219 \text{ kips}$$

Example 5-9

Find the allowable compressive load capacity for an 8-in. double-extra-strong steel pipe column which has an unbraced length of 20 ft. The ends are pin-connected, and the steel is A501 ($F_y = 36$ ksi).

Solution   For the 8-in. double-extra-strong steel pipe (AISCM, Part 1):

$$A = 21.3 \text{ in.}^2$$
$$r = 2.76 \text{ in.}$$
$$\frac{K\ell}{r} = \frac{1(20)(12)}{2.76} = 87$$
$$F_a = 14.56 \text{ ksi}$$
$$P_a = AF_a = 21.3(14.56) = 310 \text{ kips}$$

Example 5-10

Find the allowable compressive load capacity for a rectangular structural tubing, TS8 × 4 × $\frac{5}{16}$, which has an unbraced length of 13 ft. The ends are pin-connected, and the steel is A500 Grade B ($F_y = 46$ ksi).

Solution   For the TS8 × 4 × $\frac{5}{16}$ (AISCM, Part 1):

$$A = 6.86 \text{ in.}^2$$
$$r_y = 1.62 \text{ in.}$$
$$r_x = 2.80 \text{ in.}$$
$$\frac{K\ell}{r} = \frac{1(13)(12)}{1.62} = 96.3$$

F_a cannot be determined directly using the AISCM since F_a tables are available only for $F_y = 36$ ksi and 50 ksi. However, Tables 4 and 5 of Appendix A of the AISCS may be used.

From Table 5, for $F_y = 46$ ksi, $C_c = 111.6$. From Table 4, enter with the ratio

$$\frac{K\ell/r}{C_c} = \frac{96.3}{111.6} = 0.86$$

and select $C_a = 0.330$.

F_a may then be determined by

$$F_a = C_a F_y$$

$$= 0.330(46) = 15.18 \text{ ksi}$$

$$P_a = AF_a = 6.86(15.18) = 104 \text{ kips}$$

Also note that AISCS formula 1.5-1 could be used for F_a.

Example 5-11

Find the allowable compressive load capacity for a W8 × 40 with an unbraced length = 26 ft. The member is used in a wind-bracing system and is pin-connected (secondary member). Use A36 steel.

Solution   For the W8 × 40:

$$A = 11.7 \text{ in.}^2$$

$$r_y = 2.04 \text{ in.}$$

$$\frac{K\ell}{r} = \frac{1(26)(12)}{2.04} = 153$$

$$F_{as} = 7.64 \text{ ksi}$$

$$P_a = AF_{as} = 11.7(7.64) = 89.4 \text{ kips}$$

5-6 DESIGN OF AXIALLY LOADED COLUMNS

The selection of cross sections for columns is greatly facilitated by the availability of design aids. We have seen that the allowable axial stress F_a depends on the effective slenderness ratio $K\ell/r$ of the column provided. Therefore, there is no direct solution for a required area or moment of inertia.

The majority of structural steel columns are composed of W shapes, structural tubing, and steel pipes. The AISCM, Part 3, contains allowable axial load tables (which will be referred to as the "column load tables") for the popular column shapes. Allowable loads (P_a) are tabulated as a function of KL (in feet) and cover the common length ranges. The actual column length L for columns in building frames is normally taken as the floor-to-floor distance. The effective length factor K may be determined using the aids discussed in Section 5-3 of this text.

The tables may be used for analysis as well as for design. For instance, in Example 5-4, the allowable compressive load capacity of a W12 × 120 of A36 steel was computed to be 612 kips. From the AISCM, Part 3, column load table for the W12 × 120, with $KL = 1.0 \times 16$ ft., the allowable load of 611 kips may be obtained directly. (Note that the unshaded areas are for $F_y = 36$ ksi and the shaded areas are for $F_y = 50$ ksi.) The General Notes at the beginning of the load tables discuss use and limitations.

The tabular values of allowable loads are with respect to the members' minor (or weak) axis. While the column load tables are indispensable for the selection of the types of cross sections noted, if built-up sections are required (see Fig. 5-1), or a section is desired for which a column load table is not available, a trial-and-error calculation approach will have to be used.

Example 5-12

Select the lightest W shape for a column that will support an axial load P of 200 kips. The length of the column will be 20 ft and the ends may be assumed to be pinned. Use A36 steel.

Solution   Using the AISCM, Part 3, column load tables, with $KL = 1.0 \times 20$ ft and $P = 200$ kips, the following W shapes are observed to be adequate ($P_a \geq P$):

$$W\,14 \times 61 \qquad (P_a = 237 \text{ kips})$$
$$W\,12 \times 53 \qquad (P_a = 209 \text{ kips})$$
$$W\,10 \times 54 \qquad (P_a = 217 \text{ kips})$$
$$W\,8 \times 67 \qquad (P_a = 221 \text{ kips})$$

The W12 × 53 is selected since it is the lightest shape with adequate capacity.

Example 5-13

Select the lightest W10 for column AB shown in Fig. 5-9, $P = 160$ kips. The overall length (L_x) is 30 ft. The weak axis is braced at midheight ($L_y = 15$ ft). Assume pinned ends ($K = 1.0$) for both axes and A36 steel.

Solution   Assume that the weak axis (y-y axis) will control. Select the column using the AISCM, Part 3, column load tables and then check whether the assumption is correct.

$$KL_y = 15 \text{ ft} \qquad P = 160 \text{ kips}$$

Select a W10 × 39 ($P_a = 162$ kips based on weak axis buckling). For the W10 × 39, $r_x = 4.27$, $r_y = 1.98$,

$$\frac{K\ell_x}{r_x} = \frac{1.0(30)(12)}{4.27} = 84.3$$

$$\frac{K\ell_y}{r_y} = \frac{1.0(15)(12)}{1.98} = 90.9$$

The larger $K\ell/r$ controls and the assumption of the weak axis controlling was correct. A W10 × 39 will be adequate.

Under different conditions, it is possible that the strong axis will control and be the buckling axis for the column of the preceding problem. The column load tables then cannot be used directly. However, once an initial section has been selected (based on the assumption that the y-y axis controls) a very rapid analysis check can be made using the tabulated properties at the bottom of the column load tables. The procedure is as follows:

1. Divide the strong-axis effective length (KL_x) by the r_x/r_y ratio.
2. Compare with KL_y. The larger of the two values becomes the controlling KL.
3. With the controlling KL value, find P_a in the appropriate column load table.

Example 5-14

Rework Example 5-13 except that the weak axis is braced at the third points so that $L_y = 10$ ft. L_x remains at 30 ft.

Solution   As was done previously, select on the basis of the weak axis controlling:

$$KL_y = 10 \text{ ft} \qquad P = 160 \text{ kips}$$

Try a W10 × 33. $P_a = 167$ kips (based on weak axis controlling) and $r_x/r_y = 2.16$.

$$\frac{KL_x}{r_x/r_y} = \frac{30 \text{ ft}}{2.16} = 13.89 \text{ ft}$$

This is an **equivalent weak-axis length** (i.e., column length based on weak-axis buckling which results in the same capacity as does the 30-ft strong-axis buckling length). *Since 13.89 ft > L_y, the strong axis controls.* Rounding the KL of 13.89 ft to 14 ft and entering the column load table for the W10 × 33, $P_a = 142$ kips.

$$142 \text{ kips} < 160 \text{ kips} \qquad\qquad \textbf{N.G.}$$

Try W10 × 39.

$$\frac{KL_x}{r_x/r_y} = \frac{30 \text{ ft}}{2.16} = 13.89 \text{ ft} \approx 14 \text{ ft}$$

$$13.89 \text{ ft} > L_y$$

Therefore, the strong axis controls. From the column load table, $P_a = 170$ kips > 160 kips. Therefore, **use a W10 × 39.**

5-7 DOUBLE-ANGLE MEMBERS

Double-angle members are frequently used as both tension and compression members in single-plane trusses. They may be separated by the thickness of a gusset plate at each end but must be connected at intervals along their length by filler plates and welds or bolts in accordance with AISCS requirements. The AISCM contains special tables for double-angle members. The use of the tables will be illustrated with several examples.

The AISCS, Section 1.9.1.2, stipulates that double-angle struts with separators subject to axial compression may be considered fully effective in resisting the applied load provided that the ratio of each element's width to thickness does not exceed $76/\sqrt{F_y}$. When the two angles are in contact with each other, the ratio of width to thickness may not exceed $95/\sqrt{F_y}$. If these ratios are exceeded, the allowable compressive stress must be modified using a reduction factor Q_s. More information on Q_s is given in the AISCS, Appendix C. The allowable compressive stress may then be determined by the expression

$$F_a = Q_s C_a F_y$$

where Q_s = empirical reduction factor which is used when the width thickness ratio of the unstiffened elements exceeds the limiting values previously discussed; the Q_s value for double-angle members is provided in the AISCM, Part 1, tables of properties for double angles

C_a = coefficient based on the ratio $(K\ell/r)/C_c'$ and obtained from the AISCS, Appendix A, Table 4

F_y = is as previously defined

Where the value of Q_s is indicated by a dash (—) in the double-angle properties tables, the allowable compressive stress may be determined in the usual manner based on the member $K\ell/r$ and using Table 3 of the AISCS, Appendix A, for F_a.

Where Q_s is furnished with a numerical value in the double-angle tables, the procedure to determine F_a is as follows:

1. Compute maximum $K\ell/r$ for the member.
2. Compute C_c'. The expression for C_c' is given at the bottom of the double-angle tables in the AISCM, Part 1.
3. If $K\ell/r \geq C_c'$, use the standard familiar approach to obtain F_a using the AISCS, Table 3 (Appendix A).

If $K\ell/r < C_c$, compute F_a using the expression $F_a = Q_s C_a F_y$.
C_a may be obtained from Table 4 of the AISCS, Appendix A, based on ratio

$$\frac{K\ell/r}{C_c'}$$

Example 5-15

Determine the capacity of an axially loaded double-angle strut truss (main) member. The strut is composed of 2L8 × 4 × 1/2 with long legs $\frac{3}{8}$ in. back to back and is of A36 steel. The length is 20 ft. Assume that $K = 1.0$.

Solution From the double-angles properties tables of the AISCM, Part 1, the following is obtained:

$$r_x = 2.59 \text{ in.}$$

$$r_y = 1.51 \text{ in.}$$

$$A = 11.5 \text{ in.}^2$$

$$Q_s = 0.911$$

$$C_c' = \frac{126.1}{\sqrt{Q_s}} = 132$$

Following the procedure discussed previously, we have

$$\frac{K\ell}{r} = \frac{1.0(20)(12)}{1.51} = 158.9$$

$$158.9 > C_c'$$

Therefore, obtain F_a directly from the AISCS, Appendix A, Table 3-36.

$$F_a = 5.91 \text{ ksi}$$

from which

$$P_a = F_a A = 5.91(11.5) = 68.0 \text{ kips}$$

Example 5-16

Rework Example 5-15 but using a double-angle strut of length 15 ft.

Solution The properties of the double-angle strut remain the same.

$$\frac{K\ell}{r} = \frac{1.0(15)(12)}{1.51} = 119.2$$

$$119.2 < C_c'$$

Therefore, obtain F_a from

$$F_a = Q_s C_a F_y$$

From the AISCS, Appendix A, Table 4, C_a is determined using the ratio $(K\ell/r)/C_c'$:

$$\frac{K\ell/r}{C_c'} = \frac{119.2}{132} = 0.90$$

$$C_a = 0.311$$

from which

$$F_a = Q_sC_aF_y = 0.911(0.311)(36) = 10.2 \text{ ksi}$$

Therefore,

$$P_a = F_aA = 10.2(11.5) = 117 \text{ kips}$$

The allowable compressive load for double-angle members may also be determined using the AISCM load tables of Part 3. Caution should be exercised in using these tables since allowable axial loads are furnished with respect to both major and minor axes assuming a $\frac{3}{8}$-in. gusset plate thickness. The table may be used for main members with a $K\ell/r \le 200$ and for secondary members with an $\ell/r \le 120$. No reduction factors need to be applied to these values since the tabulated values reflect all AISCS requirements. The reader may wish to compare the results of the previous two analysis problems with allowable compressive loads for the double-angle strut as shown in the load tables.

The tables are of significant value for the design of double-angle members and may best be illustrated with the following examples.

Example 5-17

Design a double-angle truss main member with an unbraced length $L = 12$ ft to carry an axial load $P = 80$ kips. Use A36 steel and long legs back to back. The member is connected to a $\frac{3}{8}$-in.-thick gusset plate at each end.

Solution

$$KL = 1(12) = 12 \text{ ft}$$

This KL value applies to both the x-x and the y-y axes.

Enter the AISCM, Part 3, double-angle tables, with $KL_x = 12$ ft, $KL_y = 12$ ft, and load $P = 80$ kips. Select the most economical member that can safely carry the applied load with respect to each direction.

Angles	P_x	P_y	Weight/ft
$5 \times 3 \times \frac{1}{2}$	106	82	25.6
$5 \times 3\frac{1}{2} \times \frac{1}{2}$	113	107	27.2
$6 \times 3\frac{1}{2} \times \frac{3}{8}$	102	82	23.4
$6 \times 4 \times \frac{3}{8}$	107	97	24.6

Select the lightest (most economical) member: $2L6 \times 3\frac{1}{2} \times \frac{3}{8}$.

5-8 COLUMN BASE PLATES (AXIAL LOAD)

Columns are usually supported on concrete supports such as footings or piers. Since the steel of the column is a higher-strength material than the concrete, the column load must be spread out over the support. This is accomplished by use of a rolled-steel **base plate.** The problem is similar to that of the beam bearing plate discussed in Section 2-13 of this text.

Detail of base plate welded to a wide-flange column.

    Base plates may be square or rectangular, must be large enough to keep the actual bearing pressure under the plate below an allowable bearing pressure, and must be thick enough so that bending in the plate itself will not be critical. Two-way bending is involved since as the column pushes down on the base plate, it tends to deform into a saucer shape. The required plate thickness will be determined by considering 1-in.-wide sections of the base plate to act as cantilever beams spanning in each of two directions, fixed at the edges of a rectangle whose sides are $0.80b_f$ and $0.95d$, as shown in Fig. 5-10. In Fig. 5-10, the notation is as follows:

P = total column load (kips)

$A = B \times N$, area of plate (in.^{2})

m, n = length of cantilever from assumed critical plane of bending, for thickness determination (in.)

d = depth of the column section (in.)

b_f = flange width of the column section (in.)

F_p = allowable bearing pressure on concrete support (ksi)

F_b = allowable bending stress in plate (ksi)

f_p = actual bearing pressure on concrete support (ksi)

f_c' = compressive strength of concrete (ksi)

t_p = thickness of plate (in.)

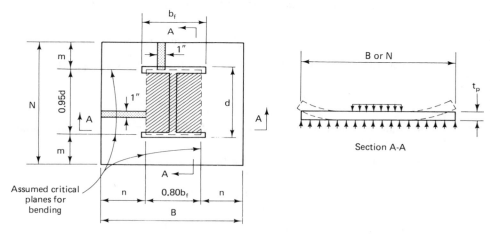

Figure 5-10   Column base plate design.

The column load P is assumed to be uniformly distributed over the top of the base plate within the described rectangle, and in turn it is assumed that the plate is sufficiently rigid and will distribute the applied load so that the pressure underneath the base plate is also uniformly distributed. For light loads the required area of the base plate will be very small and the values of m and n will be close to zero. For such cases, a theoretical analysis of the plate reveals that the highest stress in the base plate will occur at the face of the column web halfway between the insides of the flanges rather than at the edges of the rectangle of load.

The discussion in Section 2-13 of this text concerning F_b and F_p is pertinent and should be reviewed. Also from Section 2-13, the expression

$$\text{required } t_p = 2n\sqrt{\frac{f_p}{F_y}}$$

is applicable. However, since bending occurs in two directions the larger value of m or n (length of cantilever) must be substituted for n. Further, to account for the condition of a small base plate with thickness controlled by bending at the column web, a lower limit, n', is specified for the length of the cantilever m and n. The value of n' may be found in the AISCM, Part 3, Table C. The use of the table is in lieu of a theoretical plate analysis. The larger of the three values of n', m, or n must be used to determine the plate thickness.

The procedure for column base plate design is as follows:

1. Determine the required base plate area:

$$\text{required } A = \frac{P}{F_p}$$

2. Select B and N so that m and n are approximately equal (if possible); B and N to full inches.

$$B \times N \geq \text{required } A$$

3. Calculate the actual bearing pressure on the concrete support:

$$f_p = \frac{P}{BN}$$

4. Calculate m and n (refer to Fig. 5-10). Determine n' from Table C of the AISCM.
5. Calculate the required t_p using the larger of m, n, or n'.

$$\text{required } t_p = 2(m, n, \text{ or } n') \sqrt{\frac{f_p}{F_y}}$$

6. Specify the base plate: width, thickness, length.

The AISCM, Part 3, column base plate design procedure varies from the preceding. In the AISCM procedure, step 1, since F_p may be dependent on the size of the concrete support, with $F_p = 0.7f_c'$ as a maximum, an optimum concrete support size may be determined which will result in the smallest required plate area. In the AISCM procedure, step 2, a method is given which allows N and B to be calculated so that m and n will be equal. Example 5-18 provides a direct and simplified approach to the AISCM column base plate design procedure.

Example 5-18

Design a rectangular base plate for a W14 × 74 column which is to carry a load of 350 kips. Assume that the base plate will cover the full area of a concrete pier of $f_c' = 3$ ksi. Use A36 steel.

Solution   From the AISCS, Section 1.5.5,

$$F_p = 0.35 f_c' = 0.35(3) = 1.05 \text{ ksi}$$

1. The required area is

$$\text{required } A = \frac{P}{F_p} = \frac{350}{1.05} = 333 \text{ in.}^2$$

2. For the W14 × 74, $d = 14.17$ in. and $b_f = 10.07$ in. Referring to Fig. 5-10, it is seen that if B and N are selected so that the difference between the two is approximately

$$0.95d - 0.80b_f$$

$$0.95(14.17) - 0.80(10.07) = 5.4 \text{ in.}$$

then m and n will be approximately equal. Try various values of B and N so that

$$B \times N \geq 333 \text{ in.}$$

B	N	Area
10	34	340
15	23	345
16	21	336

Use $B = 16$ in. and $N = 21$ in.

3. Actual pressure is

$$f_p = \frac{P}{BN} = \frac{350}{16(21)} = 1.04 \text{ ksi}$$

4. Calculate m and n; determine n'.

$$m = \frac{N - 0.95d}{2} = \frac{21 - 0.95(14.17)}{2} = 3.77 \text{ in.}$$

$$n = \frac{B - 0.80b_f}{2} = \frac{16 - 0.80(10.07)}{2} = 3.97 \text{ in.}$$

From the AISCM, Part 3, Table C, for the W14 × 74, $n' = 4.43$ in. The latter (n') controls.

5. Calculate the required t_p.

$$\text{required } t_p = 2n' \sqrt{\frac{f_p}{F_y}} = 2(4.43) \sqrt{\frac{1.04}{36}} = 1.51 \text{ in.}$$

Refer to the AISCM, Part 1, Bars and Plates—Product Availability, for information on plate thicknesses available: use a thickness of $1\frac{5}{8}$ in.

6. **Use a base plate 16 in. × $1\frac{5}{8}$ in. × 1 ft–9 in.**

The length and width of column base plates are usually selected in multiples of full inches and their thickness in multiples of $\frac{1}{8}$ in. if the plate thickness required is between 1 and 3 in. Since good contact between the column and the plate is a necessity, the AISCS requires that rolled steel bearing plates over 2 in. but not over 4 in. in thickness must be straightened by pressing or milling. Plates over 4 in. in thickness must be milled. The bottom surface of the plates need not be milled since a layer of grout will be placed between the plate and the underlying foundation to ensure full bearing contact. Plates 2 in. or less in thickness may be used without milling provided that a satisfactory contact bearing is obtained between the plate and the column.

Steel columns and their base plates are usually anchored to the foundation by steel anchor bolts embedded in the concrete. The anchor bolts pass through the base plate in slightly oversize holes. This allows for some misalignment of the bolts without redrilling the base plate holes or taking out and resetting the anchor bolts. Angles may be used to bolt or weld the base plate to the column. If so, the anchor bolts will also pass through the angles. With the exception of base plates for larger columns, current practice is to omit the angles and shop weld the base plate to the column, thereby permitting the column and base plate to be shipped to the job site as a single unit.

PROBLEMS

Note: In these problems, the columns are main members and the AISCS applies unless specifically noted.

5-1. Two channels C12 × 25 serve as a 36-ft-long pin-ended column as shown. Find the Euler buckling load P_e. The proportional limit is 34 ksi.

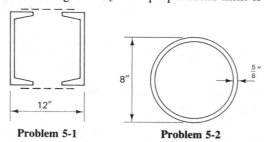

Problem 5-1   **Problem 5-2**

5-2. Calculate P_e for the pipe column shown for the following lengths. $E =$ 29,000 ksi and the proportional limit = 34 ksi

(a) 20 ft.

(b) 30 ft.

(c) 40 ft.

(d) 50 ft.

(e) 15 ft.

Draw a graph of f_e versus ℓ(ft) showing the results of the calculations above. Assume pinned ends.

5-3. A W8 $\times$ 40 is to serve as a pin-ended, 20-ft-long column. The proportional limit is 34 ksi. Find the allowable axial compressive load using the Euler formula and a factor of safety of 2.5.

5-4. Use the Euler formula and a factor of safety of 2.5 to design a W14 column to support a load of 350 kips. The length is 34 ft. The ends are pinned. The proportional limit is 34 ksi.

5-5. A solid square steel bar 2 in. $\times$ 2 in. in cross section is used as a pin-ended column. The proportional limit is 34 ksi. Find the allowable axial compressive load using the Euler formula and a factor of safety of 3.0 if the unbraced length is

(a) 4 ft.

(b) 8 ft.

5-6. A W10 $\times$ 49 is to serve as a pin-ended, 40-ft-long column. The column is braced at midheight with respect to its weak axis. The proportional limit is 34 ksi. Find the allowable axial compressive load using the Euler formula and a factor of safety of 2.5.

5-7. Find the allowable axial compressive load P_a for the following compression members. ASTM A501 steel furnishes $F_y = 36$ ksi.

Shape	K	L(ft)	Steel
(a) W16 $\times$ 50	1.3	16.0	A36
(b) HP12 $\times$ 74	1.0	14.0	A588
(c) Pipe 8 X-Strong	0.8	26.3	A501

5-8. Find the allowable axial compressive load for a W12 $\times$ 50 column of A36 steel. End conditions and the unbraced length are as follows:

(a) Fixed ends, $L = 15$ ft.

(b) Pinned-ends, $L = 15$ ft.

(c) One end pinned, one fixed, $L = 15$ ft.

(d) Pinned ends, $L = 30$ ft.

(e) Pinned ends, $L_x = 15$ ft, $L_y = 7.5$ ft.

5-9. Find the allowable axial compressive load for a W8 $\times$ 40 column of A36 steel. End conditions and the unbraced length are as follows:

(a) Pinned ends, $L = 18$ ft.

(b) Pinned ends, $L = 26$ ft.

(c) Pinned ends, $L_x = 26$ ft, $L_y = 13$ ft.

(d) Pinned ends (secondary member), $L = 26$ ft.

5-10. Find the allowable axial compressive load for a built-up column composed of a W12 × 50 with a 12 × $\frac{1}{2}$ plate bolted to each flange. The member is 30 ft long with pinned ends and is A36 steel.

5-11. Find the allowable axial compressive load for the following compression members:

    (a) W8 × 58, fixed ends, L = 16 ft–6 in., A36 steel.

    (b) W12 × 96, pinned ends, L = 20 ft, A36 steel.

    (c) W14 × 211, pinned ends, L = 20 ft, A441 steel.

    (d) W12 × 87, fixed at one end, pinned at one end, L_x = 24 ft, L_y = 12 ft, A36 steel.

    (e) W10 × 88, fixed at one end, pinned at one end. L = 24 ft–6 in., F_y = 50 ksi.

    (f) W14 × 176, fixed ends, L = 22 ft, F_y = 46 ksi.

    (g) 10-in. standard steel pipe, pinned ends, L = 22 ft, A36 steel.

    (h) W10 × 54, pinned ends, L_x = 22 ft, L_y = 11 ft, A36 steel.

    (i) Structural tubing TS6 × 4 × $\frac{3}{8}$, pinned ends, L = 14 ft, F_y = 46 ksi.

5-12. Find P_a for the built-up column shown. Use A36 steel. The length is 18 ft and the ends are fixed.

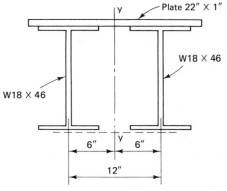

Problem 5-12

5-13. Find P_a for the built-up column shown. Use A36 steel. The length is 19 ft–6 in. and K = 1.0.

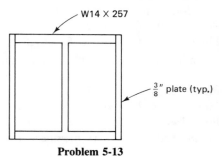

Problem 5-13

(a) Find P_a for the core W14 × 257.

(b) Find P_a for the built-up column.

5-14. Select the lightest shape indicated to serve as an axially loaded column. Use A36 steel (A500 Grade B for tubing, with F_y = 46 ksi).

	Shape	P	Ends	L(ft)
(a)	W	100	Pinned	24
(b)	W	280	Fixed	20
(c)	HP	260	Fixed/pinned	12
(d)	Tube	200	Pinned	15

5-15. Select the lightest shape indicated to serve as an axially loaded column.

	Shape	P	Ends	Length L	Steel
(a)	W	600	Fixed	15	A36
(b)	W	600	Fixed	15	A441
(c)	W	100	Fixed	$L_x = 30, L_y = 15$	A36
(d)	W	100	Pinned	26	A36
(e)	W	275	Fixed/pinned	10	A36
(f)	Std. pipe	41	Pinned	12	A501
(g)	Tube	50	Pinned	12	A500 B
(h)	W	2700	Pinned	15	A36
(i)	W	2700	Pinned	15	A441

5-16. Select the lightest W12 columns of A36 steel to support the axial loads as follows:

(a) P = 400 kips, pinned ends, L_x = 24 ft, L_y = 18 ft.

(b) P = 400 kips, pinned ends, L_x = 24 ft, L_y = 12 ft.

(c) P = 400 kips, pinned ends, L_x = 24 ft, L_y = 8 ft.

(d) P = 700 kips, pinned ends, L_x = 31 ft, L_y = 16 ft.

5-17. Design a typical interior column for the four-story braced frame shown. The

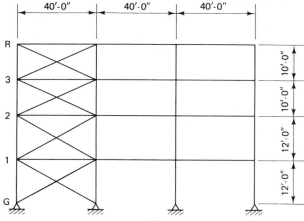

Problem 5-17

frames are 32 ft on center. Assume that $K = 1.0$. Loads may be determined
from the following: roof snow load is 45 psf, roof dead load is 30 psf, the floor
live load is 100 psf and 15 psf for partitions, and the floors are composed of
5-in.-thick reinforced concrete slabs. Include 10 psf dead load for both roof and
floor to account for the weight of beams and girders. Splice the columns at the
second-floor level. A36 steel.

5-18. An axially loaded column has its weak axis braced at the third points. $P = 600$ kips, overall length is 27 ft, $K = 1.0$, and the steel is A36. Select the lightest W shape.

5-19. Select or design the most economical column to support an axial load of 550 kips. $K = 1.3$. $L = 16$ ft. The column must fit into a 12 in. $\times$ 11 in. space. The column is to be A36 steel and a main member.

5-20. Compute the allowable axial compressive load for a double-angle strut consisting of 2L6 $\times$ 4 $\times$ 3/4 long legs back to back, straddling a $\frac{3}{8}$-in. gusset plate. The strut is a main truss member of A36 steel with an unbraced length of
(a) $L = 15$ ft.
(b) $L = 34$ ft.

5-21. Compute the allowable axial compressive load for a double-angle strut consisting of 2L7 $\times$ 4 $\times$ 3/8, long legs back to back, straddling a $\frac{3}{8}$-in. gusset plate. The strut is a main truss member of A36 steel with an unbraced length $L = 8$ ft.

5-22. Determine the maximum permissible unbraced length for a main truss compression member consisting of 2L6 $\times$ 4 $\times$ 1/2, long legs back to back, straddling a $\frac{3}{8}$-in. gusset plate. The steel is A36.

5-23. Select the lightest double-angle truss compression member, long legs back to back, straddling a $\frac{3}{8}$-in. gusset plate for the following conditions:
(a) $P = 100$ kips, A36 steel, $L = 10$ ft.
(b) $P = 32$ kips, A36 steel, $L = 19$ ft (secondary member).
(c) $P = 50$ kips, A36 steel, $L = 8$ ft.

5-24. Design the lightest base plates for the columns and loads given. Plate size to be full inches and thickness to be governed by the AISCM. Determine the *weight* of the plate in each case. Use A36 steel. $F_p = 1050$ psi. Note that the W18 $\times$ 119 is not commonly used as a column.

Column	P (kips)
(a) W12 $\times$ 45	75
(b) W14 $\times$ 132	650
(c) W18 $\times$ 119	700

6 Beam-Columns

6-1 INTRODUCTION

It is generally accepted that axially (concentrically) loaded compression members are nonexistent in actual structures and that all compression members are subjected to some amount of bending moment. The bending moment may be induced by an eccentric load as shown in Fig. 6-1(a). The interior column of Fig. 6-1(b), shown with a concentric load, will not be concentrically loaded if the live loads are not symmetrical. Bending moments may be induced in columns through continuous frame action. Since columns may be subjected to varying amounts of axial load and bending moment, two extremes may exist. If the *bending moment* approaches zero, as a limit, the member is theoretically subjected to an axial load only. The analysis and design of such a member are the same as for an axially loaded compression member as treated in Chapter 5. If the *axial load* approaches zero as a limit, the member is theoretically subjected to a bending moment only and the analysis and design are the same as for a beam (bending member) as treated in Chapter 2. A structural member that is subjected to varying amounts of both axial compression and bending moment is commonly termed a **beam-column.**

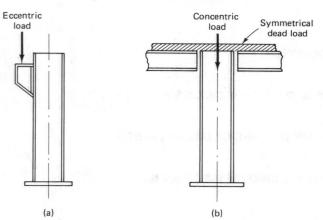

Figure 6-1 Column loadings.

The actual stresses induced in a beam-column by axial compression and bending moment are not directly additive since the combination of the two generates a secondary moment which cannot be ignored. This secondary moment results from a lateral deflection initially caused by the bending moment as shown in Fig. 6-2(a). The product of this deflection and the axial load ($P \times \Delta$, sometimes called the **P-delta moment**) causes further bending and creates secondary stresses which normally are not considered in individual beam or column analysis and design.

Neglecting the secondary moment for now, an approximate expression for the combined stresses for a short beam-column subjected to an axial load and bending moment with respect to one axis only may be expressed as

$$f_{max} = \frac{P}{A} \pm \frac{Mc}{I}$$

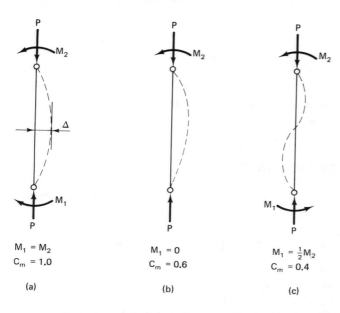

$M_1 = M_2$
$C_m = 1.0$

(a)

$M_1 = 0$
$C_m = 0.6$

(b)

$M_1 = \frac{1}{2} M_2$
$C_m = 0.4$

(c)

Figure 6-2 Values of C_m.

where f_{max} = computed maximum stress

P = axial load

A = gross cross-sectional area

M = applied moment

c = distance from the neutral axis to the extreme outside of the cross section

I = moment of inertia of the cross section about the bending neutral axis

If bending occurs with respect to both axes, the expression becomes

$$f_{max} = \frac{P}{A} \pm \left(\frac{M_x c}{I_x} \right) \pm \left(\frac{M_y c}{I_y} \right)$$

Utilizing this expression it was always a simple problem to compute approximate actual combined stresses for beam-columns. However, the results were of little significance since the allowable bending stress and the allowable compressive stress have always been appreciably different and an allowable combined stress has never been established by code.

In an effort to simplify the combined stress problem, the previous expression may be rewritten as

$$f_{max} = f_a + f_{bx} + f_{by}$$

with the negative signs neglected. Dividing both sides by f_{max}, we have

$$1 = \frac{f_a}{f_{max}} + \frac{f_{bx}}{f_{max}} + \frac{f_{by}}{f_{max}}$$

This may be further modified by substituting the applicable allowable stresses in place of the f_{max} terms:

$$1 = \frac{f_a}{F_a} + \frac{f_{bx}}{F_{bx}} + \frac{f_{by}}{F_{by}}$$

With this arrangement, if any two of the computed stresses become zero, the correct allowable stress is approached either as an axially loaded column or as a beam subjected to bending about either axis.

6-2 ANALYSIS OF BEAM-COLUMNS (AISC)

The expression developed in the preceding section is the basis for AISCS interaction formula 1.6-2. It may be rewritten as

$$\frac{f_a}{F_a} + \frac{f_{bx}}{F_{bx}} + \frac{f_{by}}{F_{by}} \le 1.0 \qquad \text{AISCS formula 1.6-2}$$

where f_a = actual axial compressive stress
f_b = actual maximum compressive bending stress
F_a = allowable axial compressive stress for axial force alone
F_b = allowable compressive bending stress for bending moment alone

It applies to members subjected to both axial compression and bending stresses when f_a/F_a is less than or equal to 0.15.

When $f_a/F_a > 0.15$ the secondary moment due to the member deflection may be of a significant magnitude. The effect of this moment may be approximated by multiplying f_{bx} and f_{by} by an amplification factor

$$\frac{1}{1 - (f_a/F'_e)}$$

where f_a is as previously defined and F'_e is the Euler stress divided by a factor of safety of 23/12 and is expressed as follows:

$$F'_e = \frac{12\pi^2 E}{23(K\ell_b/r_b)^2}$$

In this expression, K is the effective length factor in the plane of bending, ℓ_b is the actual unbraced length in the plane of bending, and r_b is the corresponding radius of gyration. Values of F'_e may be obtained from Table 9 in Appendix A of the AISCS.

Under some combinations of loading, it was found that this amplification factor overestimated the effect of the secondary moment. To compensate for this condition, the amplification factor was modified by a reduction factor C_m.

With the introduction of the two factors the AISCS interaction formula 1.6-2 was modified for the case when $f_a/F_a > 0.15$ and expressed as follows:

$$\frac{f_a}{F_a} + \frac{C_{mx}f_{bx}}{(1 - f_a/F'_{ex})F_{bx}} + \frac{C_{my}f_{by}}{(1 - f_a/F'_{ey})F_{by}} \le 1.0 \qquad \text{AISCS formula 1.6-1a}$$

where all terms are as previously defined and C_m is a coefficient defined as follows:

$$C_m = 0.6 - 0.4\left(\frac{M_1}{M_2}\right)$$

in which M_1/M_2 is the ratio of the smaller end moment to the larger end moment. M_1/M_2 is taken as positive if the moments tend to cause reverse curvature and negative if they tend to cause single curvature. Examples of C_m values are shown in Fig. 6-2.

C_m should never be less than 0.4. If frame sidesway is not prevented by adequate bracing or other means, C_m should not be taken as less than 0.85 since in this case the column ends move out of alignment causing an additional secondary moment from the axial load.

The question as to whether adequate bracing exists to prevent sidesway is difficult to answer and is usually a judgment factor. Sidesway itself may be described as a kind of deformation whereby one end of a member moves laterally with respect to the other. A simple example is a column fixed at one end and entirely free at the other (cantilever column or flagpole). Such a column will buckle as shown in Fig. 6-3. The upper end would move laterally with respect to the lower end. This lateral movement is termed **sidesway.** If the column bending moment is a result of a lateral load placed between column support points, C_m may be conservatively taken as unity.

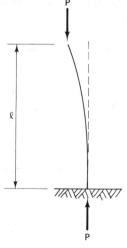

Figure 6-3   Sidesway for a flagpole-type column.

The AISCS interaction formula 1.6-1a applies where stability of a member is a problem and the critical buckling stresses are assumed to occur away from the points of bracing. To guard against overstressing at one end of a member where no buckling action is present, stresses are limited by a modified interaction expression.

$$\frac{f_a}{0.60F_y} + \frac{f_{bx}}{F_{bx}} + \frac{f_{by}}{F_{by}} \le 1.0 \qquad \text{AISCS formula 1.6-1b}$$

If only one axis of bending is involved in a problem, one of the terms will equal zero with the remaining formula still applicable.

In determining F_b for the interaction equations, the compactness of the beam-column must be established. As discussed previously in Section 2-3, the web compactness of a *beam* is based on $f_a = 0$. This applies to a beam subjected to bending only, with no axial load. With a beam-column (f_a is not zero) the web compactness must be checked utilizing AISCS equations 1.5-4a or 1.5-4b. This is simplified in the AISCM through the use of F_y''' which is defined as the theoretical maximum yield stress (ksi) based on the depth-thickness ratio of the web below which a particular shape may be considered compact for any condition of combined bending and axial stresses. F_y''' is tabulated in the Properties table of Part 1. It is determined in the same way that F_y' is determined as discussed in Section 2-3. If $F_y''' \ge F_y$, the member is compact based on the web criterion. If $F_y''' < F_y$, the member is not compact based on the web criterion and F_b cannot exceed $0.60F_y$.

In summary, to establish whether a beam-column is satisfactory, the following applies:

1. When $f_a/F_a \le 0.15$, use AISCS formula 1.6-2.
2. When $f_a/F_a > 0.15$, use both AISCS formulas 1.6-1a *and* 1.6-1b. *Both* formulas must be satisfied.

Example 6-1

An A36 steel W6 × 25 column is subjected to an eccentric load of 32 kips as shown in Figure 6-4. The column has an unbraced length of 15 ft and may be

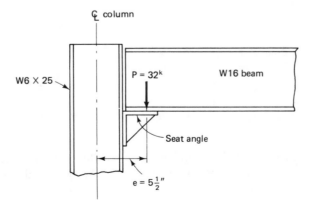

Figure 6-4 Beam-column connection.

assumed to have pinned ends. Bracing prevents sidesway. Determine if the column is adequate.

Solution   For the W6 × 25:

$$A = 7.34 \text{ in.}^2$$

$$S_x = 16.7 \text{ in.}^3$$

$$r_y = 1.52 \text{ in.}$$

$$r_x = 2.70 \text{ in.}$$

$$K = 1 \text{ (pinned ends)}$$

$$F_y''' = -$$

1. Replacing the eccentric load with a concentric load and a couple (moment):

$$P = 32 \text{ kips}$$

$$M_x = Pe = 32(5.5) = 176 \text{ in.-kips}$$

2. Calculating actual axial compressive stress:

$$f_a = \frac{P}{A} = \frac{32}{7.34} = 4.36 \text{ ksi}$$

3. F_a is a function of $K\ell/r_y$:

$$\frac{K\ell}{r_y} = \frac{1(15)(12)}{1.52} = 118.4$$

By interpolation, from the AISCS, Appendix A, Table 3-36,

$$F_a = 10.51 \text{ ksi}$$

4. Calculating actual maximum compressive bending stress, we have

$$f_{bx} = \frac{M_x}{S_x} = \frac{176}{16.7} = 10.5 \text{ ksi}$$

5. F_{bx} is a function of the actual unbraced length. Determine whether the member is adequately or inadequately braced. Actual unbraced length $L_b = 15$ ft. L_c and L_u may be obtained from the column load tables in the AISCM, Part 3:

$$L_c = 6.4 \text{ ft}$$

$$L_u = 20.0 \text{ ft}$$

Therefore, $L_c < L_b < L_u$. The W6 × 25 is compact since $F_y''' > F_y$, and the allowable bending stress is:

$$F_b = 0.6F_y = 22 \text{ ksi}$$

6. $f_a/F_a = 4.36/10.51 = 0.41 > 0.15$. Therefore, use AISCS formulas 1.6-1a and 1.6-1b.

7. Calculate C_m and F'_{ex}. Since $M_1 = 0$ and $M_2 = 176$ in.-kips and sidesway is prevented,

$$C_m = 0.6 - 0.4\left(\frac{M_1}{M_2}\right)$$

$$= 0.6$$

F'_{ex} is a function of $K\ell_b/r_b$, which in this case is $K\ell/r_x$:

$$\frac{K\ell}{r_x} = \frac{1(15)(12)}{2.70} = 66.7$$

By interpolation, from the AISCS, Appendix A, Table 9:

$$F'_{ex} = 33.58 \text{ ksi}$$

8. Checking AISCS interaction formula 1.6-1a, we have

$$\frac{f_a}{F_a} + \frac{C_{mx}f_{bx}}{(1 - f_a/F'_{ex})F_{bx}} \leq 1.0$$

$$\frac{4.36}{10.51} + \frac{0.6(10.5)}{(1 - 4.36/33.58)(22.0)} \leq 1.0$$

$$0.41 + 0.33 = 0.74 < 1.0 \qquad\qquad \textbf{O.K.}$$

9. Checking AISCS interaction formula 1.6-1b, we have

$$\frac{f_a}{0.6F_y} + \frac{f_{bx}}{F_{bx}} \leq 1.0$$

$$\frac{4.36}{22.0} + \frac{10.5}{22.0} \leq 1.0$$

$$0.20 + 0.48 = 0.68 < 1.0 \qquad\qquad \textbf{O.K.}$$

The beam-column is adequate.

Example 6-2

An A572 ($F_y = 50$ ksi) W12 × 136 column supports beams framing into it as shown in Fig. 6-5. The connections are moment connections. The column supports an axial load of 600 kips, which includes the beam reactions at its top. Due to unbalanced floor loading, moments of 80 ft-kips each are applied in opposite directions at top and bottom of columns as shown. Sidesway is prevented by a bracing system. $K_y = 1.0$ and K_x is estimated to be 0.9. Determine if the member is adequate.

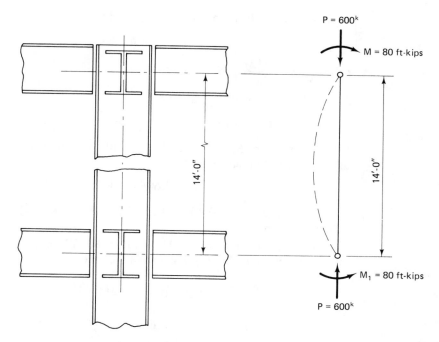

Figure 6-5   Beam-column.

Solution   For the W12 × 136: $A = 39.9$ in.^{2}

$$S_x = 186 \text{ in.}^3$$

$$r_y = 3.16 \text{ in.}$$

$$r_x = 5.58 \text{ in.}$$

$$K_x = 0.9$$

$$K_y = 1.0$$

$$F_y''' = -$$

1. $P = 600$ kips, $M = 80$ ft-kips

2. $f_a = \dfrac{P}{A} = \dfrac{600}{39.9} = 15.0$ ksi

3. F_a is a function of the largest slenderness ratio:

$$\frac{K_y \ell}{r_y} = \frac{1(14)(12)}{3.16} = 53.2$$

$$\frac{K_x \ell}{r_x} = \frac{0.9(14)(12)}{5.58} = 27.1$$

By interpolation from the AISCS, Appendix A, Table 3-50:

$$F_a = 23.85 \text{ ksi}$$

4. $f_{bx} = \dfrac{M_x}{S_x} = \dfrac{80(12)}{186} = 5.16 \text{ ksi}$

5. F_{bx} is a function of L_b, L_c, and L_u.

$$L_b = 14 \text{ ft} \qquad L_c = 11.1 \text{ ft} \qquad L_u = 38.3 \text{ ft}$$

Since $L_c < L_b < L_u$ and since W12 × 136 is compact ($F_y''' > F_y$), the allowable bending stress is

$$F_b = 0.60F_y = 30 \text{ ksi}$$

6. $f_a/F_a = 15.0/23.85 = 0.63 > 0.15$. Therefore, use AISCS formulas 1.6-1a and 1.6-1b.

7. Calculate C_m and F'_{ex}. Since $M_1 = M_2 = 80$ ft-kips and causes single curvature, the ratio M_1/M_2 is negative (see the AISCS, Section 1.6.1):

$$C_m = 0.6 - 0.4\left(\frac{M_1}{M_2}\right)$$

$$= 0.6 - 0.4(-1)$$

$$= 1.0$$

F'_{ex} is a function of $K\ell_b/r_b$, which in this case is $K_x\ell/r_x$:

$$\frac{K_x\ell}{r_x} = \frac{0.9(14)(12)}{5.58} = 27.1$$

Interpolating from the AISCS, Appendix A, Table 9, we have

$$F'_{ex} = 203.4 \text{ ksi}$$

8. Checking AISCS interaction formula 1.6-1a gives us

$$\frac{f_a}{F_a} + \frac{C_{mx}f_{bx}}{(1 - f_a/F'_{ex})F_{bx}} \le 1.0$$

$$\frac{15.0}{23.85} + \frac{1.0(5.16)}{(1 - 15.0/203.4)(30.0)} \le 1.0$$

$$0.63 + 0.19 = 0.82 < 1.0 \qquad\qquad \textbf{O.K.}$$

9. Checking AISCS interaction formula 1.6-1b yields

$$\frac{f_a}{0.6F_y} + \frac{f_{bx}}{F_{bx}} \le 1.0$$

$$\frac{15.0}{30.0} + \frac{5.16}{30.0} < 1.0$$

$$0.5 + 0.17 = 0.67 < 1.0 \qquad\qquad \textbf{O.K.}$$

The beam-column is adequate.

6-3 DESIGN OF BEAM-COLUMNS (AISC)

The use of the interaction formulas furnishes a convenient means of beam-column analysis. These may also be used for beam-column design. However, a trial section must first be selected. After the selection is made, the problem becomes one of analysis. In essence, the design process is one of trial and error since no simple design procedure exists whereby a most economical member would be selected in one quick step.

The AISCM furnishes a method of design whereby a trial section may be obtained using an equivalent axial load in conjunction with the AISCM axial load table of Part 3. The AISCM also furnishes modified interaction formulas which may be used in place of the previously discussed interaction formulas. As the modified formulas do not simplify the analysis, they will not be introduced in this text. The reader is referred to Part 3 of the AISCM.

Using the AISCM approach to determine a trial section, the reader is referred to the AISCM, Part 3, Table B. The equivalent axial load, for design purposes, is designated P_{eff}.

$$P_{eff} = P_0 + M_x m + M_y mU$$

where P_0 = actual axial load (kips)

   M_x = bending moment about the strong axis (ft-kips)

   M_y = bending moment about the weak axis (ft-kips)

   m = factor taken from the AISCM, Part 3, Table B

   U = factor taken from the AISCM, Part 3, column load tables

The procedure for selection of a trial section is as follows:

1. With the known value of KL (in feet) select a value of m from the first approximation section of Table B and assume that $U = 3$.
2. Solve for P_{eff}.
3. From the column load table in Part 3 of the AISCM, select a trial section to support P_{eff}.
4. Based on the section selected, obtain a *subsequent approximate value of m* from Table B and a U value from the column load table. Solve for P_{eff} again.

5. Select another section (if necessary) and continue the process until the values of m and U stabilize.

Using this trial section, the beam-column may then be analyzed in the manner discussed previously using the AISCM interaction formulas.

Example 6-3

Using A36 steel and the AISCS, select a wide-flange column for the conditions shown in Fig. 6-6. The column is a main member with pinned ends and sidesway prevented. Bending occurs with respect to the strong $(x\text{-}x)$ axis.

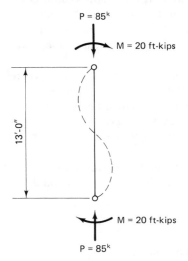

P = 85^k

M = 20 ft-kips

13'-0"

M = 20 ft-kips

P = 85^k

Figure 6-6   Beam-column.

Solution   From the AISCM, Part 3, Table B, with $KL = 13$ ft, select a value of $m = 2.25$ from the first approximation portion. Since $M_y = 0$, the expression for the effective axial load becomes

$$P_{\text{eff}} = P_0 + M_x m$$

$$= 85 + 20(2.25)$$

$$= 130 \text{ kips}$$

From the column load table of the AISCM, Part 3, select a W8 × 31.

From Table B again, select a value of $m = 2.85$ from the subsequent approximations portion.

$$P_{\text{eff}} = 85 + 20(2.85)$$

$$= 142 \text{ kips}$$

From the column load table select a W8 × 31 again. Since there is no change in the trial section, the W8 × 31 should be checked using the interaction formulas. The checking procedure is identical to that in Examples 6-1 and 6-2. AISCS formula 1.6-1a results in a value of 0.77 and formula 1.6-1b results in a value of 0.82. The W8 × 31 is therefore satisfactory. The reader may wish to verify the results of this analysis.

Example 6-4

Using A36 steel and the AISCS select a wide-flange column for the conditions shown in Fig. 6-7. Architectural requirements indicate the use of a W8, if possible. The column is a main member pinned at both ends. Bending occurs with respect to both axes. Sidesway is prevented in both directions.

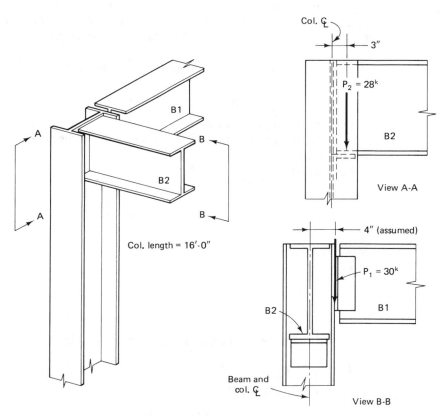

Figure 6-7   Corner beam-column.

Solution Replace the eccentric loads with concentric loads and couples (moments). A 4-in. eccentricity is assumed for strong-axis bending:

$$P = 58 \text{ kips}$$

$$M_x = 30(4) = 120 \text{ in.-kips}$$

$$M_y = 28(3) = 84 \text{ in.-kips}$$

From the AISCM, Part 3, Table B, with $KL = 16$ ft, select a value of $m = 2.2$ from the first approximation portion. Let $U = 3$; therefore,

$$P_{\text{eff}} = P_0 + M_x m + M_y mU$$

$$= 58 + \frac{120}{12}(2.2) + \frac{84}{12}(3)(2.2)$$

$$= 58 + 22 + 46.2 = 126.2 \text{ kips}$$

From the AISCM, Part 3, column load table, select a W8 × 35. From Table B again, select a value of $m = 2.6$ from the subsequent approximations portion and a value of $U = 2.59$ from the column load table.

$$P_{\text{eff}} = 58 + \frac{120}{12}(2.6) + \frac{84}{12}(2.6)(2.59)$$

$$= 58 + 26 + 47 = 131 \text{ kips}$$

From the column load table, select a W8 × 35 again. Since there is no change in the trial section, the W8 × 35 will now be checked using the interaction formulas.

For the W8 × 35: $A = 10.3 \text{ in.}^2$

$$S_x = 31.2 \text{ in.}^3$$

$$r_x = 3.51 \text{ in.}$$

$$r_y = 2.03 \text{ in.}$$

$$K = 1$$

$$S_y = 10.6 \text{ in.}^3$$

$$F_y''' = -$$

1. $P = 58$ kips, $M_x = 120$ in.-kips, $M_y = 84$ in.-kips

2. $f_a = \dfrac{P}{A} = \dfrac{58}{10.3} = 5.63 \text{ ksi}$

3. F_a is a function of $K\ell/r_y$:

$$\frac{K\ell}{r_y} = \frac{1(16)(12)}{2.03} = 94.6$$

By interpolation, from the AISCS, Appendix A, Table 3-36:

$$F_a = 13.65 \text{ ksi}$$

4. $f_{bx} = \dfrac{M_x}{S_x} = \dfrac{120}{31.2} = 3.85 \text{ ksi}$

$f_{by} = \dfrac{M_y}{S_y} = \dfrac{84}{10.6} = 7.92 \text{ ksi}$

5. F_{bx} is a function of L_b, L_c, and L_u:

$$L_b = 16 \text{ ft} \qquad L_c = 8.5 \text{ ft} \qquad L_u = 22.6 \text{ ft}$$

Since $L_c < L_b < L_u$ and since the W8 × 35 is compact ($F_y''' > F_y$), the allowable bending stress F_{bx} is

$$F_{bx} = 0.60F_y = 22.0 \text{ ksi}$$

From the AISCS, Section 1.5.1.4.3:

$$F_{by} = 0.75F_y = 27.0 \text{ ksi}$$

6. $f_a/F_a = 5.63/13.65 = 0.41 > 0.15$. Therefore, use AISCS formulas 1.6-1a and 1.6-1b.

7. Calculate C_m and F'_e. Sidesway is prevented in both directions; therefore,

$$C_m = 0.6 - 0.4\left(\frac{M_1}{M_2}\right) \geq 0.4$$

Since $M_1 = 0$,

$$C_{mx} = C_{my} = 0.6$$

Since bending occurs with respect to both axes, F'_e values must be obtained with respect to each axis.

$$\frac{K\ell}{r_y} = \frac{1(16)(12)}{2.03} = 94.6$$

$$\frac{K\ell}{r_x} = \frac{1(16)(12)}{3.51} = 54.7$$

By interpolation, from the AISCS, Appendix A, Table 9:

$$F'_{ey} = 16.69 \text{ ksi}$$

$$F'_{ex} = 49.92 \text{ ksi}$$

8. Checking AISCS interaction formula 1.6-1a gives us

$$\frac{f_a}{F_a} + \frac{C_{mx}f_{bx}}{(1 - f_a/F'_{ex})F_{bx}} + \frac{C_{my}f_{by}}{(1 - f_a/F'_{ey})F_{by}} \leq 1.0$$

$$\frac{5.63}{13.65} + \frac{0.60(3.85)}{(1 - 5.63/49.92)(22.0)} + \frac{0.6(7.92)}{(1 - 5.63/16.69)(27.0)} \leq 1.0$$

$$0.41 + 0.12 + 0.27 = 0.80 < 1.0 \qquad\qquad \textbf{O.K.}$$

9. Checking AISCS interaction formula 1.6-1b yields

$$\frac{f_a}{0.6F_y} + \frac{f_{bx}}{F_{bx}} + \frac{f_{by}}{F_{by}} \leq 1.0$$

$$\frac{5.63}{22.0} + \frac{3.85}{22.0} + \frac{7.92}{27.0} \leq 1.0$$

$$0.26 + 0.18 + 0.29 = 0.73 < 1.0 \qquad\qquad \textbf{O.K.}$$

The W8 × 35 is satisfactory.

6-4 *EFFECTIVE LENGTH FACTOR* K

In the design of steel structures one is often concerned with continuous frames of various types consisting of beams and columns that are rigidly connected. When this occurs, the columns are subjected to the combined action of compression and bending and may be categorized as beam-columns.

As discussed previously, the terms F_a and F'_e in the AISCS interaction formula are functions of slenderness ratio. This, in turn, is a function of the effective length factor K, which when multiplied by an actual length of a member will result in an effective length $(K\ell)_x$ or $(K\ell)_y$. The effective length of the member depends on the restraints against relative rotation and lateral movement (sidesway) imposed at the ends of the member.

Consider a simple portal frame as shown in Fig. 6-8. The beam is rigidly connected to the supporting columns. When sidesway (lateral movement) of the frame is effectively prevented by some means, its deformed shape under vertical load may be observed. The effective length factor K for the columns can have values that range

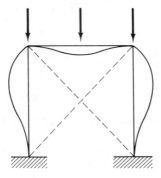

Figure 6-8 Loaded braced frame.

from 0.5 for ends fixed against rotation to 1.0 for pinned ends. When the frame depends on its own stiffness for resistance to sidesway as shown in Fig. 6-9, K will have a value larger than 1.0. Figure 6-9(a) shows the deformed shape due to vertical load. The frame will deflect to the side (sidesway) in order to equalize the moments at the tops of the columns. Sidesway may also be caused by a laterally applied force as shown in Fig. 6-9(b). As a rule, columns free to translate in a sidesway mode are appreciably weaker then columns of equal length braced against sidesway. Also of importance is the fact that the magnitude of the sidesway of a column is directly affected by the stiffness of the other members in the frame; or, the magnitude of joint rotation is directly affected by the stiffness of the members framing into the joint. The problem can easily become very complex.

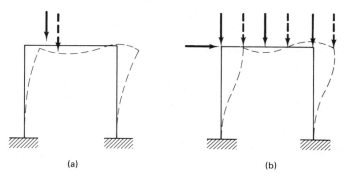

(a)   (b)

Figure 6-9 Loaded unbraced frame.

To simplify the determination of the effective length factor K, alignment charts are furnished in the AISCM, Part 3 (Figure 1), for the two cases of *sidesway prevented* and *sidesway not prevented*. The charts afford a means of obtaining more precise values for K than those offered by Table C.1.8.1 in the AISCS Commentary and discussed in Chapter 5. The use of the charts requires an evaluation of the relative stiffness of the members of the frame at each end of the column. The stiffness ratio or *relative stiffness* of the members rigidly connected at each joint may be expressed as

$$G_A = \frac{\Sigma\ (I_c/\ell_c)_A}{\Sigma\ (I_g/\ell_g)_A}$$

and

$$G_B = \frac{\Sigma\ (I_c/\ell_c)_B}{\Sigma\ (I_g/\ell_g)_B}$$

where A and B subscripts refer to the joint at which the relative stiffness is being determined; the c and g subscripts refer to *column* and *girder* or *beam*, respectively; I is the moment of inertia; and ℓ is the unsupported length of member. The I/ℓ terms are taken with respect to an axis normal to the plane of buckling under consideration.

Having determined G_A and G_B, the appropriate chart may be used to determine K. The points representing the values of G_A and G_B are connected with a straight line and the value of K is read at the intersection of this line with the central vertical K reference line.

For column ends supported by but not rigidly connected to a footing or foundation, G_B is theoretically infinity, but unless designed as a true friction-free pin may be taken as 10 for practical designs. If the column end is rigidly attached to a properly designed footing, G_B may be taken as 1.0. Figure 6-10 shows how the G values would be determined for a given column AB.

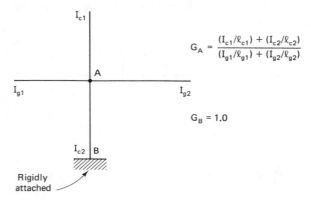

$$G_A = \frac{(I_{c1}/\ell_{c1}) + (I_{c2}/\ell_{c2})}{(I_{g1}/\ell_{g1}) + (I_{g2}/\ell_{g2})}$$

$G_B = 1.0$

Figure 6-10   Column G values.

The use of the alignment charts requires prior knowledge of the column and beam sizes. In other words, before the charts can be used, a trial design has to be made of each of the members. To start the design, it is therefore necessary to assume a reasonable value of K, choose a column section to support the axial load and moments, and then determine the actual value of K. In addition, trial beam sizes must be reasonably estimated.

Example 6-5

Compute the effective length factor K for each of the columns in the frame shown in Fig. 6-11 using AISCM alignment charts. Preliminary sizes of each member are furnished. Sidesway is *not* prevented. Webs of the wide flange shapes are in the plane of the frame.

Solution

Member	Shape	I (in.^{4})	ℓ (in.)	I/ℓ
AB	W14 × 53	541	168	3.22
BC	W14 × 53	541	144	3.76
CD	W18 × 50	800	360	2.22
BE	W21 × 62	1330	360	3.69
DE	W14 × 53	541	144	3.76
EF	W14 × 53	541	168	3.22

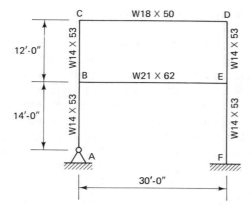

Figure 6-11 Two-story rigid frame.

G factors for each joint are determined as follows:

Joint	$\dfrac{\Sigma(I_c/\ell_c)}{\Sigma(I_g/\ell_g)}$		G
A	Pinned end	=	10.0
B	$\dfrac{3.76 + 3.22}{3.69}$	=	1.89
C	$\dfrac{3.76}{2.22}$	=	1.69
D	$\dfrac{3.76}{2.22}$	=	1.69
E	$\dfrac{3.76 + 3.22}{3.69}$	=	1.89
F	Fixed end	=	1.0

Column K factors from the chart (sidesway uninhibited):

Column	G values at Column ends		K
AB	10.0	1.89	2.08
BC	1.89	1.69	1.54
DE	1.69	1.89	1.54
EF	1.89	1.0	1.44

Because of the smaller effective length factors used for frames where sidesway is prevented, it is advisable to provide lateral support wherever possible. This may be accomplished with diagonal bracing, shear walls, attachment to an adjacent structure having adequate lateral stability, or by floor slabs or roof decks secured horizontally by walls or bracing systems.

As discussed previously, the determination of the *K*-factors utilizing the alignment charts is based on several assumptions. Two of the principal assumptions are that all columns in a story buckle simultaneously and that all column behavior is purely elastic. Either one or both of these conditions may not exist in an actual structure and, as a result, the use of the alignment charts will produce overly conservative designs.

The AISCM, Part 3, contains a design procedure to reduce the *K*-factor value by multiplying the elastic *G* value by a stiffness reduction factor. Then, utilizing the alignment charts as discussed previously, an *inelastic K*-factor is obtained. The stiffness reduction factor is obtained from Table A in Part 3 of the AISCM. For further details, see References 1 and 2 at the end of the chapter.

Example 6-6

Compute the *inelastic K*-factors for columns *BC* and *DE* of Example 6-5 as shown in Fig. 6-12. Consider behavior in the plane of the frame only. Use Figure 1 and Table A of the AISCM, Part 3, for the inelastic *K*-factor procedure. Preliminary sizes are shown. Sidesway in the plane of the frame is not prevented. Webs of the W-shape members are in the plane of the frame. Use A36 steel. Assume that the columns support loads of 245 kips as shown in Fig. 6-12.

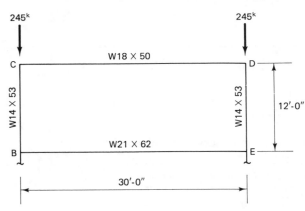

Figure 6-12   Inelastic *K*-factor determination.

Solution

1. The trial column size is W14 × 53. $A = 15.6$ in.^{2}
2. Compute f_a:

$$f_a = \frac{245}{15.6} = 15.7 \text{ ksi}$$

3. From Table A, the stiffness reduction factor f_a/F'_e is 0.621.

4. From Example 6-5, the elastic stiffness ratios are 1.89 at joints B and E (bottom) and 1.69 at joints C and D (top).
5. Calculate $G_{\text{inelastic}}$:

$$G_{\text{inelastic (top)}} = 0.621(1.69) = 1.05$$

$$G_{\text{inelastic (bottom)}} = 0.621(1.89) = 1.17$$

6. Determine K from Figure 1:

$$K = 1.35$$

This compares with a K of 1.54 as determined in Example 6-5, indicating a greater column capacity if inelastic behavior is considered.

REFERENCES

1. Joseph A. Yura, The Effective Length of Columns in Unbraced Frames, *Engineering Journal, American Institute of Steel Construction,* Vol. 8, No. 2, April 1971.
2. Robert O. Disque, Inelastic K-factor for Column Design, *Engineering Journal, American Institute of Steel Construction,* Vol. 10, No. 2, 1973.

PROBLEMS

Note: The AISCS applies in all of the following problems.

6-1. A W12 × 53 column of A36 steel supports a vertical load of 40 kips at an eccentricity of 12 in. with respect to the strong axis. The length is 12 ft and $K = 1.0$ with respect to both axes. Determine if the member is adequate.

6-2. In Problem 6-1, determine the maximum *moment* (bending about the strong axis) that the beam-column can safely support. The vertical *load* remains 40 kips.

6-3. In Problem 6-1, determine if the beam-column is adequate if the 40-kip load acts at an eccentricity of 12 in. with respect to *each* axis.

6-4. A single-story W14 × 109 of A36 steel is to support a load of 280 kips at an eccentricity of 10 in. Bending is to be about the strong axis. Sidesway is prevented. The length is 24 ft and $K = 1.0$. Determine if the member is adequate.

6-5. A single-story W10 × 45 of A36 steel supports a beam reaction of 70 kips. Bending is about the *weak* axis and eccentricity is 3 in. Sidesway is prevented. The length is 16 ft and $K = 1.0$. Determine if the member is adequate.

6-6. An A36 W10 × 39 is subjected to the loads shown. Sidesway is prevented by bracing. The member length is 24 ft and $K = 1.0$. Determine if the member is adequate.

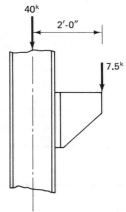

40^k

2'-0"

7.5^k

Problem 6-6

6-7. A W8 × 40 column has a length of 16 ft. The bottom is fixed and the top is pinned. Sidesway is prevented by bracing. A36 steel is used. Determine the maximum load that may be safely supported by this column assuming a strong-axis eccentricity of 6 in.

6-8. The built-up column shown supports a vertical load P of 100 kips at an eccentricity of 8 in. The length is 16 ft. Sidesway is prevented. $K = 1.0$. Determine if the column is adequate.

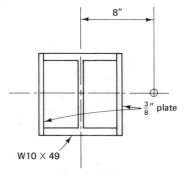

8"

$\frac{3}{8}''$ plate

W10 × 49

Problem 6-8

6-9. Select the lightest W shape of A588 steel to support an axial load of 200 kips and a moment of 50 ft-kips with respect to the strong axis. The column length is 18 ft and $K = 1.0$ in both directions. Sidesway is prevented.

6-10. Select the lightest W shape of A36 steel to support an axial load of 280 kips and a weak-axis moment of 60 ft-kips. The member is part of a frame in which sidesway is not prevented. The length is 14 ft and K has been estimated to be 1.3.

6-11. Select the lightest W12 for the beam-column shown. Bending is about the strong axis. Sidesway is not prevented and $K = 1.18$. Use A36 steel. $L = 14$ ft.

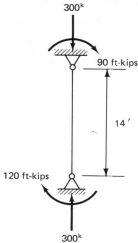

Problem 6-11

6-12. Select the lightest W shape of A572 steel ($F_y = 50$ ksi) to support an axial load of 500 kips and a lateral uniformly distributed load of 0.30 kip/ft with respect to the strong axis. The member length is 25 ft and $K = 1.0$. Sidesway is prevented.

6-13. In Problem 6-12, select the lightest W shape if the lateral load occurred with respect to the weak axis.

6-14. The corner column shown is part of a frame that is braced against sidesway and has $K = 1.0$. The length of the column is 20 ft. The column supports girts at its midheight. The girts, which support exterior wall panels, induce a maximum horizontal load of 2 kips on the column in each direction. Select the lightest W12 shape using A36 steel.

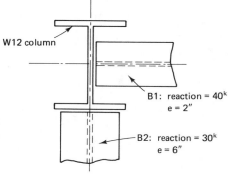

Problem 6-14

6-15. Determine the effective length factor K for each of the two columns in the rigid frame shown. The column bases are to be considered pinned. The webs of the members are in the plane of the frame. Sidesway is not prevented.

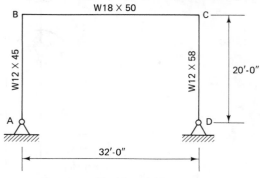

Problem 6-15

6-16. Determine the effective length factor K for each column in the frame shown. Sidesway is not prevented. The webs of the members are in the plane of the frame.

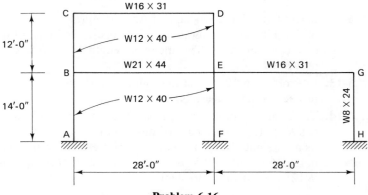

Problem 6-16

7 Bolted Connections

7-1 INTRODUCTION

The preceding chapters have covered the three fundamental structural members—bending members, tension members, and compression members—of which every structure must in part be composed, no matter how simple or how complex. A structure may be considered to be an assemblage of these various members which must be fastened together to make the finished product.

Irrespective of how scientifically or efficiently the basic structural members may have been designed, if the necessary connections are inadequate the result could be catastrophic collapse. The importance of economical and structurally adequate connections cannot be overemphasized.

Connection behavior is so complex that numerous simplifying assumptions must be made so that connection design is brought to a practical level. It is generally agreed among designers that the design of the basic structural members is simple compared to the design of the connections between those members.

The most common types of structural steel connections currently being used are bolted connections and welded connections. For many years rivets were the predominant type of fasteners in structures. However, because of their low strength, high installation costs, and other disadvantages, they have been superseded and may be considered obsolete. Despite this, the 8th Edition of the AISCM continues to include rivet data.

There are several types of bolts that can be used for connecting structural steel members. The two types generally used in structural applications are unfinished bolts and high-strength bolts. Proprietary bolts incorporating ribbed shanks, end splines, and slotted ends are also available, but in reality, these are only modifications of the high-strength bolt. Unfinished bolts are also known as machine, common, or ordinary bolts. They are designated in the AISCM as **ASTM A307 bolts**, conforming to the requirements of ASTM A307, *Specifications for Low Carbon Steel Externally and Internally Threaded Fasteners*. Permissible loads on these bolts are significantly less than those permitted on high-strength bolts. Their application should be limited to secondary members not subjected to vibrations or moving loads.

7-2 TYPES OF BOLTED CONNECTIONS

Connections serve primarily to transmit load from or to intersecting members; hence, the design of connections must be based on structural principles. This involves creating a detail that is both structurally adequate and economical as well as practical.

The simplest form of bolted connection is the ordinary **lap joint** shown in Fig. 7-1(a). Some joints in structures are of this general type, but it is not a commonly used detail due to the tendency of the connected members to deform.

A more common type of connection, the **butt joint**, is shown in Fig. 7-1(b). It

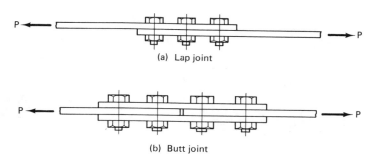

(a) Lap joint

(b) Butt joint

Figure 7-1   Types of joints.

is a type that may be used for tension member splices, in effect replacing the member at the point where it is cut. Other commonly used bolted connections may be observed in Fig. 7-2.

An understanding of the behavioral aspect of connections is important since the connections establish the support conditions of the connected members. The design of the members, which always precedes the design of the connections, must necessarily be based on assumed support conditions. There is a vast number of connection types. However, a series of relatively standard connections has been developed over the years and categorized in a behavioral sense. These connections are primarily beam-to-column and beam-to-beam building connections.

There are three AISC basic types of construction and associated design assumptions. They are as follows:

Type 1, commonly designated as *rigid frame* (continuous frame), assumes that beam-to-column connections have sufficient rigidity to hold, virtually unchanged, the original angles between intersecting members.

Type 2, commonly designated as *simple framing* (unrestrained, pin-connected), assumes that insofar as gravity loading is concerned, the ends of the beams and girders are connected for shear only, and are free to rotate under gravity load.

Type 3, commonly designated as *semi-rigid framing* (partially restrained), assumes that the connections of beams and girders possess a dependable and known moment capacity intermediate in degree between the rigidity of Type 1 and the flexibility of Type 2.

Therefore, in the design of a steel frame building, the type of construction must be established prior to the design of any of the structural members. After the structural members (beams and columns) are designed the connections must be designed consistent with the type of construction. Some of the common types of bolted building connections are shown in Fig. 7-3.

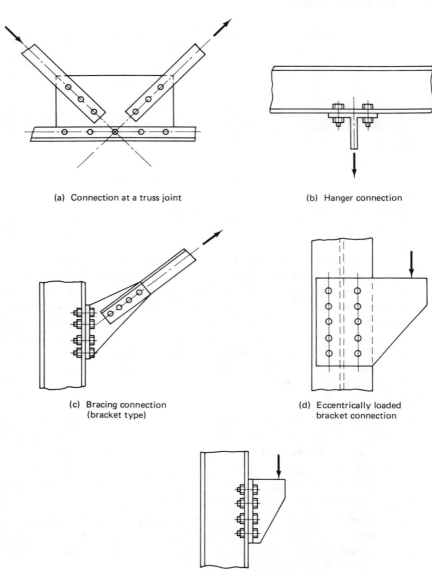

(a) Connection at a truss joint

(b) Hanger connection

(c) Bracing connection
     (bracket type)

(d) Eccentrically loaded
     bracket connection

(e) Eccentrically loaded
     connection

Figure 7-2 Common types of bolted connections.

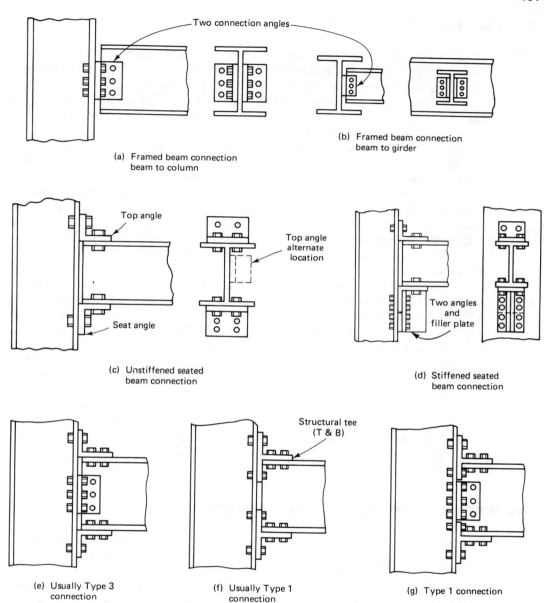

(a) Framed beam connection
      beam to column

(b) Framed beam connection
      beam to girder

(c) Unstiffened seated
      beam connection

(d) Stiffened seated
      beam connection

(e) Usually Type 3
      connection

(f) Usually Type 1
      connection

(g) Type 1 connection

Figure 7-3 Common types of bolted connections.

7-3 HIGH-STRENGTH BOLTS

The high-strength bolt is undoubtedly the most commonly used mechanical fastener for structural steel. Bolting with high-strength bolts has become the primary means of connecting steel members in the field as well as in the shop.

The two basic types of high-strength bolts are the A325 high-strength bolt and the A490 high-strength bolt. Specifications covering the chemical and mechanical requirements of the bolt *materials* are the latest ASTM A325 and ASTM A490 standards.[1,2] The A490 bolt has the higher material strength. Both fasteners may be described as heavy hex structural bolts and are used with heavy hex nuts. They have shorter thread lengths than comparable bolts used for other applications.

High-strength bolts are available in several *types*. The type specified will depend on the condition of use or performance desired, such as use at elevated temperatures or enhanced corrosion resistance or certain weathering characteristics. In general, the A325 and A490 bolts are available in diameters ranging from $\frac{1}{2}$ to $1\frac{1}{2}$ in. inclusive, with $\frac{3}{4}$ in. and $\frac{7}{8}$ in. being the most commonly used sizes. For details on types and sizes, see References 1 and 2.

With the profusion of different bolts available, it became necessary to establish a means of identification. It is required that the top of the bolt head be marked (either A325 or A490) along with a symbol identifying the manufacturer. Additional markings may be required to identify the type of bolt. The *Specification for Structural Joints Using ASTM A325 or A490 Bolts,* AISCM, Part 5, contains further details on markings and many other aspects of high-strength bolting. For simplicity, this specification will be referred to as the *SSJ*. The SSJ, together with its accompanying *Commentary,* is relatively short and should be read in its entirety. Additionally, for a concise compilation of authoritative data on the science of high-strength structural bolting, the reader is referred to Reference 3 at the end of this chapter. Part 5 of the AISCM will assist the reader in the proper identification and designation of the bolts, nuts, and washers.

7-4 INSTALLATION OF HIGH-STRENGTH BOLTS

The performance of a high-strength bolt depends on the proper installation of the bolt as well as on the type of bolt. High-strength bolts must be tightened in such a manner that the tension induced into the bolt is equal to or greater than 70% of the specified minimum tensile strength for that steel. The minimum fastener tension is prescribed in the SSJ, AISCM, Part 5. When the bolt is tightened to the minimum specified tension it produces predictable high clamping forces between the connected parts. As a result of these large clamping forces the contact surfaces are capable of transmitting loads almost entirely by friction. In addition, enough frictional resistance is produced to prevent nuts from working loose under service loads.

High-strength bolts may be tightened by any one of three recognized and accepted methods:

1. Turn-of-nut tightening
2. Calibrated wrench tightening
3. Direct tension indicator method

The objective of the tightening method is the same in all cases, namely to induce the minimum required tension into the bolts. Tightening may be accomplished by turning either the bolt head or the nut while preventing the other element from rotating. The actual tightening is almost always accomplished by using a pneumatic or electric **impact wrench.**

In the **turn-of-nut method,** enough bolts are brought to a "snug-tight" condition to ensure that the parts of the joint are brought into good contact with each other. Snug tight is defined as the tightness attained by a few impacts of an impact wrench or the full effort of a person using an ordinary spud wrench. (A spud wrench is an open-end wrench, about 15 to 18 in. long, which has the handle end formed into a tapered pin. The pin is used to align the holes of members being connected.) Bolts must then be placed in any remaining holes in the connection and also brought to snug tightness. From this snug point all bolts must be tightened additionally by some amount of turning element rotation, varying anywhere from one-third to one full turn depending on the length and diameter of the bolts. During this operation no rotation of the part *not* turned by the wrench is permissible. The precise amount of turning element rotation, as well as the need for washers when using this tightening method, is indicated in the SSJ.

In the **calibrated wrench tightening method,** calibrated wrenches are used which must be set to provide a tension at least 5% in excess of the prescribed minimum bolt tension. Wrenches must be calibrated at least once each working day for each bolt diameter being used. Calibration must be accomplished by tightening, in a device capable of indicating actual bolt tension, three typical bolts of each diameter from the bolts being installed. Despite the fact that the wrench is adjusted to stall at a designated bolt tension, another check should be made by verifying during the actual installation that the turned element rotation from snug position is not in excess of that prescribed for the turn-of-nut method. The tightening procedure for this method is identical to the turn-of-nut tightening procedure. In addition, it is necessary that the wrench be returned to touch up bolts previously tightened, which may have been loosened by the tightening of subsequent bolts, until all are tightened to the prescribed amount.

There are several tightening methods available which come under the category of **direct tension indicators.** One involves the use of a load indicator washer (Bethlehem Steel Co.). The method depends on the deformation of the washer (see Fig. 7-4) to indicate the induced bolt tension during and after tightening. The round hardened washer has a series of protrusions on one face. The washer is usually inserted

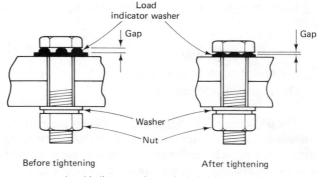

Figure 7-4 Direct tension indicator.

between the bolt head and the gripped plates with the protrusions bearing against the underside of the bolt head, leaving a gap. Upon tightening, the protrusions are partially flattened and the gap is reduced. Bolt tension is evaluated by measurement of the gap closure. When the gap is reduced to a prescribed dimension, the bolt has been properly tightened. Indicators are produced for both A325 and A490 bolts and are identified appropriately to avoid any errors. Installation by this method still requires the use of wrenches. However, the calibration of the wrench, or establishing the nut rotation from a snug-tight condition, is not necessary when using the indicator. In addition, the use of the indicator furnishes simplified and economical inspection which may be performed by one person using a simple metal feeler gage.

Another innovation in the category of the direct tension indicator method is the modification of a high-strength bolt by adding a spline to the end of the bolt. This bolt has been termed the **Load Indicator Bolt** by Bethlehem Steel Co. In combination with a special wrench, the installation process has been simplified and reduced to a one-person operation. The principle is quite simple (see Fig. 7-5). The wrench, by virtue of its construction, grasps both the nut and the spline, applying a clockwise turning force to the nut and a counterclockwise force to the spline. When the fastener assembly reaches the proper torque, the bolt tension is in excess of the specified minimum

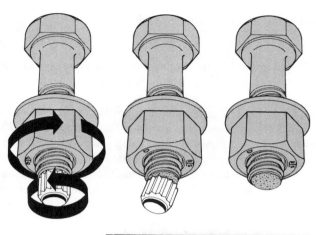

Figure 7-5 Load indicator bolt. (Copyright, Bethlehem Steel Corporation. Reproduced by permission.)

Framed beam connection which is shop bolted to the beam web using load-indicator bolts. The bolts have not yet been tightened. Note also the cope in the beam which is made to allow clearance for the flange of the supporting member.

tension and the wrench will twist off the splined end. Inspection is quick; if the spline has been twisted off, the bolt tension is adequate. At this writing the Bethlehem Load Indicator Bolt only meets the specifications for the A325 high-strength bolt. These may be reused (if approved by the engineer) as a conventional high-strength bolt by using the alternate methods of tightening.

With respect to the reuse of all high-strength bolts, it is stipulated in the SSJ that A490 bolts and galvanized A325 bolts shall not be reused. Other A325 bolts may be reused once or twice if approved by the engineer responsible. However, a properly installed high-strength bolt will have experienced some deformation. Proper reinstallation of the bolt will be difficult if not impossible. According to the SSJ, retightening previously tightened bolts which may have been loosened by the tightening of adjacent bolts should not be considered as a reuse.

7-5 *STRENGTH AND BEHAVIOR OF HIGH-STRENGTH BOLTED CONNECTIONS*

In determining the strength of high-strength bolted connections, one must consider the aspects of *shear, bearing,* and *tension* with regard to both the fasteners and the connected materials.

In most structural connections the bolt is required to prevent the movement of connected material in a direction perpendicular to the length of the bolt, as in Fig. 7-6. In such cases the bolt is said to be loaded in **shear.** In the lap joint shown, the bolt has a tendency to shear off along the single contact plane of the two plates. Since the bolt is resisting the tendency of the plates to slide past one another along the contact surface and is being sheared on a single plane, the bolt is said to be in **single shear.**

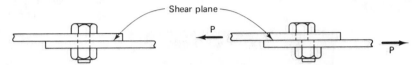

Figure 7-6   Bolt in single shear.

In a butt joint such as that shown in Fig. 7-7, there are two contact planes; therefore, the bolt is offering resistance along two planes and is said to be in **double shear.**

It is easy to visualize the plates slipping in the direction of the applied forces until they bear against the bolt. This is called a **bearing-type connection.** If, on the other hand, there is sufficient friction between the connected parts to prevent the slip from occurring, the connection is considered to be a **friction-type connection.** Friction-type connections should be used in structures subject to vibrations from dynamic loadings such as bridges, as well as for members that will be subjected to load reversals.

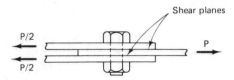

Figure 7-7   Bolt in double shear.

The load-carrying capacity, or working strength, of a bolt in single shear is equal to the product of the cross-sectional area of its shank and an allowable shear stress.

$$r_v = A_b F_v$$

where r_v = allowable shear for one bolt (kips)
A_b = cross-sectional area of one bolt (in.^{2})
F_v = allowable shear stress (ksi)

The allowable shear stress depends on the type of high-strength bolt, the type of connection (friction or bearing), and the type of hole. The hole may be either a *standard hole, oversized,* or *slotted.* Oversized and slotted holes are used to facilitate erection and are defined in Section 3 (Bolted Parts) of the SSJ. Values for allowable shear stress F_v may be observed in Table 2 of the SSJ.

When a bolt is subjected to more than one plane of shear, such as in double shear (Fig. 7-7), the allowable shear for the one bolt will be r_v multiplied by the number of shear planes.

While the bolts in a connection may be adequate to transmit the applied load in shear, the connection will fail unless the *material joined* is capable of transmitting the load *into* the bolts. This capacity is a function of the bearing (or crushing) strength of the connected material, as shown in Fig. 7-8. The true distribution of the bearing pressure on the material around the perimeter of the hole is unknown; therefore, the resisting contact area has been taken as the nominal diameter of the bolt multiplied by the thickness of the connected material. This assumes a uniform pressure acting over a rectangular area.

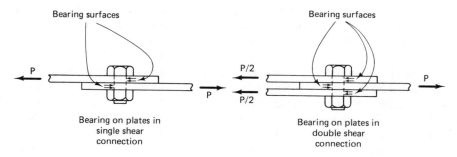

Bearing surfaces

Bearing surfaces

Bearing on plates in single shear connection

Bearing on plates in double shear connection

Figure 7-8   Bearing pressures.

The strength of one bolt in bearing may be expressed as

$$r_v = dtF_p$$

where r_v = allowable bearing for one bolt (kips)
  d = nominal bolt diameter (in.)
  t = thickness of plate or connected part (in.)
  F_p = allowable bearing stress (ksi)

The allowable bearing stress F_p is taken as the smaller value of

$$F_p = \frac{LF_u}{2d} \qquad \text{or} \qquad F_p = 1.5F_u$$

where F_p = allowable bearing stress (ksi)
  d = nominal bolt diameter (in.)
  F_u = lowest specified minimum tensile strength of the connected parts (ksi)

L = distance in inches measured in the line of force from the centerline of a bolt to (a) the nearest edge of an adjacent bolt, or to (b) the end of the connected part toward which the force is directed, as shown in Fig. 7-9

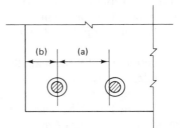

Figure 7-9   Explanation of L.

In the friction-type connection it is assumed that the load is transmitted from one connected part to another entirely by the friction that results from the high tension in the bolt. The theory behind the friction-type connection is that *no slippage* occurs between the connected parts and that the bolts are not actually loaded in shear or bearing. However, *for design purposes* it is assumed that the bolts *are* in shear and bearing and therefore, allowable stress values are furnished. Even though no slippage is expected to occur, the bearing consideration will result in an acceptable design should the remotely possible slippage take place. The allowable shear stress values in Table 2 are modified in Table 2a of the SSJ, which takes into consideration the surface condition of the contact area between the connected parts. The most commonly used categories are Classes A, B, and C, which are the uncoated categories. For these classes, the contact surfaces must be free of oil, paint, lacquer, or other coatings.

In the bearing-type connection it is accepted that the bolt is actually in shear and that the load is transmitted by the shearing resistance of the bolt as well as bearing of the connected parts on the bolt. The frictional resistance between the connected parts is of no concern nor is the surface condition of the connected parts. Bearing-type connections may be designed with the bolt threads in, or out of, the shear plane. The allowable shearing stresses will reflect the selected condition. Depending on the type of connection and location of the threads, the AISCM uses the following bolt designations:

Bolt designation	Type of application
A325F, A490F	Friction-type connection
A325N, A490N	Bearing-type connection, threads in the shear plane
A325X, A490X	Bearing-type connection, threads excluded from the shear plane

In summary, the bolts in each type of connection are tensioned in exactly the same manner and to the same magnitude of tensile force as determined by the size and material. The difference lies in the allowable stresses used in the analysis or design. A friction-type connection will generally be composed of more fasteners than it would have been if it had been designed as a bearing-type connection. Another important difference concerns the contact surface between the connected parts, as discussed previously. For both types of connection, the contact surfaces must be brought into solid contact, and no loose mill scale, burrs, dirt, or other foreign material is permitted.

A high-strength bolted connection, where the bolts are subjected to pure tensile loads, is shown in Fig. 7.2(b). This is a hanger-type connection. The permissible tension in a bolt may be taken as the product of its nominal cross-sectional area and its allowable unit tensile stress. This may be expressed as

$$r_t = A_b F_t$$

where r_t = allowable tension for one bolt (kips)
A_b = nominal cross-sectional area of one bolt (in.²)
F_t = allowable unit tensile stress (ksi)

The allowable tensile stresses are furnished in Table 1.5.2.1 of the AISCS and Table 1-A in Part 4 of the AISCM. Despite the fact that tensile stresses exist in high-strength bolts before any external tensile load is applied, it has been found that the external tensile load does not substantially affect the stresses in the bolt until the externally applied load exceeds the tension initially induced into the bolt by one of the tightening methods discussed previously. Therefore, the bolt is permitted to develop its full allowable tensile strength to resist the load.

In the design of hanger-type connections, prying action must be considered. In Fig. 7-10, this is the tendency of the flanges of the structural tee to act as cantilever beams under the action of the downward load P and the upward force of the restraining bolt. Actually, prying forces are present to some extent in nearly all connections employing bolts loaded in tension. The effect of the prying action is to increase the

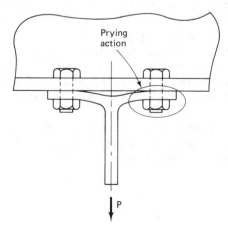

Figure 7-10 Prying action.

tension in the bolts. The magnitude of this increased tensile load is a function of the supporting member, bolt type and size, and the connection geometry. A design example considering prying forces will be presented later in this chapter.

Connections subjected simultaneously to shear and tension loads require special analysis to assure conformance to specification provisions for values of F_v and F_t. Connections for this type of combined loading occur frequently at the end of diagonal bracing members, as shown in Fig. 7-2(c). Interaction formulas permit the computation of a new allowable tensile stress F_t which must exceed the actual tensile stress induced by the applied loads. The new F_t is based on f_v, which is the actual bolt shear stress induced by the applied loads.

According to the AISCS, Table 1.6.3, the interaction expressions may be taken as follows:

Bolt		*Expression for F_t (ksi)*
A325N	———————————	$55 - 1.8f_v \leq 44$
A325X	———————————	$55 - 1.4f_v \leq 44$
A490N	———————————	$68 - 1.8f_v \leq 54$
A490X	———————————	$68 - 1.4f_v \leq 54$

In all cases the shear stress f_v cannot exceed the allowable shear stresses as furnished in Table 2 of the SSJ or in Table I-D of the AISCM, Part 4. In friction-type connections (A325F and A490F) the allowable shear stress, as furnished in these tables, must be multiplied by a reduction factor

$$1 - \frac{f_t A_b}{T_b}$$

where f_t is the average tensile stress due to a direct load applied to all the bolts in a connection, and T_b is the initial tensile load applied in the tightening process as specified in Table 3 of the SSJ. A_b is as defined previously. In this expression, f_t cannot exceed the allowable tensile stress furnished in Table 1.5.2.1 of the AISCS. In addition, the reader is referred to Sections 1.5.2 and 1.6.3 of the AISCS.

The design and detailing of bolted connections require proper bolt spacing as well as proper *edge distances* in accordance with the AISCS. Edge distance may be defined as the distance from the *center of the hole* to an edge of a connected part. This is particularly important as the allowable bearing stress is based on bolt spacing and edge distances.

The **distance between centers** of standard, oversized, or slotted fastener holes (bolt spacing) must not be less than $2.67d$, where d is the nominal diameter of the bolt. A distance of $3d$ is preferred. However, bolt spacing along a line of transmitted force cannot be less than

$$\frac{2P}{F_u t} + \frac{d}{2}$$

for *standard* holes, where

P = force transmitted by one bolt to the critical connected part (kips)
F_u = lowest specified minimum tensile strength of the critical connected part (ksi)
t = thickness of the critical connected part (in.)

For *oversized and slotted holes* this distance is increased by an increment C_1 given in Table 1.16.4.2 of the AISCS. This provides the same clear distance between holes as for standard holes.

The **edge distance** must not be less than that shown in Table 1.16.5.1 of the AISCS. However, the edge distance along a line of transmitted force, in the direction of the force, cannot be less than

$$\frac{2P}{F_u t}$$

where P, F_u, and t are as defined previously. For oversized and slotted holes this distance is increased by an increment C_2 given in Table 1.16.5.4 of the AISCS. This provides the same clear distance from the edge of the hole as for a standard hole.

At end connections bolted to the web of a beam and designed for beam shear reaction only, the distance from the center of the nearest standard hole to the end of the beam web must not be less than

$$\frac{2P_R}{F_u t}$$

where P_R is the beam reaction (kips) and F_u and t are as defined previously. Note that this special edge distance is actually perpendicular to the force. The maximum distance from the center of a bolt hole to the nearest edge of any member must be 12 times the thickness of the connected part under consideration, but not exceed 6 in.

In the design or analysis of a connection, it may be necessary to check the tensile capacity of the connected member itself since the original design of the tension member was based on an assumed connection. For simplicity in the following example problems, the allowable load on the connections as governed by fastener strength will be denoted as P. The tensile capacity of the connected member itself will be denoted P_t, following the convention of Chapter 4 of this text. Naturally, the final tensile capacity (or allowable load) of the member may be controlled either by the member itself or by the connection.

Example 7-1

Compute the tensile capacity, P, for the single shear lap joint shown in Fig. 7-11. The plates are A36 steel and the high-strength bolts are $\frac{7}{8}$-in.-diameter A325F (friction-type connection) in standard holes.

Solution   *Note*: All table references are to the AISCM, Part 4.

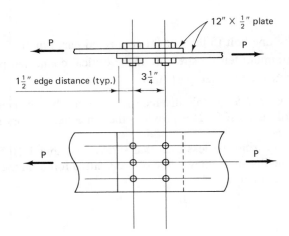

Figure 7-11   Single shear lap joint.

Bolt shear: Allowable shear stress (Table I-D):

$$F_v = 17.5 \text{ ksi}$$

Allowable shear load per bolt (A_b from Table I-D):

$$r_v = A_b F_v$$
$$= 0.6013(17.5) = 10.5 \text{ kips}$$

Note that the allowable shear per bolt could also be obtained directly without computations from Table I-D.

Allowable shear load per connection (six bolts):

$$P = 10.5(6) = 63.0 \text{ kips}$$

Bearing (on $\frac{1}{2}$-in.-thick plate):
Allowable bearing stress F_p equals the smaller of

$$F_p = 1.5F_u = 1.5(58) = 87.0 \text{ ksi}$$

or, based on bolt spacing,

$$F_p = \frac{LF_u}{2d} = \frac{(3.25 - 0.875/2)(58)}{2(0.875)} = 93.2 \text{ ksi}$$

or, based on edge distance:

$$F_p = \frac{LF_u}{2d} = \frac{1.5(58)}{2(0.875)} = 49.7 \text{ ksi}$$

Therefore, use $F_p = 49.7$ ksi.

Allowable bearing load per bolt:

$$r_v = dtF_p = 0.875(0.5)(49.7) = 21.8 \text{ kips}$$

Note that the allowable bearing load per bolt could also be obtained directly without computations from Table I-E.

Allowable bearing load per connection:

$$P = 21.8(6) = 130.8 \text{ kips}$$

Tension: This is a check of the tensile capacity of the plates (bolts are *not* in tension) and is based on the principles of Sections 4-2 and 4-3.

$$A_n = A_g - A_{\text{holes}} = \left(12 \times \frac{1}{2}\right) - 3\left(\frac{7}{8} + \frac{1}{8}\right)\frac{1}{2} = 4.50 \text{ in.}^2$$

Based on A_g ($F_t = 0.60\,F_y = 22$ ksi):

$$P_t = A_g F_t = \left(12 \times \frac{1}{2}\right)(22) = 132 \text{ kips}$$

Based on A_n [$F_t = 0.50\,F_u = 0.50(58) = 29.0$ ksi; shear lag consideration not applicable]:

$$P_t = A_n F_t = 4.50(29.0) = 130.5 \text{ kips}$$

Therefore, bolt shear governs the strength of the lap joint. The joint has a tensile capacity of 63.0 kips.

Example 7-2

Rework Example 7-1 assuming that the bolts are $\frac{7}{8}$-in.-diameter A325X (bearing-type connection with threads excluded from the shear plane) in standard holes.

Solution   *Note:* All table references are to the AISCM, Part 4.

Bolt shear: The allowable shear stress F_v from (Table I-D) is 30 ksi. Therefore, the allowable shear load per bolt:

$$r_v = A_b F_v$$

$$= 0.6013(30) = 18 \text{ kips}$$

Note that the allowable shear load per bolt could also be obtained directly, without computations from Table I-D.

Allowable shear load per connection (six bolts):

$$P = 18(6) = 108 \text{ kips}$$

Bearing (on $\frac{1}{2}$-in.-thick plate):   Same as for friction-type joint of Example 7-1:

$$P = 130.8 \text{ kips}$$

Tension: Same as for friction-type joint of Example 7-1:

$$P_t = 130.5 \text{ kips}$$

Therefore, the tensile capacity = 108 kips.

Example 7-3

A single-angle tension member in a roof truss is attached to a $\frac{3}{8}$-in.-thick gusset plate with A325 $\frac{3}{4}$-in.-diameter high-strength bolts (A325N) in standard holes, as shown in Fig. 7-12. The gusset plate and the angle are A36 steel ($F_u = 58$ ksi). The tension member is to support a 42-kip load. Determine if the member and the connection are adequate.

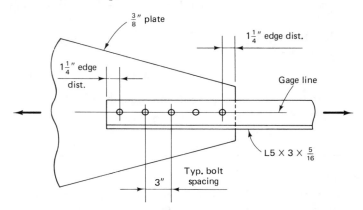

Figure 7-12   Single-angle tension member.

Solution   *Note:* All table references are to the AISCM, Part 4.

Bolt shear (single shear)   See Table I-D.

$$P = 9.3(5) = 46.5 \text{ kips}$$

Bearing (on thinnest material: $\frac{5}{16}$-in.-thick angle):   From Table I-E, based on 3-in. bolt spacing, the capacity of one bolt is 20.4 kips. Based on $1\frac{1}{4}$-in. edge distance, the capacity of one bolt is 11.3 kips. Therefore,

$$P = 11.3(5) = 56.5 \text{ kips}$$

Tension: In the design of this truss, the angle tension member was designed based on an assumed end connection of one gage line and one hole per cross section. The actual connection agreed with the assumption. Under these conditions, it would not be necessary to check the angle in tension. However to be complete, we will determine the tensile capacity of the single angle. Calculate the net area (A_n):

$$A_n = A_g - A_{holes}$$

$$= 2.40 - 0.875(0.3125) = 2.13 \text{ in.}^2$$

For calculation of effective net area A_e, $C_t = 0.85$ (see Table 4-1 of this text). Based on A_e, $[F_t = 0.50F_u = 0.50(58.0) = 29.0 \text{ ksi}]$:

$$P_t = A_e F_t = C_t A_n F_t = 0.85(2.13)(29.0) = 52.5 \text{ kips}$$

Based on A_g $(F_t = 0.60F_y = 22 \text{ ksi})$:

$$P_t = A_g F_t = 2.40(22) = 52.8 \text{ kips}$$

The tensile capacity of the angle is taken as the lowest of the three capacities calculated. Bolt shear, $P = 46.5$ kips, controls. This is greater than the actual load of 42 kips. Therefore, the connection and the member are O.K.

Example 7-4

Compute the tensile capacity P for the double shear butt joint shown in Fig. 7-13. The plates are A36 steel with an $F_u = 58$ ksi. The high-strength bolts are $\frac{7}{8}$-in.-diameter A325 in standard holes. Assume that the connection is

(a) Friction-type (A325F).
(b) Bearing-type, with threads excluded from the shear plane (A325X).
(c) Bearing-type, with threads in the shear plane (A325N).

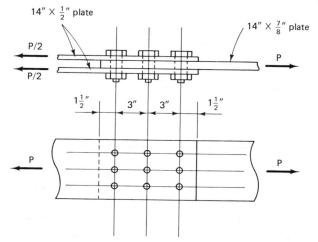

Figure 7-13 Double shear butt joint.

Solution *Note:* All table references are to the AISCM, Part 4.

(a) *A325F connection.*   Bolt shear (DS):

$$P = 21(9) = 189 \text{ kips (Table I-D)}$$

Bearing (on $\frac{7}{8}$-in. plate):

$$P = 38.1(9) = 342.9 \text{ kips (Table I-E)}$$

Net area tension:

$$P_t = A_n F_t = (14 - 3)(0.875)(29) = 279 \text{ kips}$$

Gross area tension:

$$P_t = A_g F_t = 14(0.875)(22) = 270 \text{ kips}$$

Tensile capacity = 189 kips.

(b) *A325X connection.*   Bolt shear (DS):

$$P = 36.1(9) = 324.9 \text{ kips (Table I-D)}$$

Bearing (on $\frac{7}{8}$-in. plate):

$$P = 38.1(9) = 342.9 \text{ kips (Table I-E)}$$

Tension:

$$P_t = 270 \text{ kips (same as A325F connection)}$$

Tensile capacity = 270 kips.

(c) *A325N connection.*   Bolt shear (DS):

$$P = 25.3(9) = 227.7 \text{ kips (Table I-D)}$$

Bearing (on $\frac{7}{8}$-in. plate):

$$P = 38.1(9) = 342.9 \text{ kips (Table I-E)}$$

Tension:

$$P_t = 270 \text{ kips (same as A325F connection)}$$

Tensile capacity = 228 kips.

Example 7-5

A tension member made up of a pair of angles is connected to a column with eight $\frac{3}{4}$-in.-diameter A325 high-strength bolts in standard holes as shown in Fig. 7-14. Note that the 70.7-kip forces are components of the 100-kip force. All structural steel is A36. Determine if the connection to the column is satisfactory. (Assume that the connection between the angles and the structural tee is satisfactory.) Consider

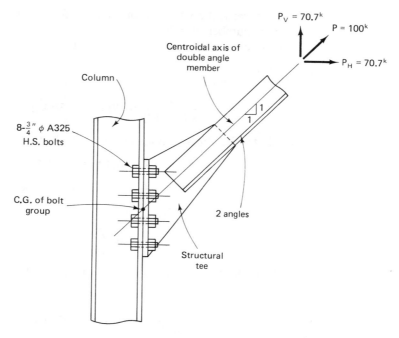

Figure 7-14   Bracing connection.

(a) A bearing-type connection with threads excluded from the shear plane (A325X).

(b) A friction-type connection (A325F).

Solution   In this connection, the bolts are subjected to combined shear and tension. Note that the connection is detailed so that the centroidal axis of the tension member intersects the center of the bolt group. This will eliminate eccentricity and each bolt is considered equally loaded. The problem of combined stresses and the associated allowable stresses is discussed in the AISCS, Section 1.6. In this solution, actual stresses will be compared with allowable stresses.

(a) *A325X connection.*   Allowable bolt shear stress:

$$F_v = 30 \text{ ksi (Table I-D, AISCM, Part 4)}$$

Actual bolt shear stress:

$$f_v = \frac{P_V}{nA_b}$$

where P_V = vertical component of the axial force in the tension member
n = number of bolts
A_b = cross-sectional area of one bolt

Therefore,

$$f_v = \frac{70.7}{8(0.4418)} = 20 \text{ ksi} < 30 \text{ ksi} \qquad \textbf{O.K.}$$

Allowable tensile stress in bolts (AISCS, Table 1.6.3):

$$F_t = 55 - 1.4f_v \text{ (not to exceed 44 ksi)}$$

$$= 55 - 1.4(20) = 27.0 \text{ ksi} < 44 \text{ ksi} \qquad \textbf{O.K.}$$

Actual tensile stress in bolts:

$$f_t = \frac{P_H}{nA_b}$$

where P_H is the horizontal component of the axial force in the tension member. Thus

$$f_t = \frac{70.7}{8(0.4418)} = 20 \text{ ksi} < 27.0 \text{ ksi} \qquad \textbf{O.K.}$$

The connection is satisfactory.

(b) *A325F connection.*   Actual tensile stress in bolts:

$$f_t = \frac{70.7}{8(0.4418)} = 20 \text{ ksi}$$

From Table I-A in Part 4 of the AISCM, the allowable tensile stress in the bolts is

$$F_t = 44 \text{ ksi} > 20 \text{ ksi} \qquad \textbf{O.K.}$$

Allowable bolt shear stress, where bolts are subjected to both shear and tension (AISCS, Section 1.6.3):

$$F_v = 17.5\left(1 - \frac{f_t A_b}{T_b}\right)$$

$$= 17.5\left[1 - \frac{20(0.4418)}{28}\right]$$

$$= 11.98 \text{ ksi}$$

T_b, the minimum bolt tension applied in the tightening process, may be found in the AISCS, Table 1.23.5, or in the SSJ, Table 3.

Actual bolt shear stress:

$$f_v = \frac{70.7}{8(0.4418)} = 20 \text{ ksi}$$

Since $f_v > F_v$, the connection is *not satisfactory* as a friction type.

Example 7-6

A truss member made up of two C10 × 15.3 channels is connected to a $\frac{3}{8}$-in.-thick gusset plate as shown in Fig. 7-15. All structural steel is A36 ($F_u = 58$ ksi). Determine the number of $\frac{3}{4}$-in.-diameter high-strength bolts in standard holes required to develop the full tensile capacity of the channels. Assume two gage lines. Design the connection and establish pitch and edge distance. Consider

(a) A friction-type connection (A325F).
(b) A bearing-type connection with threads in the shear plane (A325N).
(c) A bearing-type connection with threads excluded from the shear plane (A325X).

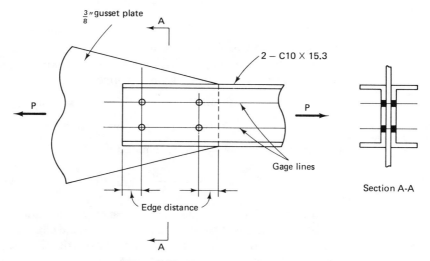

Figure 7-15   Truss member connection.

Solution   *Note:* Table references are to the AISCM, Part 4.

The *tensile capacity* of the two channels assuming two gage lines and two holes per cross-sectional plane per channel (refer to Chapter 4 of this text):

$$A_n = A_g - A_h$$
$$= 2(4.49) - 4(0.875)(0.24) = 8.14 \text{ in.}^2$$

Based on A_e, assuming that there will be at least three bolts per gage line so that $C_t = 0.85$, and using $F_t = 0.50F_u = 29.0$ ksi:

$$P_t = A_e F_t = C_t A_n F_t = 0.85(8.14)(29.0) = 201 \text{ kips}$$

Based on A_g, ($F_t = 0.60F_y = 22$ ksi):

$$P_t = A_g F_t = 2(4.49)(22) = 198 \text{ kips}$$

Therefore, the tensile capacity for which the connection will be designed is 198 kips as controlled by the A_g consideration.

(a) *Design the A325F friction-type connection.* The capacity of one bolt in double shear is 15.5 kips (Table I-D). The capacity of one bolt in bearing on $\frac{3}{8}$-in.-thick plate (Table 1-E) is

$$24.5 \text{ kips (assuming 3-in. bolt spacing)}$$

$$16.3 \text{ kips (assuming } 1\tfrac{1}{2}\text{-in. edge distance)}$$

Therefore, the number of bolts required is

$$n = \frac{198}{15.5} = 12.8 \text{ bolts}$$

Use 14 bolts with a pitch of 3 in. and an edge distance of $1\tfrac{1}{2}$ in.

(b) *Design the A325N bearing-type connection* with threads in the shear plane. The capacity of one bolt in double shear is 18.6 kips (Table I-D). The capacity of one bolt in bearing on $\frac{3}{8}$-in.-thick plate (Table I-E) is

$$24.5 \text{ kips (assuming 3-in. bolt spacing)}$$

$$16.3 \text{ kips (assuming } 1\tfrac{1}{2}\text{-in. edge distance)}$$

Therefore, the number of bolts required is

$$n = \frac{198}{16.3} = 12.2 \text{ bolts}$$

Use 14 bolts. Since bearing (edge distance consideration) controls, the number of bolts could be reduced by increasing the edge distance, thereby increasing the bolt capacity in bearing (based on edge distance) and permitting double shear capacity to control.

If double shear controlled, the number of bolts required would be

$$n = \frac{198}{18.6} = 10.7$$

This would require 12 bolts instead of 14. Therefore, increase the edge distance. The *allowable bearing load per bolt P* is

$$P = tdF_p$$

From the SSJ, Table 2:

$$F_p = \frac{LF_u}{2d}$$

where L is the edge distance (see Fig. 7-9). Therefore,

$$P = \frac{tdLF_u}{2d}$$

and noting that the load per bolt P is $198/12$ (kips), the minimum required edge distance L is (refer to the AISCS, Section 1.16.5)

$$\frac{2dP}{tdF_u} = \frac{2P}{F_u t} = \frac{2(198)}{12(58)(0.375)} = 1.52 \text{ in.}$$

Use 12 bolts with a pitch of 3 in. and edge distance of $1\frac{3}{4}$ in. Note that in part (a), since bolt shear controlled, an increase of the edge distance would have been no benefit.

(c) *Design the A325X bearing-type connection* with threads excluded from the shear plane. The capacity of one bolt in double shear is 26.5 kips (Table I-D). The capacity of one bolt in bearing on $\frac{3}{8}$-in.-thick plate (Table I-E) is

$$24.5 \text{ kips (assuming 3-in. bolt spacing)}$$

$$16.3 \text{ kips (assuming } 1\frac{1}{2}\text{-in. edge distance)}$$

Increase the edge distance so that 24.5 kips per bolt will control. Then number of bolts required is

$$n = \frac{198}{24.5} = 8.08 \text{ bolts}$$

Use 10 bolts with a pitch of 3 in.

$$\text{edge distance required} = \frac{2(198)}{10(58)(0.375)} = 1.82 \text{ in.}$$

Use an edge distance of 2 in.

Example 7-7

Determine the number of $\frac{3}{4}$-in.-diameter high-strength bolts (A325) in standard holes required for the bracing connection shown in Fig. 7-16. Note that the 60-kip and 80-kip forces are components of the 100-kip force. All structural steel is A36 ($F_u = 58$ ksi). Design as a bearing-type connection with threads excluded from the shear plane (A325X). Assume that the double-angle tension member

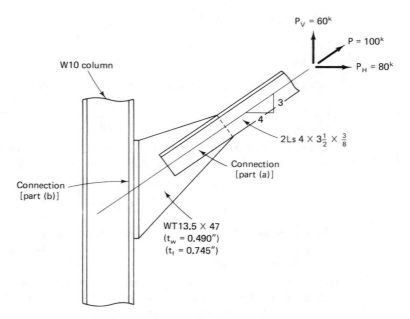

Figure 7-16   Bracing connection design.

is adequate. Consider

(a) Tension member-to-structural tee connection.
(b) Structural tee-to-column connection.

Solution   *Note:* All table references are to the AISCM, Part 4, unless noted otherwise.

(a) This is the connection of the angles to the structural tee (A325X). (Bolts are in double shear. The bearing is on the web of the structural tee.) The capacity of one bolt in double shear (Table I-D) is 26.5 kips. From Table I-E, the capacity of one bolt in bearing on the web of the structural tee ($t_w = 0.490$ in.) is

$$0.490(74.3) = 36.4 \text{ kips (assuming 3-in. bolt spacing)}$$

$$0.490(43.5) = 21.3 \text{ kips (assuming } 1\tfrac{1}{2}\text{-in. edge distance)}$$

Increase the edge distance so that double shear will control (26.5 kips per bolt). The number of bolts required is

$$n = \frac{100}{26.5} = 3.8 \text{ bolts}$$

Use four bolts with a 3-in. pitch. The new required edge distance is

$$\frac{2P}{F_u t} = \frac{2(100)}{4(58)(0.490)} = 1.76 \text{ in.}$$

Use an edge distance of 2 in.

(b) This is the connection of the structural tee to the flange of the column. Bolts are in combined shear and tension (A325X). The design of this connection is a trial-and-error procedure. Assume eight bolts, with four in each of two vertical rows and assume that the tensile load passes through the center of gravity of the bolt group.

Allowable bolt shear stress:

$$F_v = 30 \text{ ksi (Table I-D)}$$

Allowable tensile stress in bolts:

$$F_t = 44 \text{ ksi (Table I-A)}$$

Actual bolt shear stress (single shear):

$$f_v = \frac{P_V}{nA_b} = \frac{60}{8(0.4418)} = 17.0 \text{ ksi} < 30 \text{ ksi} \qquad \textbf{O.K.}$$

Allowable tensile stress in bolts (AISCS, Table 1.6.3):

$$F_t = 55 - 1.4 f_v \text{ (not to exceed 44)}$$

$$= 55 - 1.4(17) = 31.2 \text{ ksi} < 44 \text{ ksi} \qquad \textbf{O.K.}$$

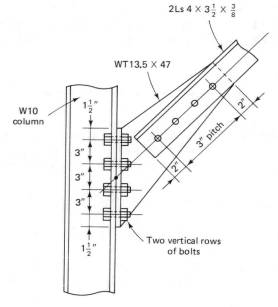

Figure 7-17 Design sketch.

Actual tensile stress in bolts:

$$f_t = \frac{P_H}{nA_b} = \frac{80}{8(0.4418)} = 22.6 \text{ ksi} < 31.2 \text{ ksi} \qquad \textbf{O.K.}$$

Use eight bolts, as shown in Fig. 7-17. The prying force on these bolts should also be considered (reference is made to the AISCM, Part 4, for example design calculations).

7-6 FRAMED BEAM CONNECTIONS

Framed beam connections are probably the most commonly used type of beam-to-column and beam-to-girder connections. These connections are portrayed in Fig. 7-3. This type of connection is used in Type 2 construction, where lateral loads (such as wind) are neglected in the design or where other systems in the structure will resist these loads. The framed connection is categorized as a simple beam connection. It is unrestrained and flexible and it permits freedom of rotation at the supports. In reality a certain amount of capacity to resist moment is developed by these connections but it is disregarded and the connections are usually designed to resist shear only.

The analysis of a framed beam connection using angles on the beam web follows. In addition to the possible failure modes discussed previously (bolt shear and bearing as governed by bolt spacing and edge distance), other types of failure modes are considered in framed connection analysis. These are (1) net area shear for the angles, and (2) block shear in the webs of coped beams. Net area shear is based on an allowable shear stress of $0.3F_u$. Block shear is a combined shear and tension tear-out type failure of the beam web. It is discussed in the AISCS Commentary, Section 1.5.1.2.

Example 7-8

Compute the load-carrying capacity of the framed connection shown in Fig. 7-18. The structural steel is A36 ($F_u = 58$ ksi). The bolts are $\frac{3}{4}$-in.-diameter high-strength bolts in standard holes. The connection is an A325F friction-type connection.

Solution   *Note:* All table references are to the AISCM, Part 4.

A. Consider bolts through the web of the W18 × 50. These bolts are in double shear.
   1. The capacity of the connection based on *bolt shear* may be calculated to be 46.4 kips or it may be read directly from Table II-A.

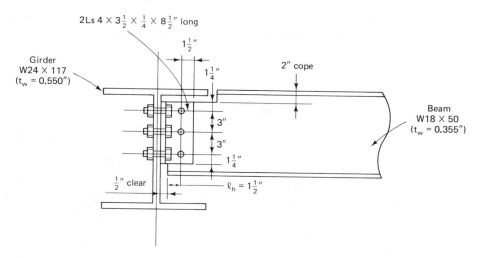

Figure 7-18   Framed beam connection analysis.

2. The capacity of the connection based on *bearing on the web* (0.355 in. thick) is:

   Considering a $1\frac{1}{4}$-in. vertical edge distance, from Table I-E:

   $$0.355(36.3)(3) = 38.7 \text{ kips}$$

   Considering 3-in. bolt spacing, from Table I-E:

   $$0.355(65.3)(3) = 69.5 \text{ kips}$$

   Considering a $1\frac{1}{2}$-in. horizontal edge distance, from Table I-F:

   $$0.355(43.5)(3) = 46.3 \text{ kips}$$

3. The capacity of the connection based on *bearing on the angles* (total thickness is $\frac{1}{2}$ in.) is:

   Considering a $1\frac{1}{4}$-in. vertical edge distance from Table I-E:

   $$3(18.1) = 54.3 \text{ kips}$$

   Considering 3-in. bolt spacing, from Table I-E:

   $$3(32.6) = 97.8 \text{ kips}$$

   Considering a $1\frac{1}{2}$-in. horizontal edge distance, from Table I-F:

   $$0.50(43.5)(3) = 65.3 \text{ kips}$$

4. The capacity of the connection based on *shear on the net area* of the connecting angles may be calculated (using $F_v = 0.30F_u$) or it may be read directly from Table II-C as 52.7 kips.

5. The capacity of the connection based on *web tear out* (block shear) with

$\ell_v = 1\frac{1}{4}$ in. and $\ell_h = 1\frac{1}{2}$ in. may be determined by calculation or through the use of coefficients from Table I-G:

$$C_1 = 1.13$$

$$C_2 = 0.99$$

$$\text{connection capacity} = (C_1 + C_2)F_u t_w$$

$$= (1.13 + 0.99)(58)(0.355)$$

$$= 43.7 \text{ kips}$$

The controlling value for this part of the connection is 38.7 kips. Some of the steps shown could have been eliminated and this portion significantly abbreviated; however, for purposes of illustration, all steps are included.

B. Consider the bolts through the supporting member (W24 × 117). These bolts are in single shear. The following items should be checked:
1. Capacity in single shear from Table II-A: 46.4 kips.
2. Capacity in bearing on the 0.550-in.-thick web of the supporting member, based on 3-in. bolt spacing, from Table I-E:

$$0.550(65.3)(6) = 215 \text{ kips}$$

3. Capacity in bearing on the $\frac{1}{4}$-thick angles:
   Based on a $1\frac{1}{4}$-in. vertical edge distance from Table I-E:

$$6(9.1) = 54.6 \text{ kips}$$

Based on 3-in. bolt spacing from Table I-E:

$$6(16.3) = 97.8 \text{ kips}$$

4. Capacity based on shear on the net area of the connection angles is the same as for the other part of the connection: 52.7 kips.
5. Web tear-out is not applicable for this part of the connection.

Hence the controlling value for the total connection is 38.7 kips. Again, for part B of the calculations, some of the steps could have been eliminated and the problem shortened. A shortened solution is furnished in Example 7-9.

Example 7-9

Compute the load-carrying capacity of the simple beam connection shown in Fig. 7-19. The structural steel is A36 ($F_u = 58$ ksi). The bolts are 1-in.-diameter high-strength bolts in standard holes. The connection is an A325F friction-type connection.

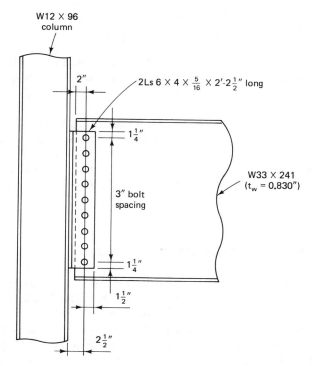

Figure 7-19   Framed beam connection.

Solution   *Note:* All table references are to the AISCM, Part 4.

A. Consider bolts through the W33 × 241 web. These bolts are in double shear. Capacities are as follows:
   1. Double shear: 247 kips (Table II-A).
   2. Bearing on the total angle thickness of $\frac{5}{8}$ in. is more critical than on the web thickness of 0.830 in. Of the three possible critical edge distances and bolt spacing, the vertical edge distance of $1\frac{1}{4}$ in. in the angles is most critical. From Table I-E, we see that the $1\frac{1}{4}$-in. vertical edge distance will control overall. Therefore, the capacity is

   $$22.7(9) = 204.3 \text{ kips}$$

   3. Shear on net area of connection angles: 184 kips (Table II-C).
      Since the beam is not coped, web tear-out (block shear) does not apply.

B. Consider bolts through the column flange (flange $t_f = 0.900$ in.). These bolts are in single shear. Capacities are as follows:
   1. Single shear: 247 kips (Table II-A).
   2. Bearing on angle thickness of $\frac{5}{16}$ in. controls over bearing on flange thickness of 0.900 in. From Table I-E, vertical edge distance of $1\frac{1}{4}$ in.

controls over bolt spacing of 3 in. Therefore, the capacity is

$$11.3(2)(9) = 203.4 \text{ kips}$$

3. Shear on the net area of the connection angles is the same as for part A of this connection = 184 kips. Also, web tear-out does not apply to this part of the connection.

Therefore, the load-carrying capacity of this connection is 184 kips.

The design of this type of connection is usually made by the structural detailer as opposed to the structural designer. Hence the design of a typical connection is furnished in Chapter 12.

7-7 UNSTIFFENED SEATED BEAM CONNECTIONS

An alternative type of "simple beam" connection that theoretically will behave as does the framed beam connection with respect to end rotation is the unstiffened seated beam connection. The seated connection has an angle under the beam which is usually shop-connected to the supporting member. The supporting member may be a column or a girder. In addition, there is another angle (generally on top of the beam) which is field-connected to the supporting member. The top angle may be placed at an optional location as shown in Fig. 7-3 if space is restricted. The design of an unstiffened seated connection is generally accomplished by the use of tables in the AISCM. A sample design problem using these tables is furnished in Chapter 12 of this text. A behavioral approach to the design problem is furnished in Example 7-10.

In the design of a seated connection, it is assumed that all of the end reaction from the beam is delivered to the seat angle. The top angle is added to provide lateral support and stability at the top flange of the beam. This angle is assumed to carry no load. Therefore, it can be relatively small and can have as few as two bolts in each of its legs. The common size for a top angle would be L$3\frac{1}{2} \times 3\frac{1}{2} \times \frac{1}{4}$ or L$4 \times 3 \times \frac{1}{4}$. The size is generally based on judgment and practical considerations. There are two main considerations in the design of the main supporting seat angle. First, the seat, or outstanding leg, must be sufficiently long to satisfy web crippling requirements in the beam. Second, it must be sufficiently thick to support the beam reaction without exceeding the allowable bending stress.

The minimum length of the outstanding leg of the seat angle may be obtained from the web crippling expressions of Section 2-12 of this book:

$$N = \frac{R}{t_w(0.75F_y)} - k$$

The required thickness of the seat angle may be obtained by designing the outstanding leg as a cantilever to support the end of the beam. The critical section for bending moment in the angle is assumed to be at the toe of the fillet, which may be taken as $(t + \frac{3}{8})$ from the back of the angle vertical leg. Also, an average distance of $\frac{1}{2}$ in. is assumed between the end of the beam and the face of the support, as shown in Fig. 7-20.

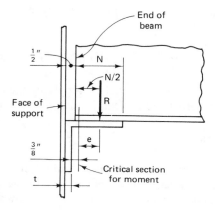

Figure 7-20   Seat angle analysis.

Therefore, the distance from the applied load R to the critical moment section may be computed from

$$e = \frac{N}{2} + \frac{1}{2} - t - \frac{3}{8}$$

$$= \frac{N}{2} + \frac{1}{8} - t$$

and the bending moment M is

$$M = Re = R\left(\frac{N}{2} + \frac{1}{8} - t\right)$$

This is the applied moment and must not exceed the moment capacity of the outstanding leg of the angle, which may be calculated as

$$M_R = F_b S$$

$$= F_b\left(\frac{bt^2}{6}\right)$$

where F_b = allowable bending stress, which may be taken as $0.75F_y$ (ksi)

b = width of beam seat (in a direction normal to the beam it supports) (in.)

t = thickness of angle (in.)

S = section modulus of the outstanding leg of the seat angle (rectangular shape) (in.^{3})

Since most steel used for connections is A36, F_b for the angle would be

$$0.75F_y = 0.75(36) = 27 \text{ ksi}$$

and the expression for moment capacity would become

$$M_R = 27\left(\frac{bt^2}{6}\right) = 4.5\ bt^2\ (\text{in.-kips})$$

Equating the applied moment and moment capacity to each other and rearranging terms, we have

$$4.5bt^2 + Rt = \frac{R}{8}(4N + 1)$$

where R is in kips and all other terms in inches. Applying the quadratic equation and discarding the minus sign before the radical as having no significance gives us

$$\text{required } t = \frac{-R + \sqrt{R^2 + 2.25Rb(4N + 1)}}{9.0(b)}$$

This equation provides a direct solution for the required thickness of the seat angle.

The number of bolts necessary for the connection of the *vertical leg* of the seat angle to the *supporting member* may be determined by allowable shear and bearing values. Despite the eccentricity of the reaction, it is assumed that no additional stresses are developed in the bolts.

This type of connection is generally used where beam reactions are relatively small. Therefore, a 4-in. *outstanding* leg with a minimum of two bolts is usually sufficient for the connection between the *beam and seat angle*. Corresponding slotted holes (for erection purposes) are generally provided in the beam flange.

Example 7-10

Design a seated beam connection for a W18 × 55 beam. The supporting column is a W10 × 60, as shown in Fig. 7-21. All structural steel is A36. The beam reaction is 31.4 kips. Use $\frac{3}{4}$-in.-diameter A325N high-strength bolts in standard holes.

Solution   *Note:* All table references are to the AISCM, Part 4, unless noted otherwise.

1. The required bearing length is

$$N = \frac{R}{0.75F_y t_w} - k$$

$$= \frac{31.4}{27(0.390)} - 1\frac{5}{16} = 1.67 \text{ in.}$$

Therefore, use a 4-in. horizontal leg. This will allow sufficient space for the bolt holes.

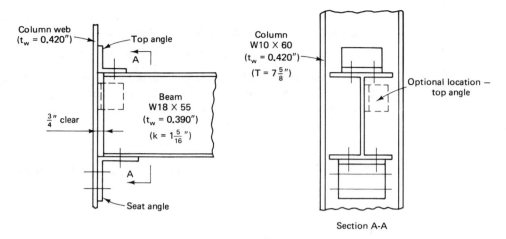

Figure 7-21   Seated beam connection.

2. Based on the T dimension for the W10 × 60, where T is the distance on the web between toes of the fillets (see the AISCM sketches, Part I, W shapes), use a 6-in. length of seat angle. For required thickness calculations, this will be the b dimension.

   To be consistent with the AISCM tables, the actual furnished bearing length should be used for calculations rather than the required bearing length. This will be

$$N = 4 - \frac{3}{4} = 3\frac{1}{4} \text{ in.}$$

As mentioned previously, the nominal beam setback is $\frac{1}{2}$ in. However, to provide for possible mill underrun in beam length, which would create a greater eccentricity, a $\frac{3}{4}$-in. setback is assumed.

$$\text{required } t = \frac{-R + \sqrt{R^2 + 2.25Rb(4N + 1)}}{9b}$$

$$= \frac{-31.4 + \sqrt{31.4^2 + 2.25(31.4)(6)[4(3.25) + 1]}}{9(6)}$$

$$= \frac{-31.4 + \sqrt{985.96 + 5510.7 + 423.9}}{54}$$

$$= \frac{-31.4 + \sqrt{6920.56}}{54} = \frac{-31.4 + 83.2}{54} = 0.96 \text{ in.}$$

Therefore, use an angle thickness of 1 in.

3. The connection for the vertical leg of the seat angle will be designed assuming that beams frame into the column web from both sides.

$$\text{total reaction} = 31.4(2) = 62.8 \text{ kips}$$

The capacity of one bolt in double shear is 18.6 kips, from Table I-D. For bearing capacity, assume a 3-in. bolt spacing. Therefore, the capacity of one bolt in bearing on the 0.420-in.-thick column web may be determined. From Table I-E, the bearing capacity is

$$0.420(65.3) = 27.4 \text{ kips per bolt}$$

Therefore, double shear controls and the number of bolts required is

$$\frac{62.8}{18.6} = 3.4 \text{ bolts}$$

Use four bolts in a pattern such as depicted in the Type B seat angle in the AISCM, Part 4, Table V. From Table V-D, it is seen that an L8 × 4 must be used in order to obtain the required 1 in. thickness. Therefore, for this seated beam connection, use one L8 × 4 × 1 × 6 in. with four $\frac{3}{4}$-in. high-strength bolts in the vertical leg (Type B pattern).

When end reactions are relatively large, the required thickness of the seat angle becomes excessive and it becomes necessary to use a *stiffened* seated beam connection as shown in the diagrams accompanying Table VII of the AISCM, Part 4. This type of connection is similar to the unstiffened type, except that in this type stiffener angles are placed under the outstanding leg of the seat angle and fitted to bear on the underside, thereby relieving the bending in the horizontal leg. The design is usually accomplished through the use of the standard tables in the AISCM, Part 4, Table VII.

7-8 END-PLATE SHEAR CONNECTIONS

A relatively recent type of flexible connection that has evolved is the end-plate shear connection. This type is generally economical and has been used successfully where beam end reactions are relatively light. It consists of a rectangular plate welded to the end of the *beam web* and generally bolted to the supporting member. Fabrication of this type of connection requires close control in cutting the beam to length as well as in squaring the beam ends. For adequate end rotation capacity the end-plate thickness should range from $\frac{1}{4}$ to $\frac{3}{8}$ in. inclusive, the plate depth is limited to the beam T dimension and only two lines of bolts are permitted. The gage, g, between lines of bolts should be between $3\frac{1}{2}$ and $5\frac{1}{2}$ in. and the edge distance should be $1\frac{1}{4}$ in.

The design of this type of connection is usually accomplished through the use of the standard tables in the AISCM, Part 4, Table IX. At this point, we will assume that the end plate is adequately welded to the end of the beam. The welded connection will be considered in Chapter 8. The remaining problem is to design the bolted connection to the supporting member. The design assumes no eccentricity and only considers shear and bearing.

Example 7-11

Design the end-plate shear connection as shown in Table IX, Part 4 of the AISCM, for a W18 × 60 beam framing to the flange of a W8 × 31 column. All structural steel is A36 (F_u = 58 ksi). The end reaction is 40 kips. Use $\frac{3}{4}$-in.-diameter A325N bolts (threads in the shear plane) in standard holes. Neglect welding of plate to end of beam. Use $\frac{5}{16}$-in.-thick plate and assume that the welding is adequate.

Solution   *Note:* All table references are to the AISCM, Part 4.

The capacity of one bolt in single shear is 9.3 kips, from Table I-D. For bearing capacity, assume an edge distance of $1\frac{1}{4}$ in. and a 3-in. bolt spacing. The $\frac{5}{16}$-in. plate will control since it is thinner than the 0.435-in. flange thickness of the column. The capacity in bearing is governed by the $1\frac{1}{4}$-in. edge distance and is 11.3 kips per bolt from Table I-E. Therefore, shear controls and the required number of bolts is

$$n = \frac{40.0}{9.3} = 4.30 \text{ bolts} \qquad (\text{use six bolts})$$

The *required* plate length (vertically) is

$$2(3) + 2(1.25) = 8.5 \text{ in.}$$

The *allowable* plate length is the beam T dimension, which is $15\frac{1}{2}$ in. from AISCM, Part 1. Therefore, for this end-plate connection, use a plate $8 \times 8\frac{1}{2} \times \frac{5}{16}$ with six $\frac{3}{4}$-in.-diameter A325N high-strength bolts. The connection is shown in Fig. 7-22.

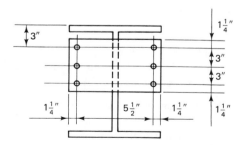

Figure 7-22   End-plate connection.

7-9 SEMI-RIGID CONNECTIONS

The framed beam, seated beam and end-plate shear connections discussed previously are designed primarily to transfer shear only. These connections are used in Type 2 construction. In essence, these are the connections that permit end rotation, thereby making the bending members behave as simply supported members. When members are to be continuous or when resistance to wind or other lateral forces is to be provided by the joints, connections must be used which offer predictable moment resistance.

Connections for Type 1 and Type 3 construction are categorized as rigid and semirigid connections, respectively. Various types are shown in Fig. 7-3. One of the simplest semirigid connection types is shown in Fig. 7-3(e). The web angles are designed to resist the shear as in the framed beam connection. The top and bottom angles (sometimes called **clip angles**) are designed to resist the moment. A modification of this connection is the use of a pair of structural tees instead of the top and bottom angles. The tendency of the end of the beam to rotate is resisted by a couple produced by a horizontal tensile force on the top flange and a horizontal compressive force on the bottom flange. The magnitude of these horizontal forces may be determined by dividing the end moment by the nominal depth of the beam

$$T = C = \frac{M}{d}$$

Assuming the use of angles on top and bottom of the beam, the horizontal force can be developed by transferring the force in the beam flange to the angles by single shear in the connecting bolts. This force is then transmitted by bending in the angles to the column flange by tension in the top connecting bolts and bearing between the column flange and the bottom clip angle. The design for the end moment involves determining how many bolts are required through top and bottom angles as well as the size and thickness of the angles themselves. In the classical approach to this design problem, the angle thickness may be determined by assuming some bending behavior of the angle and applying the beam flexure formula. The following example demonstrates this approach. Example 7-13 will demonstrate the current AISC approach.

Example 7-12

Design a connection of the type shown in Fig. 7-23 to resist a moment of 50 ft-kips and a shear (reaction) of 35 kips. Use $\frac{3}{4}$-in.-diameter A325N high-strength bolts in standard holes. All structural steel is A36 ($F_u = 58$ ksi).

Solution   *Note:* All table references are to the AISCM, Part 4.

1. The connection on the beam web to resist the shear has been designed using the AISCM and consists of 2L5 × 3 1/2 × 5/16 × 8 1/2 in. with three rows of bolts.

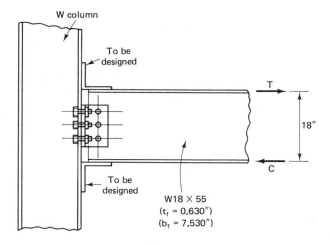

Figure 7-23   Semi-rigid connection.

2. Consider the bolts through the *beam* flanges and the top and bottom angles. The capacity of one bolt in single shear is 9.3 kips, from Table I-D. The capacity of one bolt in bearing (assume an angle thickness of $\frac{5}{8}$ in. or more) is 22.7 kips (based on a $1\frac{1}{4}$-in. edge distance), from Table I-E. The force in each flange (tension in the top and compression in the bottom) is calculated as

$$T = C = \frac{50(12)}{18} = 33.3 \text{ kips}$$

Therefore, the number of bolts required is

$$n = \frac{33.3}{9.3} = 3.6 \text{ bolts} \qquad (\text{use four bolts})$$

3. Consider the bolts through the *column* flanges and the top and bottom angles. The capacity of one bolt in tension is 19.4 kips, from Table I-A. Conservatively use the applied force $T = C = 33.3$ kips. The force is actually less since the distance between the bolts is in excess of 18 in. The number of bolts required is

$$n = \frac{33.3}{19.4} = 1.72 \text{ bolts} \qquad (\text{use two bolts})$$

4. Using an L7 × 4 to accommodate the required bolts and assuming a $\frac{3}{4}$-in. angle thickness, the design will be accomplished based on an assumed point of contraflexure located as shown in Fig. 7-24(b). This point is a point of zero moment, a point at which only shear exists. Therefore, the calculation of moment at the assumed points of equal maximum moment, shown in Fig. 7-24(c), becomes possible. Assume that the angle is 8 in.

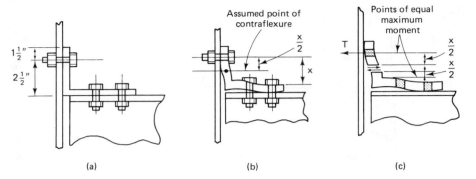

Figure 7-24 Top angle analysis.

long (to match approximately the width of the beam) and compute the thickness required based on an allowable bending stress F_b of $0.75F_y$.

The distance from the bolts in tension to the top of the horizontal leg of the angle is [see Fig. 7-24(b)]

$$x = 2.5 - 0.75 = 1.75 \text{ in.}$$

Determine the tensile force using the actual vertical distance between forces T and C of 23 in.:

$$T = C = \frac{50(12)}{23} = 26.1 \text{ kips}$$

Therefore, the moment at the assumed critical sections is

$$M = 26.1\left(\frac{1.75}{2}\right) = 22.8 \text{ in.-kips}$$

Solving for the required thickness, recall that for a rectangular shape

$$S = \frac{bt^2}{6}$$

and that

$$\text{required } S = \frac{M}{F_b}$$

Therefore,

$$\text{Required } t = \sqrt{\frac{6(M)}{F_b(b)}}$$

$$= \sqrt{\frac{6(22.8)}{0.75(36)(8)}} = 0.80 \text{ in. } > 0.75 \text{ in.} \qquad \textbf{N.G.}$$

Recalculate, assuming an L8 × 4 × 1 × 8 in.:

$$x = 2.5 - 1 = 1.5 \text{ in.}$$

$$T = C = 26.1 \text{ kips (unchanged)}$$

$$M = 26.1\left(\frac{1.50}{2}\right) = 19.6 \text{ in.-kips}$$

$$\text{required } t = \sqrt{\frac{6(19.6)}{0.75(36)(8)}} = 0.74 \text{ in.}$$

Therefore, use top and bottom angles L8 × 4 × 1 × 8 in.

The preceding example does not take into account *prying action*. The actual distribution of stress in the clip angle is very complex. The stiffness of the clip angle, rather than its bending strength, is usually more critical since large deformation would be detrimental. The AISCM provides an empirical design method which will produce conservative results. It is contained in Part 4 and is entitled "Hanger Type Connections." The following example will demonstrate the method. Reference should be made to the AISCM.

Example 7-13

Rework Example 7-12 using the AISCM empirical design method.

Solution   *Note:* The steps of the AISCM **Design Procedure** will be followed.

1. This refers to the bolts into the column flange. Two bolts are required, from step 3 of Example 7-12.

$$p = \frac{8}{2} = 4 \text{ in.}$$

$$T = \frac{26.1}{2} = 13.0 \text{ kips}$$

2. Load per linear inch:

$$\frac{26.1}{8} = 3.26 \text{ kips/in.}$$

Assuming a $\frac{5}{8}$-in. angle thickness:

$$\text{estimated } b = 2.5 - 0.63 = 1.87 \text{ in.}$$

From the Preliminary Selection Table, try an angle thickness $t = \frac{5}{8}$ in. (0.625 in.) and tentatively select an L7 × 4 × $\frac{5}{8}$. From Fig. 7-25:

$$t = 0.625 \text{ in.}$$

$$b = 2.5 - 0.625 = 1.875 \text{ in.}$$

$$a = 4 - 2.5 = 1.5 \text{ in.}$$

$$b' = 2.5 - 0.375 - 0.625 = 1.50 \text{ in.}$$

$$a' = 1.5 + 0.375 = 1.875 \text{ in.}$$

$$d' = \frac{13}{16} \text{ in.} = 0.813 \text{ in.}$$

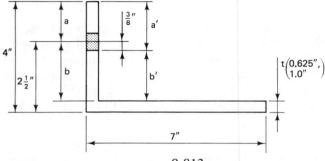

Figure 7-25 Clip angle design.

3. $\delta = 1 - \dfrac{0.813}{4} = 0.797$

4. $M = \dfrac{4(0.625)^2(36)}{8} = 7.031$ in.-kips

5. $\alpha = \dfrac{[13.0(1.875)/7.031] - 1}{0.797} = 3.095 > 1.0$ (use $\alpha = 1.0$)

6. $B_c = 13.0\left[1 + \dfrac{0.797(1.0)}{1 + 0.797(1.0)}\left(\dfrac{1.50}{1.875}\right)\right] = 17.61$ kips

7. Required $t = \left[\dfrac{8(17.61)(1.875)(1.50)}{4(36)(1.875 + 0.797(1.0)(1.875 + 1.50))}\right]^{1/2}$

$$= 0.776 \text{ in.}$$

$$0.776 > \frac{5}{8} \qquad \text{select thicker angle}$$

Try L8 × 4 × 1 × 8 in.

$$t = 1 \text{ in.}$$

$$p = 4 \text{ in.}$$

$$T = 13.0 \text{ kips}$$

1-in. angle thickness:

$$b = 1.5 \text{ in.}$$

$$a = 1.5 \text{ in.}$$

$$b' = 2.5 - 1 - 0.375 = 1.125 \text{ in.}$$

$$a' = 4 - 1 - 1.125 = 1.875 \text{ in.}$$

$$d' = 0.813 \text{ in.}$$

$$\delta = 0.797$$

$$M = \frac{4(1)^2(36)}{8} = 18.0 \text{ in.-kips}$$

$$\alpha = \frac{[13.0(1.5)/18] - 1}{0.797} = 0.105 < 1.0 \qquad (\text{use } \alpha = 0.105)$$

$$B_c = 13\left[1 + \frac{0.797(0.105)}{1 + (0.797)(0.105)}\left(\frac{1.125}{1.875}\right)\right] = 13.60 \text{ kips}$$

$$\text{required } t = \left[\frac{8(13.60)(1.875)(1.125)}{4(36)\{1.875 + 0.797[0.105(1.875 + 1.125)]\}}\right]^{1/2} = 0.87 \text{ in.}$$

$$0.87 < 1.0 \qquad\qquad\qquad\qquad\qquad\qquad\qquad\qquad \textbf{O.K.}$$

Use L8 × 4 × 1 × 8 in.

7-10 *ECCENTRICALLY LOADED BOLTED CONNECTIONS*

When bolt groups are loaded by some external load that does not act through the center of gravity of the group, the load is said to be **eccentric.** It will tend to cause a relative rotation and translation of the connected parts, and the individual bolts will have unequal loads induced in them.

If the eccentrically applied load lies in the plane of the connection as shown in Fig. 7-26, and elastic behavior is assumed, the bolt group may be analyzed by resolving the eccentric load P into a concentric load acting through the centroid of the bolt group and a torsional moment M (where $M = Pe$). The moment acts with respect to the centroid of the bolt group as a center of rotation. Hence the forces acting on the bolts will be made up of two components: Q_v due to the axial effect of the eccentric load and Q_m due to the torsional moment effect as shown in Fig. 7-27.

Q_v will be the same for all the bolts and may be taken as the load P divided by the number of bolts (n):

$$Q_v = \frac{P}{n}$$

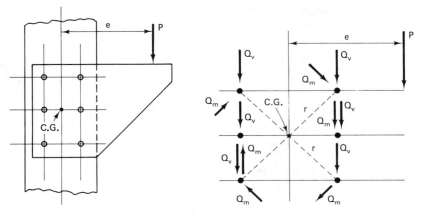

Figure 7-26   Eccentrically loaded
connection.

Figure 7-27   Bolt group forces.

Q_m will vary with the distance r from the center of gravity of the bolt group to the bolt and will act in a direction normal to a line from the bolt to the center of gravity. Therefore, the connection must be designed so that the resultant of these two components acting on any bolt does not exceed the maximum permissible bolt capacity as determined by shear or bearing.

The torsional load Q_m may be determined by applying the classic torsional stress formula for circular members to the bolt group:

$$f_v = \frac{Mr}{J}$$

where f_v = shear stress in any bolt
M = torsional moment (Pe)
r = radial distance from center of gravity of bolt group to any bolt
J = polar moment of inertia = $\Sigma\, Ar^2$

In the expression for polar moment of inertia (taken from any engineering mechanics text), A represents the cross-sectional area of one bolt. Since all bolts in any given connection will have the same cross-sectional area, and since r^2 may be expressed as x and y coordinates,

$$J = \sum A(x^2 + y^2)$$

which may be written as

$$J = A \sum(x^2 + y^2) = A\left(\sum x^2 + \sum y^2\right)$$

The torsional stress formula then becomes

$$f_v = \frac{Mr}{A(\Sigma\, x^2 + \Sigma\, y^2)}$$

If we multiply both sides of the equation by A, we obtain the torsional load Q_m on any bolt:

$$f_v A = Q_m = \frac{Mr}{\Sigma\, x^2 + \Sigma\, y^2}$$

If we assume r to be a *unit distance* from the center of gravity, the expression becomes

$$Q_{mu} = \frac{M}{\Sigma\, x^2 + \Sigma\, y^2}$$

This represents a force or load acting at a unit distance from the center of gravity. Knowing this force we can compute any force acting on any bolt normal to a radial line from the center of gravity to the bolt by multiplying Q_{mu} by the radial distance r to that bolt.

It is generally more convenient to resolve the Q_m force into vertical and horizontal components and then vectorially add the Q_v force to obtain the resultant force on the bolt. Usually, the bolt most remote from the group center of gravity and on the load side will be subjected to the critical (greatest) load and will determine the adequacy of the connection.

Example 7-14

For the eccentrically loaded bolted connection shown in Fig. 7-28, determine the force acting on the most critical bolt. Assume $\frac{3}{4}$-in.-diameter A325N bolts in standard holes. The applied load is 10,000 lb.

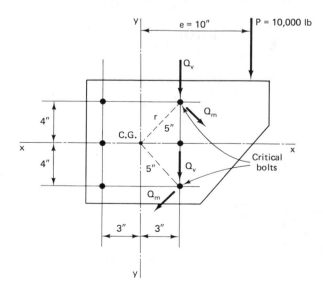

Figure 7-28 Eccentrically loaded bracket.

Solution

1. The torsional moment is

$$M = Pe = 10,000(10) = 100,000 \text{ in.-lb}$$

2. The polar moment of inertia (divided by A) is

$$\sum x^2 = 6(3)^2 = 54 \text{ in.}^2$$

$$\sum y^2 = 4(4)^2 = 64 \text{ in.}^2$$

$$\frac{J}{A} = \sum x^2 + \sum y^2 = 118 \text{ in.}^2$$

3. The torsional load (Q_{mu}) acting at a unit distance from the center of gravity of the bolt group is

$$Q_{mu} = \frac{M}{\sum x^2 + \sum y^2} = \frac{100,000}{118} = 847 \text{ lb/in.}$$

   The torsional load on the critical bolt is

$$Q_m = 847(5) = 4235 \text{ lb}$$

4. Horizontal component of $Q_m = 4235(4/5) = 3388$ lb
   Vertical component of $Q_m = 4235(3/5) = 2541$ lb

5. The force on the bolt due to the 10,000-lb load applied at the center of gravity is

$$Q_v = \frac{P}{n} = \frac{10,000}{6} = 1667 \text{ lb}$$

   The forces are depicted in Fig. 7-29.

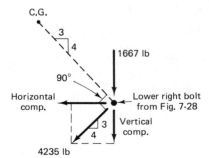

Figure 7-29 Forces on critical bolt.

6. The resultant force R on the critical bolt is determined with reference to Fig. 7-30.

$$R^2 = (1667 + 2541)^2 + 3388^2$$

$$R = 5402 \text{ lb}$$

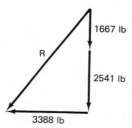

1667 lb

R

2541 lb

3388 lb

Figure 7-30   Vector diagram.

Example 7-15

Find the maximum load P that can be supported by the bracket shown in Fig. 7-31. Column and bracket are A36 steel. Use $\frac{7}{8}$-in.-diameter A325F high-strength bolts. Assume that the column flange and bracket are thick enough so that single shear in the bolts will control.

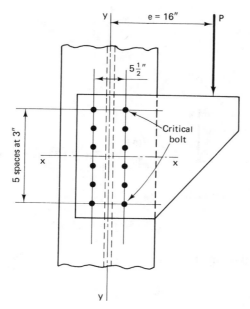

y

e = 16"

P

$5\frac{1}{2}''$

Critical bolt

5 spaces at 3"

x

x

y

Figure 7-31   Bracket analysis.

Solution

1. The torsional moment is

$$M = Pe = 16P \text{ in.-lb} \quad (\text{when } P \text{ is in pounds})$$

2. The polar moment of inertia (divided by A) is

$$\sum x^2 = 12(2.75)^2 = 90.75 \text{ in.}^2$$

$$\sum y^2 = 4(1.5)^2 + 4(4.5)^2 + 4(7.5)^2 = 315 \text{ in.}^2$$

$$\frac{J}{A} = \sum x^2 + \sum y^2 = 405.8 \text{ in.}^2$$

3. The torsional load (Q_{mu}) at unit distance from the center of gravity of the bolt group is

$$Q_{mu} = \frac{M}{\sum x^2 + \sum y^2} = \frac{16P}{405.8}$$

The torsional load on the critical bolt is

$$Q_m = \frac{16P(7.99)}{405.8} = 0.315P$$

where the radius of 7.99 in. is determined from Fig. 7-32.

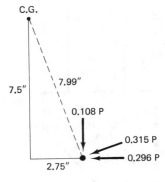

Figure 7-32 Torsional loads on critical bolts.

4. The horizontal component of Q_m is

$$\frac{7.5}{7.99}(0.315P) = 0.296P$$

and the vertical component of Q_m is

$$\frac{2.75}{7.99}(0.315P) = 0.108P$$

5. The force on the bolt due to P load applied at the center of gravity is

$$Q_v = \frac{P}{n} = \frac{P}{12} = 0.083P$$

6. The resultant force R on the critical bolt cannot exceed the capacity of one bolt in single shear: 10.5 kips (AISCM, Part 4, Table I-D). The vector diagram for the determination of R is shown in Fig. 7-33.

$$R^2 = (0.191P)^2 + (0.296P)^2$$

$$R = 0.352P$$

from which

$$0.352P = 10,500 \text{ lb}$$

$$P = 29,830 \text{ lb}$$

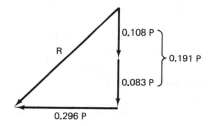

$$0.108\,P \left.\vphantom{\begin{array}{c}a\\b\end{array}}\right\} \\ 0.083\,P$$

R

0.191 P

0.296 P

Figure 7-33   Vector diagram.

The two examples for eccentrically loaded bolt groups are based on a so-called **elastic method.** This method, although providing a simplified and conservative approach, does not result in designs having a consistent factor of safety. As a result, a **modified elastic method** was introduced in 1963 by the AISC. In this method, two empirical expressions were introduced which in effect reduced the actual eccentricity to an effective eccentricity. This modified method provided for more economical designs; however, the factor of safety was still found to lack uniformity when compared to the *ultimate strength* of a connection. As a result, the **ultimate-strength method** was developed and is the current method used in the AISCM 8th edition tables for "Eccentric Loads on Fastener Groups."

As a means of comparison, Example 7-15 will be recalculated using the modified elastic method and the ultimate-strength method.

Example 7-16

Rework Example 7-15 using a *modified elastic method* approach. Refer to the AISCM, Part 4, "Eccentric Loads on Fastener Groups."

Solution

1. The torsional moment Pe is based on an effective eccentricity (e_{eff}) for two or more rows of bolts:

$$e_{\text{eff}} = e_{\text{act}} - \frac{1+n}{2}$$

where n is the number of bolts in one row, and e_{act} is the actual eccentricity.

$$e_{\text{eff}} = 16 - \frac{1+6}{2} = 12.5 \text{ in.}$$

$$M = P(12.5)$$

2. As in Example 7-15,

$$\sum x^2 + \sum y^2 = 405.8 \text{ in.}^2$$

3. The torsional load Q_{mu} at unit distance from the center of gravity of the bolt group is

$$Q_{mu} = \frac{12.5P}{405.8}$$

and the torsional load on the critical bolt is

$$Q_m = \frac{12.5P(7.99)}{405.8} = 0.246P$$

4. The horizontal component of Q_m is

$$\frac{7.5}{7.99}(0.246P) = 0.231P$$

and the vertical component of Q_m is

$$\frac{2.75}{7.99}(0.246P) = 0.085P$$

5. As in Example 7-15,

$$Q_v = 0.083P$$

6. As in Example 7-15, the capacity of one bolt in single shear is 10.5 kips. The vector diagram for R is shown in Fig. 7-34.

$$R^2 = (0.168P)^2 + (0.231P)^2$$

$$R = 0.286P$$

from which

$$0.286P = 10,500 \text{ lb}$$

$$P = 36,700 \text{ lb}$$

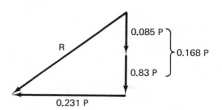

Figure 7-34   Vector diagram.

Since 36,700 lb > 29,830 lb, it can be observed that the original elastic method is appreciably conservative.

Example 7-17

Rework Example 7-15 using an ultimate-strength method.

Solution   Refer to the discussion of the method in the AISCM, Part 4, and utilize Tables X–XVIII. When entering Table XII, use the following values:

$n = 6$ (number of bolts in one vertical row)

$r_v = 10.5$ kips (permissible load on one bolt in single shear)

$\ell = e = 16$ in. (actual eccentricity)

$D = 5\frac{1}{2}$ in. (distance between gage lines)

$b = 3$ in. (bolt pitch)

From Table XII, the coefficient C is 3.55. Therefore, the allowable load is calculated as

$$P = Cr_v = 3.55(10,500) = 37,275 \text{ lb}$$

The three methods are discussed further in the AISCM on the pages preceding the tables. The use of the tables expedites and simplifies design and analysis of eccentrically loaded bolted connections. In fact, with the use of the tables, the process of design is probably best accomplished by assuming a number and arrangement of bolts and then checking the group capacity. Revising and rechecking may also be quickly accomplished.

When an eccentrically applied load lies outside the plane of the connection as shown in Fig. 7-35, the eccentric load tends to separate the bracket from the column flange at the top and press the bracket against the flange at the bottom. Therefore, the bolts are subjected to a varying and decreasing tensile force from the top down to the neutral axis. The bolt is also placed in shear by the vertical effect of the eccentric load.

Simplifying assumptions have been made with respect to analyses of this type of connection and are based on an assumption that the tension induced in any bolt as

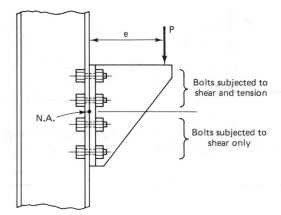

Figure 7-35   Eccentrically loaded connection.

a result of the applied eccentric load will not be in excess of the initial tension induced by tightening.

As assumption is made that the neutral axis lies at the middepth or center of gravity of the bolt group. Further, it is assumed that the tensile stress distribution above the neutral axis varies linearly from zero at the neutral axis to a maximum at the bolt farthest from the neutral axis. From the flexure formula

$$f_t = \frac{Mc}{I} = \frac{Pec}{I}$$

where P = applied eccentric load
e = eccentricity of load from face of column
c = distance from the neutral axis to the center of the most distant bolt
I = moment of inertia of the bolt areas

In addition, the shear stress is

$$f_v = \frac{P}{nA_b}$$

where n is the total number of bolts, and A_b is the cross-sectional area of one bolt.

As discussed previously, the allowable tensile stress must be modified by an interaction formula which is based on an actual bolt shear stress f_v, which, in turn, cannot be in excess of an allowable shear stress (see the AISCS, Table 1.6.3, and the SSJ, Table 2). In friction-type connections the allowable shear stress must be modified by an expression based on an actual tensile stress, which, in turn, cannot exceed the allowable tensile stress as shown in the AISCS, Table 1.5.2.1, or the AISCM, Part 4, Table I-A. The modification factor is from the AISCS, Section 1.6.3.

Example 7-18

For the connection shown in Fig. 7-35, the bracket is a structural tee. $P = 25$ kips and $e = 12$ in. Determine if the connection is satisfactory assuming $\frac{3}{4}$-in.-diameter A325N high-strength bolts in two vertical rows. The bolts are in standard holes spaced 3 in. vertically. Assume that the column and bracket are adequate.

Solution

1. The moment of inertia of the bolts about the neutral axis is

$$I = \sum (A_b d^2)$$

$$I = 0.4418[4(1.5)^2 + 4(4.5)^2] = 39.8 \text{ in.}^4$$

2. The actual tensile stress in the top bolt is

$$f_t = \frac{Pec}{I} = \frac{25[12(4.5)]}{39.8} = 33.9 \text{ ksi}$$

3. The allowable shear stress F_v is 21 ksi (from the AISCS, Table 1.5.2.1). The actual shear stress in the bolts is

$$f_v = \frac{25}{8(0.4418)} = 7.07 \text{ ksi} < 21 \text{ ksi} \qquad \textbf{O.K.}$$

4. The allowable tensile stress (from the AISCS, Table 1.6.3) is

$$F_t = 55 - 1.8f_v = 55 - 1.8(7.07) = 42.3 \text{ ksi} < 44 \text{ ksi} \qquad \textbf{O.K.}$$

Since the actual tensile stress is less than the allowable tensile stress, the connection is satisfactory.

REFERENCES

1. *Standard Specification for High-Strength Bolts for Structural Steel Joints,* ANSI/ASTM A325, The American Society for Testing and Materials, 1916 Race Street, Philadelphia, PA 19103.

2. *Standard Specification for Quenched and Tempered Alloy Steel Bolts for Structural Steel Joints,* ANSI/ASTM A490, The American Society for Testing and Materials, 1916 Race Street, Philadelphia, PA 19103.

3. *High-Strength Bolting for Structural Joints,* Bethlehem Steel Corp., Bethlehem, PA 18016.

PROBLEMS

7-1. Compute the tensile capacity P for the single shear lap joint shown. The plates are A36 steel, the high-strength bolts are $\frac{3}{4}$-in.-diameter A325 in standard holes. Assume that the connection is
(a) A325F.
(b) A325X.
(c) A325N.

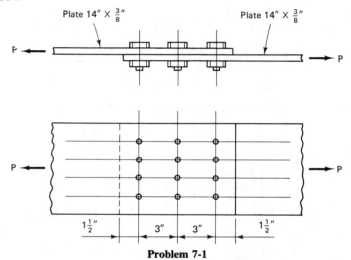

Problem 7-1

7-2. Compute the tensile capacity P for the connection shown. The gusset plate and the angle are A36 steel. The high-strength bolts are $\frac{3}{4}$-in.-diameter A325 in

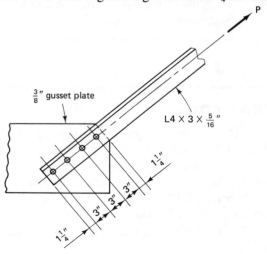

Problem 7-2

standard holes. Assume the connection is

(a) A325X.

(b) A325N.

7-3. Find the tensile capacity P for the double shear butt joint shown. Bolts are $\frac{3}{4}$-in.-diameter high-strength bolts in standard holes. All structural steel is A36.

(a) A325F.

(b) A325N.

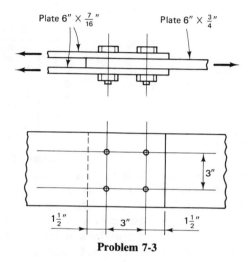

Plate 6″ × $\frac{7}{16}$″   Plate 6″ × $\frac{3}{4}$″

3″

$1\frac{1}{2}$″   3″ $1\frac{1}{2}$″

Problem 7-3

7-4. Compute the tensile capacity P for the double shear butt joint shown. The plates are A36 steel with $F_u = 58$ ksi. The high-strength bolts are $\frac{3}{4}$-in.-diameter in

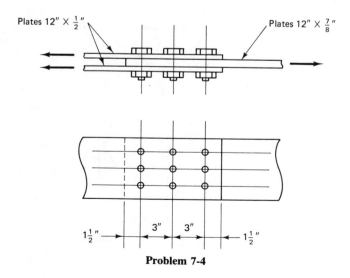

Plates 12″ × $\frac{1}{2}$″   Plates 12″ × $\frac{7}{8}$″

$1\frac{1}{2}$″   3″   3″ $1\frac{1}{2}$″

Problem 7-4

standard holes. Assume that the connection is

(a) A325F.

(b) A325X.

(c) A325N.

7-5. Compute the tensile capacity for the double-angle tension member shown. Fasteners are $\frac{7}{8}$-in.-diameter A325X bolts in standard holes. All structural steel is A36. Use standard gage lines in the angles.

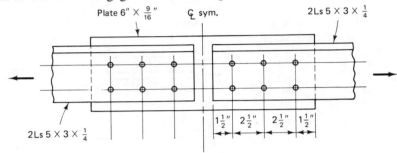

Plate 6" × $\frac{9}{16}$"   ₵ sym.   2 Ls 5 × 3 × $\frac{1}{4}$

2 Ls 5 × 3 × $\frac{1}{4}$

$1\frac{1}{2}$" $2\frac{1}{2}$" $2\frac{1}{2}$" $1\frac{1}{2}$"

Problem 7-5

7-6. A tension member made up of a pair of angles is connected to a column with eight $\frac{7}{8}$-in.-diameter A325 high-strength bolts in standard holes similar to that shown in Fig. 7-14. All structural steel is A36. The angles are oriented 30° with the horizontal and subjected to an axial load of 120 kips. Assume that the connection between the angles and the tee is satisfactory and that the angles are satisfactory for the applied load. Determine if the connection to the column is satisfactory. Consider

(a) A325F bolts.

(b) A325X bolts.

7-7. Determine the capacity of the bracket-to-column connection shown. All structural steel is A36. Bolts (total of 10) are $\frac{7}{8}$-in.-diameter A325X in standard holes.

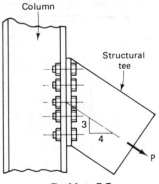

Column

Structural tee

3

4

P

Problem 7-7

7-8.   A truss member made up of two C12 × 25 channels is connected to a $\frac{3}{4}$-in.-thick gusset plate, as shown. Structural steel is A36 steel. Determine the number of $\frac{7}{8}$-in.-diameter A490F high-strength bolts required to develop the full tensile capacity of the channels.

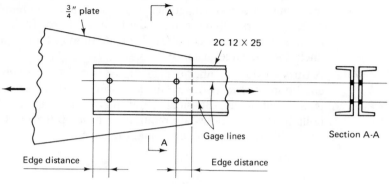

Problem 7-8

7-9.   A single shear lap joint is made up of two 14 × $\frac{3}{4}$ in. A36 plates. The plates are to be connected with A490 high-strength bolts in standard holes. Determine the number of bolts required, the pitch, and the edge distance to transfer a tensile load of 150 kips (assume three gage lines) if the connection is designed as
(a) A490F ($\frac{3}{4}$ in. diameter).
(b) A490X ($\frac{3}{4}$ in. diameter).
(c) A490N ($\frac{3}{4}$ in. diameter).

7-10. Determine the number of $\frac{3}{4}$-in.-diameter A325F high-strength bolts required to

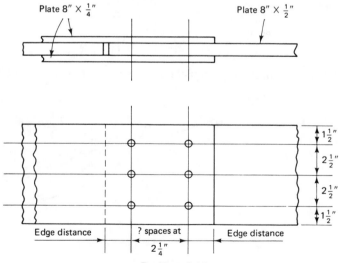

Problem 7-10

develop the full tensile capacity of the member that contains the butt joint shown. Determine the edge distance to be used. All structural steel is A36.

7-11. A roof truss tension member is made up of 2L6 × 4 × 1/2 (long legs, back to back) of A36 steel and is subjected to an axial load of 180 kips. If the angles frame into a $\frac{3}{8}$-in.-thick gusset plate, determine the number of $\frac{7}{8}$-in.-diameter A325 high-strength bolts in standard holes required for its end connection using a friction-type connection (A325F). Assuming two gage lines, establish the pitch and the edge distance.

7-12. A truss tension member is made up of two C10 × 30 channels laced together and connected to two $\frac{3}{8}$-in.-thick gusset plates, as shown. The channels and gusset plates are A36 steel. How many $\frac{7}{8}$-in.-diameter A325X high-strength bolts in standard holes are required to develop the full tensile capacity of the channels? Establish the pitch and the edge distance.

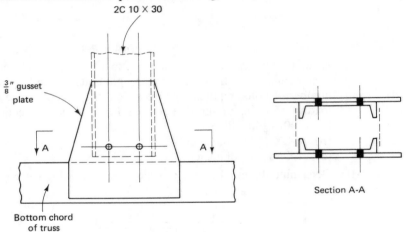

Problem 7-12

7-13. The single-angle tension member shown, part of a bracing system, is to support

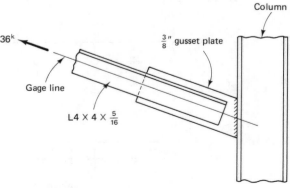

Problem 7-13

a design tensile load of 36 kips. The $\frac{3}{8}$-in.-thick gusset plate will be shop-welded to the column and field-bolted to the angle. Structural steel is A36.

    (a) Using $\frac{3}{4}$-in.-diameter A325N bolts, design the connection. Include number of bolts, gage line location, edge distances, and bolt spacing.

    (b) Design the gusset plate. (Neglect weld design.)

7-14. In Problem 7-7, assume that the applied tensile load is 80 kips. Determine the number of $\frac{3}{4}$-in.-diameter A325F bolts required for the bracket-to-column connection.

7-15. Refer to Fig. 7-19. Assume that the beam is a W36 × 170 and the bolts are $\frac{7}{8}$-in.-diameter A325X. Compute the load-carrying capacity of the connection.

7-16. Compute the load-carrying capacity of the framed connection shown. The structural steel is A36. The bolts are $\frac{3}{4}$-in.-diameter A325N bolts in standard holes.

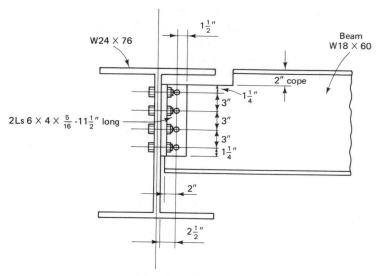

Problem 7-16

7-17. Refer to Fig. 7-20 and assume the following: the beam is a W21 × 57, the column is a W12 × 40, the seat angle is an L7 × 4 × $\frac{3}{4}$ × 8 in. (4-in. leg is horizontal) connected to the column web with four $\frac{7}{8}$-in.-diameter A490X bolts, and the setback is $\frac{1}{2}$ in. Only one beam frames to the column at this point. All structural steel is A36. Determine the maximum beam reaction that can be transmitted by this connection.

7-18. Refer to Fig. 7-21. Design the seated beam connection assuming that the beam is a W16 × 77 and the column is a W8 × 35. All structural steel is A36. Use $\frac{7}{8}$-in.-diameter A325X bolts. The beam reaction is 35 kips.

7-19. Rework Problem 7-18 except that the supporting column is a W12 × 65 and the reaction is 40 kips.

7-20. Design an end-plate shear connection for a W14 × 38 beam framing to the flange of a W8 × 35 column. All structural steel is A36. The end reaction is 38 kips. Use $\frac{3}{4}$-in.-diameter A325N high-strength bolts in standard holes. Neglect the welding of the plate to the end of the beam. Use $\frac{5}{16}$-in.-thick plate and assume that the welding is adequate.

7-21. Design the top and bottom clip angles and bolts for a semirigid connection similar to that shown in Fig. 7-23. Assume that the beam is a W21 × 62, the reaction is 60 kips, and the end moment to be transmitted is 75 ft-kips. Neglect the connection on the beam web. Bolts are $\frac{7}{8}$-in.-diameter A325N in standard holes. All structural steel is A36. Neglect prying action and use the classical approach of assuming a point of contraflexure in the angle.

7-22. Rework Problem 7-21 using the AISCM empirical design method.

7-23. For the eccentrically loaded bolted connection shown, determine the force acting on the most critical bolt. Assume $\frac{3}{4}$-in.-diameter A325N high-strength bolts in standard holes. The applied load is 10,000 lb. Use
   (a) The elastic method.
   (b) The modified elastic method.
   (c) The ultimate strength method.

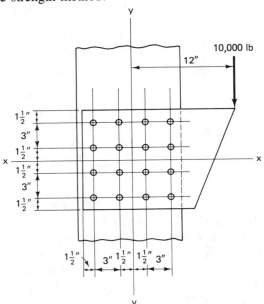

Problem 7-23

7-24. For the eccentrically loaded bolted connection shown, determine the maximum allowable load that can be supported by the $\frac{3}{8}$-in. plate connected to the W8 × 24 column. The plate and column are A36 steel. Use $\frac{3}{4}$-in. A325F high-strength bolts in standard holes. Assume that the plate and column are adequate. Use

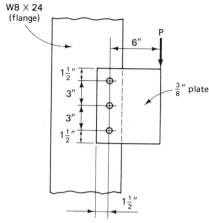

Problem 7-24

(a) The elastic method.

(b) The modified elastic method.

(c) The ultimate strength method.

7-25. Determine if the connection shown is adequate. The angle thickness is $\frac{5}{16}$ in. and both angle and column are A36 steel. Bolts are $\frac{7}{8}$-in. A490F high-strength bolts in standard holes. Assume that the angle and column are adequate. Use

(a) The elastic method.

(b) The modified elastic method.

(c) The ultimate strength method.

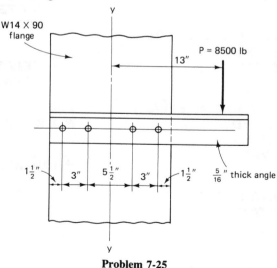

Problem 7-25

7-26. Refer to Fig. 7-35. Determine if the connection is satisfactory if $P = 20$ kips, $e = 15$ in., and the bolts are ¾-in.-diameter A325F, in two rows at 3 in. pitch.

8 Welded Connections

8-1 INTRODUCTION

Welding is a process in which two pieces of metal are fused together by heat to form a joint. In structural welding this is usually accompanied by the addition of filler metal from an electrode. Structural welds are usually made either by the manual shielded metal-arc process or by the submerged arc process.

The **manual shielded metal-arc welding process,** commonly called **stick welding,** is designed primarily for manual application and is used in both the shop and the field. An electric arc is formed between the end of a coated metal electrode and the steel components to be welded. This arc generates heat of an approximate temperature of 6500°F, which in turn melts a small area of the base metal. The tip of the electrode also melts and this metal is forcibly propelled across the arc. The small pool of molten metal that is formed is called a **crater.** As the electrode is moved along the joint, the crater follows it, solidifying rapidly as the temperature of the pool behind it drops below the melting point. Figure 8-1 illustrates this process. During the welding process, as the electrode coating decomposes, it forms a gas shield to prevent absorption of impurities from the atmosphere. In addition, the coating contains a material (commonly called **flux**) which will prevent or dissolve oxides and other undesirable substances in the molten metal, or which will facilitate removal of these substances from the molten metal.

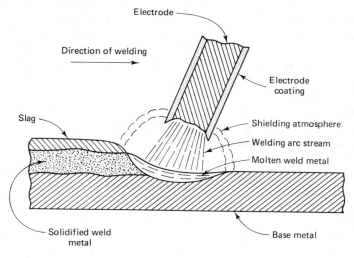

Figure 8-1   Manual shielded metal arc welding process.

The **submerged arc welding process** is primarily a shop-welding process performed by either an automatic or a semiautomatic method. The principle is similar to manual shielded metal-arc welding, but a bare metal electrode is used instead of a coated electrode. Loose flux is supplied separately in granular form and is placed over the joint to be welded. The electrode is pushed through the flux and as the arc is formed, part of the flux melts to form a shield that coats the molten metal. This welding process is faster and results in deeper weld penetration. In the automatic process, an electrically controlled machine supplies the flux and metal electrode through separate nozzles as it moves along a track. Figure 8-2 illustrates this process.

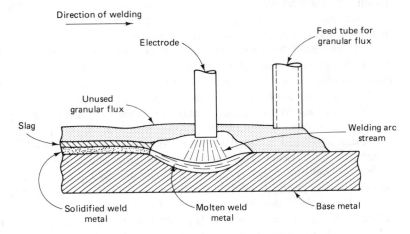

Figure 8-2 Submerged arc welding process.

The chemical and mechanical properties of the deposited weld metal (electrodes) should be as similar as possible to those of the base metal. Hence a variety of electrodes are needed to satisfy the requirements of the various steels. As a result, the American Welding Society (AWS) in cooperation with the ASTM has established an electrode numbering system which classifies welding electrodes. The system utilizes a prefix letter, E, which indicates an electrode, followed by four or five digits. In the manual shielded metal-arc welding process, the first two (or three) digits designate the minimum tensile strength (in ksi) of the deposited metal. The third (or fourth) digit indicates the welding position in which the electrode is capable of making sound welds. These positions are illustrated in Fig. 8-9. A 1 in the electrode numbering system indicates all positions: flat, horizontal, vertical, and overhead; 2 indicates flat and horizontal; and a 3 indicates the flat position only. The fourth (or fifth) digit refers to the current supply and the type of coating on the electrode. As an example, an E7014 electrode indicates an electrode with a minimum tensile strength of 70 ksi, usable in all positions with either ac or dc current and with an iron powder added to the electrode covering so that the arc can easily be maintained. Initially, for structural design, the item of significance is the minimum tensile strength of the electrode material, since the designer is concerned that the weld metal be of adequate strength with respect to the base metal. The 7th edition of the AISCM (Table 1.5.3) establishes the proper electrode–base metal combinations. The table stipulates that E70XX electrodes should be used with A36 steel. The minimum yield strength of the E70 electrode is 60 ksi, which is approximately 65% higher than the minimum yield strength of A36 steel. Thus weld metal is invariably stronger than the metals it connects.

In the submerged arc process the electrode numbering system is somewhat different, as it includes a combination of flux and electrode designations such as F7X-E7XX. The first portion of the designation is pertinent to the flux and the second portion to the electrode. F represents flux and the first digit after F represents the tensile strength requirement of the resulting weld. The second digit indicates the impact strength requirement. The second portion E7XX indicates an E for electrode with the first digit after the E indicating the minimum tensile strength of the weld metal. The last two digits classify the electrode. For this process the 7th edition of the AISCM stipulates that an F7X-EXXX flux–electrode combination should be used with A36 steel. The most widely used electrode for structural work at present is the E70 since it is compatible with all grades of steel with a yield stress F_y up to 60 ksi.

8-2 TYPES OF WELDS AND JOINTS

The two types of welds that predominate in structural applications are the fillet weld and the groove weld. Other structural welds are the plug weld and slot weld which generally are used only under circumstances where the fillet welds lack adequate load-carrying capacity. These four weld types may be observed in Fig. 8-3. The most commonly used weld for structural connections is the fillet weld. Groove welds may require extensive edge preparation as well as precise fabrication and, as a result, are more costly.

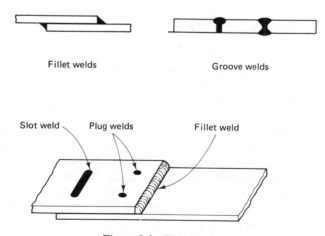

Fillet welds   Groove welds

Slot weld   Plug welds   Fillet weld

Figure 8-3   Weld types.

In any given welded structure the adjoining members may be situated with respect to each other in several ways. These joints may be categorized as butt, tee, corner, lap, and edge. They are illustrated in Fig. 8-4.

Fillet welds are welds of theoretically triangular cross section joining two surfaces approximately at right angles to each other in lap, tee, and corner joints. The

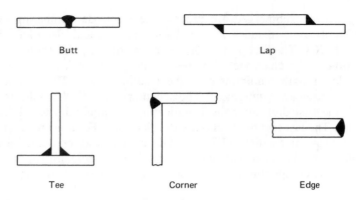

Butt

Lap

Tee

Corner

Edge

Figure 8-4 Joint types.

cross section of a typical fillet weld is a right triangle with equal legs. Figure 8-5 illustrates a typical fillet weld together with its pertinent nomenclature. The leg size designates the size of the weld. The **root** is the vertex of the triangle, or the point at which the legs intersect.

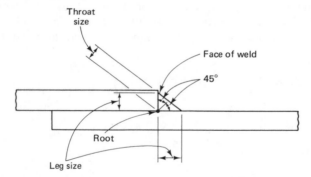

Figure 8-5 Typical fillet weld.

The **face of weld** is a theoretical plane, since weld faces will be either convex or concave, as shown in Fig. 8-6. The convex fillet weld is the more desirable of the two since the convex weld has less tendency to crack as a result of shrinking while cooling. The distance from the theoretical face of weld to the root is called the **throat size.** Variations of this fillet weld are permitted and may be necessary. Leg sizes may be unequal. If the pieces to be joined do not intersect at right angles, the welds are considered to be **skewed fillets,** as shown in Fig. 8-7. If the intersection is not within the angular limits shown in Fig. 8-7, the welds are considered to be **groove welds.**

Groove welds are welds made in a groove between adjacent ends, edges, or surfaces of two parts to be joined in a butt, tee, or corner joint. The weld configuration in these joints can be made in various ways. A welded butt joint can be made square, double-square, single-bevel, double-bevel, single V, double V, single J, double J, single U, or double U, as illustrated in Fig. 8-8. With the exception of the square

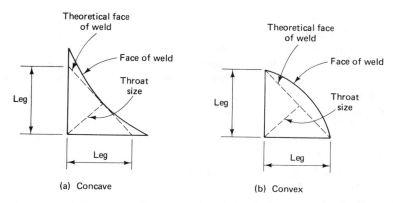

(a) Concave   (b) Convex

Figure 8-6 Fillet welds.

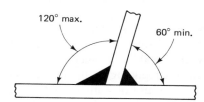

Figure 8-7 Fillet weld limitations.

	Single	Double
Square groove		
Bevel groove		
Vee groove		
J groove		
U groove		

Figure 8-8 Groove welds.

groove weld, some edge preparation is required for either one or both of the members to be connected.

Groove welds are further classified as either complete penetration or partial penetration welds. A **complete penetration weld** is one that achieves fusion of weld and base metal throughout the depth of the joint. It is made by welding from both sides of the joint, or from one side to a backing bar. The throat dimension of a full penetration groove weld is considered to be the full thickness of the thinner part joined, exclusive of weld **reinforcement.** Reinforcement is added weld metal over and above the thickness of the welded material.

       Partial penetration groove welds are used when load requirements do not require full penetration or when welding must be done from one side of a joint only without the use of a backing bar. For more detailed information regarding groove preparation, welding processes and positions, the reader is referred to the AISCM, Part 4, Welded Joints, as well as the *Structural Welding Code* of the AWS.[1] Where possible, groove welds should be avoided because of their high cost compared to fillet welds. Where groove welds are necessary, the type required should be simply stated either as a complete joint penetration weld or a partial joint penetration weld. This permits the fabricator to use the most economical groove weld for the particular situation and equipment. With respect to groove weld design, if the proper electrode is used with the parent metals, allowable stresses in the weld will be the same as those for the parent material.

       Plug and slot welds are used in lap joints (see Fig. 8-3). Round holes or slotted holes are punched or otherwise formed passing through one of the members to be joined (before assembly). Weld metal is deposited in the openings. The openings may be partially or completely filled depending on the thickness of the punched material. A variation of the slot weld is the use of a fillet weld in the slotted hole.

       The AWS has established certain joints used in structural welding as **prequalified.** These may be observed in Part 4 of the AISCM. The joints may be made by manual shielded metal arc or submerged arc welding and used without the need for performing welding procedure qualification tests, provided that the electrodes and flux conform to AWS specifications and the fabricator meets certain workmanship standards. Prequalified joints are made possible because such joints and the procedures for making them have established a long history of satisfactory performance. Joints that are not prequalified under AWS codes and specifications may be used in structural welding but they must first pass a procedure qualification test as prescribed by the AWS. This is usually costly. Therefore, whenever possible, only prequalified joints should be used.

       Welds are also classified as being flat, horizontal, vertical, and overhead. These four positions are illustrated in Fig. 8-9. The position of the joint when welding is

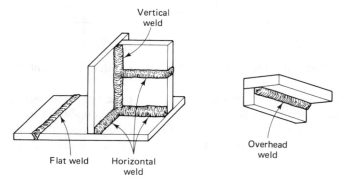

Vertical weld

Flat weld   Horizontal weld

Overhead weld

Figure 8-9   Weld types.

performed is of economic significance. The flat weld is the most economical and the overhead weld the most expensive. Therefore, the flat position is preferred in all types of welding.

8-3 STRENGTH AND BEHAVIOR OF FILLET WELDED CONNECTIONS

A fillet weld is a surface weld and its shape or size is not restricted by the shape and size of a groove. Hence it is necessary to establish the size and length of the fillets to avoid overwelding or underwelding. Since tests have shown that fillet welds are stronger in tension and compression than they are in shear, the controlling fillet weld stresses are considered to be shear stresses acting on an effective (theoretical) throat area. This throat area establishes the strength of a fillet weld and is defined as the shortest distance from the root of the joint to the theoretical face of weld (as shown in Fig. 8-5). In a fillet with equal leg sizes, where the cross-sectional shape of the weld is theoretically a 45° right triangle, the effective throat distance is

$$\sin 45° \times \text{leg size} = 0.707 \times \text{leg size}$$

If weld metal exists outside the theoretical right triangle, this additional weld metal is considered to be *reinforcement* and is assumed to add no strength.

The strength of a fillet depends on the direction of the applied load, which may be parallel or perpendicular to the axis of the weld. In parallel loading, the applied load is transferred parallel to the weld from one leg face to the other as shown in Fig. 8-10(a). The minimum resisting area in the fillet occurs at the throat and is equal to 0.707 times the leg size. (This assumes equal leg sizes for the weld, which is generally the case.) The strength of the weld is computed by multiplying the allowable shear stress of the weld by the throat area.

Tests indicate that a fillet weld loaded perpendicular to the fillet [transverse loading as shown in Fig. 8-10(b)] is approximately one-third stronger than when loaded in a parallel direction. However, the AISCS does not permit this fact to be considered when designing welds. The strengths of all fillets are based on the values calculated for loads applied in a parallel direction. The fillet weld loaded in a perpendicular direction has greater strength due to the fact that the failure plane develops at an angle other than 45°; hence the resisting area of the fillet is greater than the throat area, which is perpendicular to the theoretical face of the weld. In addition, transverse fillet welds are more uniformly stressed than the parallel loaded fillet welds.

The allowable shear stress for the weld metal is (AISCS, Table 1.5.3)

$$F_v = 0.3F_u$$

where F_u is the specified minimum tensile strength of the electrode. Therefore, the strength of a fillet weld per linear inch of weld is

$$P = F_v(0.707)(\text{leg size})$$

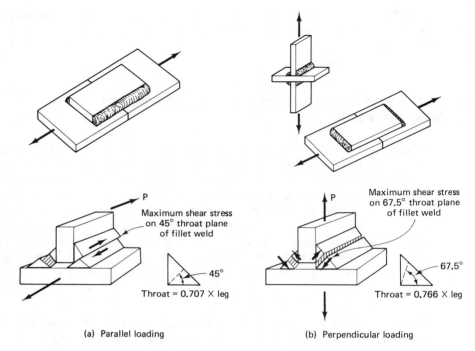

(a) Parallel loading   (b) Perpendicular loading

Figure 8-10   Fillet weld loading.

or

$$P = 0.3F_u(0.707)(\text{leg size}) = 0.212F_u(\text{leg size})$$

We will now calculate strength per inch for a fillet weld having a leg size of $\frac{1}{16}$ in. (which is a hypothetical value since a minimum weld size according to the AISCS is $\frac{1}{8}$ in.). The strength or load-carrying capacity of other leg-size fillet welds can then be obtained by multiplying by the number of sixteenths in the leg size. For an E70XX electrode ($F_u = 70$ ksi),

$$P = 0.212(70)\left(\frac{1}{16}\right) = 0.928 \text{ kip/in.}$$

This value is generally rounded off to 0.925 kip/in. Using 0.925 kip/in. as a basic value, the strength of other sizes of fillet welds may be computed and tabulated. For example, the strength of a $\frac{3}{16}$-in. fillet weld would be

$$0.925(3) = 2.78 \text{ kips/in.}$$

A similar approach could be used for E60XX electrodes, where $F_u = 60$ ksi.

When the submerged arc process is used, greater heat input produces a deeper weld penetration and as a result the AISCS stipulates that the effective throat distance for welds larger than $\frac{3}{8}$ in. may be taken equal to the theoretical throat plus 0.11 in.

In addition, for welds of $\frac{3}{8}$ in. or less, the strength of the weld is based on leg size rather than throat distance. These values may be observed in Table 8-1.

TABLE 8-1   STRENGTH OF WELDS (kips per linear inch)

Weld size (in.)	E70XX SMAW[a]	E60XX SMAW	E70XX SAW[b]	E60XX SAW
$\frac{1}{16}$	0.925	0.795	1.31	1.13
$\frac{1}{8}$	1.85	1.59	2.63	2.25
$\frac{3}{16}$	2.78	2.39	3.94	3.38
$\frac{1}{4}$	3.70	3.18	5.25	4.50
$\frac{5}{16}$	4.63	3.98	6.56	5.63
$\frac{3}{8}$	5.55	4.77	7.88	6.75
$\frac{7}{16}$	6.48	5.57	8.81	7.55
$\frac{1}{2}$	7.40	6.36	9.73	8.34
$\frac{9}{16}$	8.33	7.16	10.66	9.14
$\frac{5}{8}$	9.25	7.95	11.59	9.93
$\frac{11}{16}$	10.18	8.75	12.52	10.73
$\frac{3}{4}$	11.10	9.54	13.45	11.52
$\frac{13}{16}$	12.03	10.34	14.37	12.32
$\frac{7}{8}$	12.95	11.13	15.30	13.12

[a] Shielded metal-arc welding.
[b] Submerged arc welding.

In addition to the strength criteria, the AISCS furnishes design requirements with respect to minimum and maximum sizes and length of fillet welds. Minimum leg sizes for various thicknesses of members being joined are shown in the AISCS, Table 1.17.2A. Note that the minimum size of a fillet weld allowed in structural work is $\frac{1}{8}$ in. Also, the minimum size is based on the *thicker* of the two parts being joined except that the weld size need not exceed the thickness of the thinner part. The minimum size limitation is based on the fact that the heat generated in depositing a small weld is not enough to heat a much thicker member beyond the immediate vicinity of the weld. As a result, the weld cools rapidly, with subsequent cracks developing.

The *maximum size* of fillet welds against the edges of connected parts of a joint is limited so that the weld does not overstress the parts it connects. This means that the fillet weld capacity cannot exceed the capacity of the connected material either in tension or shear. The maximum permissible leg size is $\frac{1}{16}$ in. less than the thickness of the material for material thickness of $\frac{1}{4}$ in. or over. Along edges of material less than $\frac{1}{4}$ in. thick, the weld leg size may equal the thickness of the material (see Fig. 8-11).

Within the constraints of the minimum and maximum criteria for fillet welds, economy can best be achieved by using welds that require a minimum amount of metal and can be deposited in the least amount of time. As shown previously, the strength of a fillet weld is directly proportional to its size; however, the volume of deposited

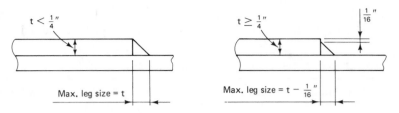

Figure 8-11   Maximum leg sizes for fillet welds.

metal, and hence the cost of the weld, increases as the square of the weld size. Hence it is generally preferred to use a long small-leg-size weld rather than a short large-leg-size weld. In addition, a weld deposited in a single pass by the welder is cheaper than welds made in multiple passes. The largest weld that can be made by a welder in a single pass is a $\frac{5}{16}$-in. fillet weld. Multiple passes require appreciably more time and weld metal and as a result are more expensive.

In addition, the AISCS, Section 1.17.4, imposes limitations on the lengths of fillet welds. The minimum effective length of a fillet weld must not be less than four times the nominal size or else the size of the weld must be considered not to exceed one-fourth of its effective length. This also applies to intermittent fillet welds (see Fig. 8-3) with the added requirement that each weld length not be less than $1\frac{1}{2}$ in. If longitudinal fillet welds are used alone (without transverse welds) in end connections of flat bar tension members, the length of each fillet weld cannot be less than the perpendicular distance between them. Nor can the transverse spacing of longitudinal fillet welds used in end connections exceed 8 in. unless specific design provisions are utilized.

Side or end fillet welds terminating at ends or sides, respectively, of parts or members should, whenever practicable, be returned continuously around the corners for a distance not less than two times the nominal size of the weld. This weld detail is called an **end return** (see Fig. 8-14). The AISCS, Section 1.17.7, states that the effective length of fillet welds includes the length of the end returns used.

Where lap joints are used the minimum amount of lap should be five times the thickness of the thinner part joined but not less than 1 in. Lap joints joining plates or bars subjected to an axial load should be fillet welded along the end of both lapped parts (see the AISCS, Section 1.17.6).

Example 8-1

Determine the allowable tensile load that may be applied to the connection shown in Fig. 8-12. The steel is A36 and the electrode used was E70 (manual shielded metal-arc welding). The weld is a $\frac{7}{16}$-in. fillet weld.

Solution   The total length of $\frac{7}{16}$-in. weld is 16 in. From Table 8-1, the capacity of a $\frac{7}{16}$-in. weld is 6.48 kips/in.

$$\text{weld capacity} = 6.48(16) = 103.7 \text{ kips}$$

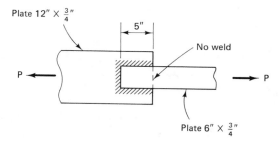

Plate 12" × $\frac{3}{4}$"

5"

No weld

P

P

Plate 6" × $\frac{3}{4}$"

Figure 8-12   Welded lap joint.

The tensile capacity of the plate (using $F_t = 22$ ksi) is

$$P_t = 6(0.75)(22) = 99 \text{ kips}$$

Therefore, the allowable tensile load is 99 kips.

Example 8-2

Determine the allowable tensile load that may be applied to the connection shown in Fig. 8-13. The steel is A36 and the electrode used was E70. The fillet weld is $\frac{5}{16}$ in. and the shielded metal-arc welding (SMAW) process was used.

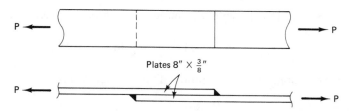

P

P

Plates 8" × $\frac{3}{8}$"

P

P

Figure 8-13   Transverse welded lap joint.

Solution   The length of $\frac{5}{16}$-in. weld is 16 in. The capacity of a $\frac{5}{16}$-in. weld per linear inch, from Table 8-1, is 4.63 kips/in.

$$\text{weld capacity} = 4.63(16) = 74.1 \text{ kips}$$

The tensile capacity of the plate (using $F_t = 22$ ksi) is

$$P_t = 8(0.375)(22) = 66 \text{ kips}$$

Therefore, the allowable tensile load is 66 kips.

Example 8-3

Design longitudinal fillet welds to develop the tensile capacity of the plate shown in Fig. 8-14. The steel is A36 and the electrode is E70 (SMAW).

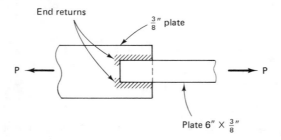

Figure 8-14 Parallel loaded welded lap joint.

Solution   The tensile capacity of the plate is

$$P_t = 6(0.375)(22) = 49.5 \text{ kips}$$

$$\text{maximum weld size} = \frac{3}{8} - \frac{1}{16} = \frac{5}{16} \text{ in.}$$

(see AISCS, Section 1.17.3). From Table 8-1, the capacity of a $\frac{5}{16}$-in. weld per linear inch is 4.63 kips/in. The length of the weld required is

$$\frac{49.5}{4.63} = 10.7 \text{ in.}$$

Use end returns with a minimum length of 2 × (leg size):

$$2\left(\frac{5}{16}\right) = \frac{5}{8} \text{ in.} \qquad (\text{use 1 in.})$$

The length of the longitudinal welds required is

$$10.7 - 1 - 1 = 8.7 \text{ in.}$$

$$\frac{8.7}{2} = 4.35 \text{ in. each side of plate}$$

However, the AISCS stipulates that the minimum length of longitudinal fillet welds must not be less than the perpendicular distance between them. Therefore, use a minimum length of 6 in. on each side of the plate.

The welds connecting the plates in Example 8-3 could also have included an end transverse weld together with the longitudinal welds. With this type of end connection, the criterion for the minimum length of longitudinal fillet weld used in Example 8-3 does not apply (see the AISCS, Section 1.17.4). In the special case where single or double angles subject to static tensile loads are welded to plates, as in a welded truss, the AISCS permits their connections to be designed using procedures similar to those of the previous examples.

Example 8-4

Design an end connection using longitudinal welds and an end transverse weld to develop the full tensile capacity of the angle (see Fig. 8-15). Use A36 steel and E70 electrodes (SMAW). The angle is an L6 $\times$ 4 $\times$ $\frac{7}{16}$ with the long leg connected to the plate.

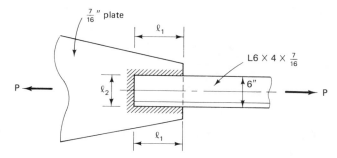

Figure 8-15 Welded connection for angle.

Solution The tensile capacity of the angle (using $F_t = 22$ ksi) is

$$P_t = 4.18(22) = 92 \text{ kips}$$

$$\text{maximum weld size} = \frac{7}{16} - \frac{1}{16} = \frac{3}{8} \text{ in.}$$

From Table 8-1, the capacity of a $\frac{3}{8}$-in. weld per linear inch is 5.55 kips/in. The length of the weld required is

$$\frac{92}{5.55} = 16.6 \text{ in.}$$

End weld ℓ_2 has length of 6 in. Therefore, the side welds must furnish

$$16.6 - 6 = 10.6 \text{ in.} \qquad (\text{say 11 in.})$$

Use a length ℓ_1 of $5\frac{1}{2}$ in. for each side of the angle.

Where the angle tension member is subject to repeated variation of stress, such as stress reversal that may occur with moving loads, the placement of the welds must conform to the distribution of the angle area, with the centroid of the resisting welds collinear with the centroidal axis of the angle. In effect, the welding pattern will not be symmetrical if the member is not symmetrical.

Example 8-5

Design an end connection using longitudinal welds and an end transverse weld to develop the full tensile capacity of the angle shown in Fig. 8-16. The member

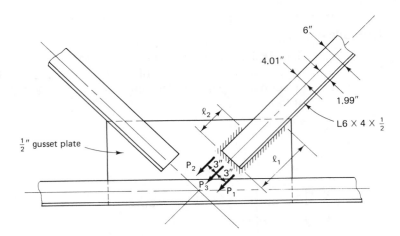

Figure 8-16 Welded truss joint.

is subjected to repeated stress variations. Use A36 steel and an E70 electrode (SMAW). The angle is an L6 × 4 × $\frac{1}{2}$ with the long leg connected to a gusset plate.

Solution   The tensile capacity of the angle is

$$P_t = 4.75(22) = 104.5 \text{ kips}$$

$$\text{maximum weld size} = \frac{1}{2} - \frac{1}{16} = \frac{7}{16} \text{ in.}$$

The capacity of a $\frac{7}{16}$-in. weld per linear inch is 6.48 kips (Table 8-1). Different-size welds could be used along the back and toe of the angle; however, for both economy purposes and practical reasons the weld will be kept the same size. In addition, a single-pass weld ($\frac{5}{16}$ in. maximum) could be used, but would require a greater length of weld. Therefore, a $\frac{7}{16}$-in. fillet weld will be used for this connection.

Since the force in the angle is assumed to act along its centroidal axis, the centroid of the resisting welds must be collinear to eliminate any eccentricity.

The resistance of the transverse end weld is

$$P_3 = 6(6.48) = 38.9 \text{ kips}$$

P_3 acts along the centerline of the attached leg.

Taking moments about line ℓ_1, resisting force P_2 may be determined:

$$104.5(1.99) = P_2(6) + 38.9(3)$$

$$P_2 = 15.2 \text{ kips}$$

P_1 may be determined by a summation of forces parallel to the length of the angle:

$$P_1 = 104.5 - 38.9 - 15.2$$

$$= 50.4 \text{ kips}$$

The length of longitudinal weld required is based on the weld capacity per linear inch, 6.48 kips.

$$\ell_2 = \frac{15.2}{6.48} = 2.3 \text{ in.} \qquad (\text{use 3 in.})$$

$$\ell_1 = \frac{50.4}{6.48} = 7.8 \text{ in.} \qquad (\text{use 8 in.})$$

8-4 STRENGTH AND BEHAVIOR OF PLUG AND SLOT WELDED CONNECTIONS

Where inadequate length of fillet welds exists to resist the applied load in a given connection, additional strength may be furnished through the use of plug or slot welds. The AISCS states that plug or slot welds may be used to transmit shear in a lap joint or to prevent buckling of lapped parts and to join component parts of built-up members.

The terms **plug weld** and **slot weld** are used with reference to circular holes, or slotted holes with circular ends, that are filled with weld metal completely or to such depth as prescribed by the AISCS. The effective area for such welds is assumed to be the nominal area of the hole or slot in the plane of the contact surfaces of the elements being joined.

Plug and slot welds are not used to a very great extent for strength. But plug welds, especially, often serve a useful purpose in stitching together elements of a member. In addition, joints or connections made with these welds have exhibited poor fatigue resistance.

Fillet welds in holes or slots also may be used to transmit shear in lap joints; however, they are not to be considered plug or slot welds. Since large plug and slot welds may exhibit excessive shrinkage, it is usually more desirable to use fillet welds in large holes or slots. Also, special care and special procedures are necessary for sound plug or slot welds, whereas the making of a fillet weld in a hole or slot is a normal procedure.

The AISCS requirements for plug and slot welds may be summarized as follows:

1. The width of slot or diameter of hole cannot be less than material thickness plus $\frac{5}{16}$ in. (rounded to the next greater odd $\frac{1}{16}$ in.) and cannot exceed 2.25 times the thickness of the weld.

2. For material up to $\frac{5}{8}$ in. thick, the weld thickness must equal the material thickness.

3. For material greater than $\frac{5}{8}$ in. thick, the weld thickness may not be less than $\frac{1}{2}$ times the material thickness or less than $\frac{5}{8}$ in.

4. The maximum length of slot is 10 times the weld thickness.

5. Maximum center-to-center spacing of plug welds is four times the hole diameter.

6. Minimum spacing of lines of slot welds transverse to their length is four times the width of the slot.

7. Minimum center-to-center spacing in a longitudinal direction on any line is two times the length of the slot.

8. The ends of the slot must be semicircular or be rounded to a radius not less than the thickness of the material (except for ends that extend to the edge of the material).

Example 8-6

Design the connection of a C10 × 30 to a $\frac{3}{8}$-in. gusset plate, shown in Fig. 8-17, to develop the full tensile capacity of the channel. Welding is not permitted on the back of the channel. All steel is A36. Use E70 electrode (SMAW). The maximum length of lap on the gusset plate is 10 in. (space limitations).

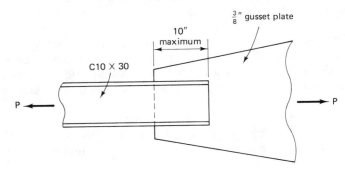

Figure 8-17   Channel connection.

Solution

1. The tensile capacity of the C10 × 30 is

$$P_t = AF_t$$

$$= 8.22(22) = 194 \text{ kips}$$

2. The web thickness of the channel is 0.673 in.; therefore, a $\frac{9}{16}$-in. fillet weld could be used. However, for economy reasons, a $\frac{5}{16}$-in. weld will be tried. This is the maximum size for a single-pass weld. The capacity of a $\frac{5}{16}$-in. weld per linear inch is 4.63 kips, from Table 8-1.

3. The length of weld required is

$$\frac{194}{4.63} = 41.9 \text{ in.}$$

4. The length available for welding is

$$10(2) + 10 = 30 \text{ in.}$$

Since there is insufficient length available, a slot weld, as shown in Fig. 8-18, will be designed.

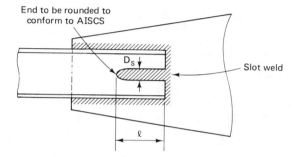

Figure 8-18   Slot weld symbols.

5. For the slot weld, use a weld thickness equal to the thickness of the channel web.

6. The minimum width of slot is

$$\text{minimum } D_s = 0.673 + \frac{5}{16} = 0.986 \text{ in.}$$

The maximum width of slot is

$$\text{maximum } D_s = 2.25(0.673) = 1.51 \text{ in.}$$

Use a slot width $D_s = 1\frac{1}{16}$ in.

7. The total length of the longitudinal and transverse end fillet weld is reduced by the width of the slot. The length of the $\frac{5}{16}$-in. fillet weld is

$$\ell = 30 - 1\frac{1}{16} = 28\frac{15}{16} \text{ in.}$$

$$= 28.94 \text{ in.}$$

8. The capacity of the $\frac{5}{16}$-in. fillet weld is

$$P = 28.94(4.63) = 134 \text{ kips}$$

9. The load to be resisted by the slot weld is

$$194 - 134 = 60 \text{ kips}$$

10. The required length of the slot weld ℓ may be determined from the relationships

$$P = AF_v = D_s \ell F_v$$

where P = load to be resisted by the slot weld

           A = area of the slot in the plane of the contact surfaces

           F_v = allowable shear stress in weld (see Section 8-3 of this text or the AISCS, Table 1.5.3).

D_s and ℓ are as shown in Fig. 8-18 ($A = D_s \ell$). Substituting yields

$$60 = \left(1\frac{1}{16}\right)(\ell)(0.3)(70)$$

from which

$$\text{required } \ell = 2.69 \text{ in.} \qquad (\text{use } \ell = 3 \text{ in.})$$

11. The maximum length of slot

$$\ell = 10(0.673) = 6.73 \text{ in.} > 3 \text{ in.} \qquad \textbf{O.K.}$$

Use a 3 in. $\times$ $1\frac{1}{16}$ slot weld.

8-5 END-PLATE SHEAR CONNECTIONS

This type of connection is discussed in Section 7-8 of this text. It consists of a rectangular plate welded to the end of a beam web and bolted to the supporting member. Example 7-11 furnishes the design for the bolted portion of the connection. The design of the welded portion of the connection will be treated here.

    The end-plate length L is always made less than the beam depth so that all the welding will be on the beam web. The plate is welded to the beam web with fillet welds as shown in Fig. 8-19. The welds should not be returned across the web at the top or bottom of the end plates and the effective weld length should be taken as equal to the plate length minus twice the weld size. The weld size to the beam web should be such that the weld shear capacity per linear inch does not exceed the beam web shear capacity per linear inch. This portion of the design assumes no eccentricity.

Example 8-7

    See Example 7-11 for all data. Design the welded portion of the connection using an E70 electrode (SMAW).

Solution

1. The plate length L was established as $8\frac{1}{2}$ in.

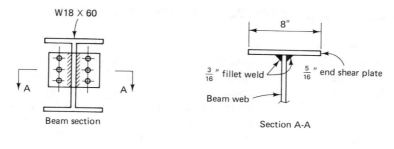

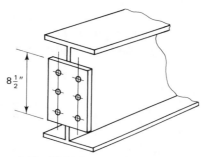

Figure 8-19   Weld for end-plate shear connection.

2. Assuming a $\frac{3}{16}$-in. weld on each side of the beam web, the available effective length of one weld is

$$8\frac{1}{2} - 2\left(\frac{3}{16}\right) = 8.13 \text{ in.}$$

3. The total weld capacity $= 2.78(8.13)(2)$

$$= 45.2 \text{ kips} > 40.0 \text{ kips (end reaction)}   \textbf{O.K.}$$

4. Check to ensure that the shear capacity of the welds (per linear inch) does not exceed the shear capacity of the web (per linear inch). The shear capacity of the welds is

$$2(2.78) = 5.56 \text{ kips/in.}$$

The shear capacity of the web (A36 steel) is

$$F_v(t_w) = 0.40F_y t_w$$

$$= 0.40(36)(0.415) = 5.98 \text{ kips/in.}$$

$$5.56 \text{ kips/in.} < 5.98 \text{ kips/in.}   \textbf{O.K.}$$

Use a $\frac{3}{16}$-in. fillet weld.

8-6 ECCENTRICALLY LOADED WELDED CONNECTIONS

In Section 7-10 of the text, eccentrically loaded bolted connections were analyzed and designed. The analysis and design of an eccentrically loaded welded connection is approached in a similar way. Note in Fig. 8-20 that the eccentrically applied load P lies in the plane of the connection. P may be resolved into a concentric load–moment combination. The concentric load acts through the centroid of the weld configuration and the torsional moment ($M = Pe$) is with respect to the same centroid as a center of rotation. Therefore, the forces acting on the welds will be made up of two components: P_v due to the axial effect of the eccentric load, and P_m due to the torsional moment effect, as shown in Fig. 8-21. The axial effect produces a load of P/ℓ per inch of weld, where ℓ is the total length of weld. This load acts in a direction parallel to P and will be the same for each linear inch of weld.

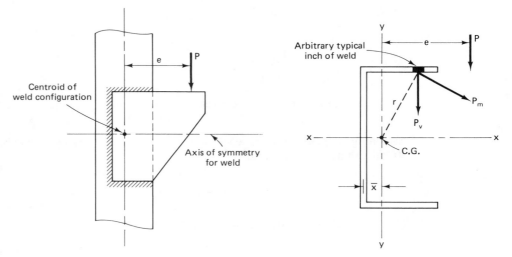

Figure 8-20   Eccentrically loaded
welded connection.

Figure 8-21   Forces in welds.

P_m will vary with the distance r from the centroid of the weld configuration to that element of weld being considered and will act in a direction normal to the line that connects the centroid with that weld element. Therefore, the connection must be so designed or analyzed that the resultant of these two components acting at any point of the weld does not exceed the weld capacity.

The torsional load P_m may be determined by applying the classic torsional stress formula to the weld configuration:

$$f_v = \frac{Mr}{J}$$

where f_v = unit shearing stress in the weld
     M = torsional moment (Pe)

r = radial distance from center of gravity of the weld configuration to any point of the weld being considered

J = polar moment of inertia of the weld

For purposes of design, each weld element may be assumed to be a line coincident with the root (see Fig. 8-5) of the fillet weld. Hence the weld may be considered to have location and length only. Therefore, the computed unit stress in the torsional stress formula becomes a force per unit length (kips/in.) rather than a force per unit area (kips/in.^{2}). This force per linear inch may be designated P_m. It should also be noted that for units and assumptions to be consistent, the polar moment of inertia will be in units of in.^{3} rather than in.^{4}. This is based on the fact that the weld only has length, thereby removing one dimension from I_x and I_y. Moment-of-inertia formulas are shown in Fig. 8-22 for a length of weld. For this type of problem it is probably easier to use the polar moment of inertia in the form

$$J = I_x + I_y$$

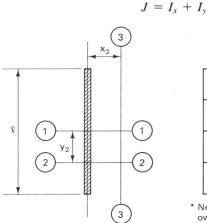

Moment of inertia	Formula
I_{1-1}	$\frac{1}{12}\ell^3$
I_{2-2}	$\frac{1}{12}\ell^3 + \ell y_2{}^2$
I_{3-3}	$*\ell x_3{}^2$

* Neglects moment of inertia about its own centroidal axis (which = 0)

Figure 8-22   Moment of inertia formulas.

Example 8-8

Determine the size of the fillet weld required to resist a load of 20 kips on the bracket shown in Fig. 8-23. The steel is A36 and the welding is to be performed using E70 electrodes (SMAW).

Solution

1. Assuming the weld to be a line, the center of gravity of the weld configuration may be obtained by taking a summation of moments of lengths of weld about the 12-in. side.

$$2(6)(3) + 12(0) = (6 + 6 + 12)(\bar{x})$$

$$\bar{x} = 1.50 \text{ in.}$$

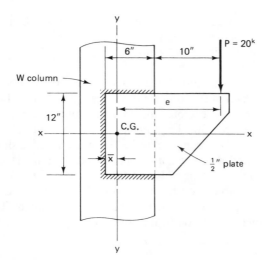

Figure 8-23 Welded bracket connection.

2. For calculation of the polar moment of inertia, note the reference axes through the weld centroid. See Fig. 8-22 for formulas.

$$J = I_x + I_y$$

$$I_x = \left(\frac{1}{12}\right)(12)^3 + 2(6)(6)^2 = 576 \text{ in.}^3$$

$$I_y = 2\left(\frac{1}{12}\right)(6)^3 + 2(6)(1.5)^2 + 12(1.5)^2$$

$$= 90 \text{ in.}^3$$

$$J = 576 + 90 = 666 \text{ in.}^3$$

3. The torsional moment M is

$$M = Pe = 20(14.5) = 290 \text{ in.-kips}$$

4. The force on the weld due to the torsional moment is

$$P_m = \frac{Mr}{J}$$

Since the most stressed parts of the weld are those which are the greatest distance from the weld center of gravity, the largest r value should be used. In this example, with reference to Fig. 8-24,

$$\text{maximum } r = \sqrt{6^2 + 4.5^2} = 7.5 \text{ in.}$$

Therefore,

$$P_m = \frac{290(7.5)}{666} = 3.266 \text{ kips/in.}$$

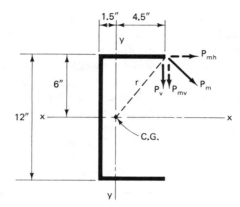

Figure 8-24   Weld geometry and forces.

5. Calculating the vertical and horizontal components of P_m, we have

$$P_{mh} = \frac{6.0}{7.5}(3.266) = 2.61 \text{ kips/in.}$$

$$P_{mv} = \frac{4.5}{7.5}(3.266) = 1.96 \text{ kips/in.}$$

6. The force on the weld due to axial effect of the eccentric load is

$$P_v = \frac{P}{\ell} = \frac{20}{24} = 0.83 \text{ kips/in.}$$

7. Adding the forces vectorially and determining the resultant force (see Fig. 8-25) gives us

$$R = \sqrt{2.61^2 + (1.96 + 0.83)^2}$$

$$= 3.82 \text{ kips/in.}$$

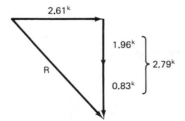

Figure 8-25   Force resolution.

8. Since the capacity of a $\frac{1}{16}$-in. weld for an E70 electrode is 0.925 kip/in. (Table 8-1), the fillet leg size required is (in number of sixteenths of an inch)

$$\frac{3.82}{0.925} = 4.13$$

Use a $\frac{5}{16}$-in. fillet weld.

The solution used in Example 8-8 is based on an elastic method which assumes that each element of weld will resist an equal share of the components of the load and a proportional share of the eccentric moment portion of the load (dependent on the element's distance from the center of gravity of the weld configuration). This method, although providing a simplified and conservative approach, does not result in a consistent factor of safety. As a result, an ultimate strength method is provided in the AISCM, 8th edition, and is the recommended approach. The elastic method is still acceptable for conditions where the AISCM tables do not apply.

As a means of comparison, Example 8-8 will be recalculated using the AISCM ultimate strength method.

Example 8-9

Rework Example 8-8 using the AISCM ultimate strength method. This is simplified through the use of the AISCM, Part 4, Table XXIII. Referring to AISCM nomenclature for Tables XIX–XXVI (Part 4), we find

$$\ell = 12 \text{ in.} = \text{length of vertical weld}$$

$$k\ell = 6 \text{ in.} = \text{length of horizontal weld}$$

$$A = 16 \text{ in.} = \text{distance from vertical weld to } P$$

Solution

$$k = \frac{k\ell}{\ell} = \frac{6}{12} = 0.5 \qquad A = 16 \text{ in.}$$

Enter Table XXIII with $k = 0.5$ and obtain $x = 0.125$:

$$x\ell = 0.125(12) = 1.50 \text{ in.}$$

(Check with Example 8-8. This is the distance from the vertical weld to the center of gravity of the weld group.)

$$a\ell = A - x\ell = 16 - 1.50 = 14.5 \text{ in.}$$

$$a = \frac{a\ell}{\ell} = \frac{14.5}{12} = 1.21$$

Interpolating between $a = 1.20$ and $a = 1.40$ for $k = 0.5$ yields

$$C = 0.532$$

$$C_1 = 1.0 \text{ for E70 electrode}$$

Required minimum size of weld in sixteenths of an inch

$$D = \frac{P}{CC_1\ell} = \frac{20}{0.532(1.0)(12)}$$

$$= 3.13 \text{ (sixteenths)}$$

Use a $\frac{1}{4}$-in. fillet weld.

8-7 UNSTIFFENED WELDED SEATED BEAM CONNECTIONS

This type of connection was discussed in Section 7-7 of this text. Bolts were used to connect the vertical leg of the seat angle to the supporting member. We will now discuss the use of welds for the connection of the vertical leg of the seat angle.

The required bearing length for the beam on the horizontal leg of the seat angle, the length of the angle, and the thickness of the angle are all determined in the same manner as that used for the unstiffened bolted seated beam connection. The attachment of seat angle to the supporting member is achieved by welds along the vertical edges of the angle. The welds are considered as eccentrically loaded with the load applied outside the plane of the welds.

The eccentric load is resolved into an axial load and a moment. The axial load is assumed to be uniformly distributed over the total length of weld. The weld stress (shearing stress) resulting from the axial load may be computed from

$$f_v = \frac{P}{\text{total weld length}}$$

where P = beam reaction
f_v = actual shearing stress

The effect of the moment is to create a bending stress distribution varying linearly from zero at the weld neutral axis, which is assumed to occur at middepth of weld. The maximum bending stress (tension) in the weld is assumed to occur at the top end of the vertical weld.

The combined effect of the shear stress (a result of the axial load) and the bending stress (a result of the moment) will be the vectorial sum of the two. Since the stresses act at right angles to each other, the resultant stress may be computed from

$$f_r = \sqrt{f_h^2 + f_v^2}$$

where f_r = resultant stress
f_h = maximum bending stress at top edge of weld
f_v = actual shear stress

Example 8-10

Design an unstiffened seated beam connection for a W18 × 55 beam to a W10 × 60 column web as shown in Fig. 8-26. All structural steel is A36. The beam reaction is 31.4 kips. Use an E70 electrode (SMAW).

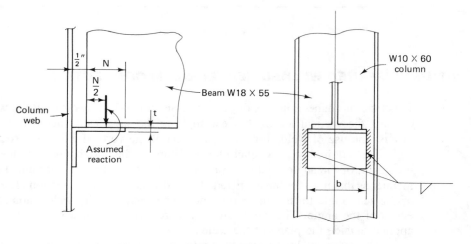

Figure 8-26   Welded unstiffened seated beam connection.

Solution   See Example 7-10 for the determination of the minimum required bearing length N, length of angle b, and thickness of angle t. These values were computed to be: minimum $N = 1.67$ in., $b = 6$ in., and $t = 1$ in. A 4-in. horizontal leg was selected with actual $N = 3.25$ in. (for calculation purposes).

  Both the length of the weld and the size of the weld are to be determined. If one of these quantities is assumed, the other may be calculated to suit the strength requirement.

  The welds are subjected to both bending moment and direct (axial) load. The bending moment referenced to the face of the column web is

$$M = 31.4\left(0.75 + \frac{N}{2}\right)$$

$$= 31.4\left(0.75 + \frac{3.25}{2}\right)$$

$$= 31.4(2.38) = 74.7 \text{ in.-kips}$$

Since there are two vertical lines of weld, the bending moment per line of weld is

$$\frac{74.7}{2} = 37.4 \text{ in.-kips}$$

The bending stress distribution in these vertical welds is based on the assumption that the neutral axis occurs at the middepth of the weld. For each line of weld, the horizontal stress due to the bending moment will be based on an assumed 8-in. vertical leg (therefore, an 8-in. length of weld).

$$f_h = \frac{Mc}{I} = \frac{37.4(4)}{8^3/12} = 3.51 \text{ kips/in.}$$

The vertical shear stress due to the direct load is

$$f_v = \frac{31.4}{16} = 1.96 \text{ kips/in.}$$

The resultant stress may be obtained from

$$f_r = \sqrt{f_h^2 + f_v^2}$$
$$= \sqrt{3.51^2 + 1.96^2} = 4.02 \text{ kips/in.}$$

The weld size required (number of sixteenths) is

$$D = \frac{4.02}{0.925} = 4.35$$

Use $\frac{5}{16}$-in. welds, 8 in. long, on each side of the L8 $\times$ 4 $\times$ 1 seat angle.

8-8 WELDED FRAMED BEAM CONNECTIONS

Bolted framed beam connections are discussed in Chapter 7 of this text. In this section, fillet welds will be used as the fastening technique rather than high-strength bolts. As

Typical framed connection between a beam and the flange of a column. The angles have been shop-welded to the beam web and field bolted to the column. Note the bolted connection for the open web steel joist (see Chapter 9).

in the bolted connection, angles are used on the beam web to transmit the end reaction. The connection may be categorized as a simple beam connection.

The angles may be shop-welded to the beam web and subsequently field-welded to the supporting member as shown in Fig. 8-27. A combination-type connection is often utilized where the angles are shop-welded to the beam web but field-connected to the supporting member with high-strength bolts.

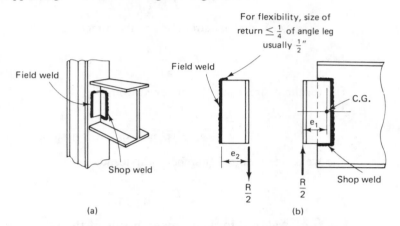

Figure 8-27   Welded framed beam connection.

The bolted framed beam connection is designed to resist shear only and all eccentricity is neglected. But in the welded framed beam connection, the eccentricity of the end reaction with respect to the welds *is* taken into consideration.

Each angle is subject to a vertical shear equal to $R/2$, where R is the beam reaction. In Fig. 8-28(a) and (b), this force may be observed as eccentric to both the

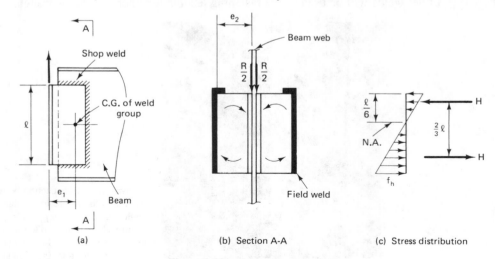

Figure 8-28   Connection behavior.

shop and field welds, thus subjecting the welds to both an axial load and a moment. The moment may be expressed as

$$\frac{R}{2}(e)$$

where e is the eccentricity as shown in Fig. 8-28 and may be either e_1 or e_2.

The design and/or analysis of the shop weld is basically the same as for the eccentrically loaded welded connection of Section 8-6 of this text. However, the field welds are subject to a rotational effect that forces the top portion of the web angles against the beam web and the bottom portion of the angles is pushed apart, as indicated in Fig. 8-28(b). Hence the resistance to this rotation is the bearing of the angles on the beam web together with a horizontal shear stress in the field weld. It is commonly assumed that the neutral axis is located at one-sixth of the length of the angles from the top and that a triangular stress distribution exists, as shown in Fig. 8-28(c). The resisting moment becomes $H(\frac{2}{3}\ell)$, which must be equal to the applied moment $(R/2)(e_2)$. Equating the two in terms of the horizontal shear stress gives us

$$\frac{R}{2}(e_2) = \frac{1}{2}(f_h)\left(\frac{5}{6}\ell\right)\left(\frac{2}{3}\ell\right)$$

from which

$$f_h = \frac{R(e_2)}{0.56(\ell^2)}$$

The vertical shear f_v in the field weld is equal to $R/2$ divided by the length of the weld. The vertical shear stress and horizontal shear stress may then be combined vectorially to determine the maximum shearing stress:

$$f_r = \sqrt{f_h^2 + f_v^2}$$

Example 8-11

Design the shop and field welds for the W21 × 73 framed beam connection shown in Fig. 8-29. Use an E70 electrode (SMAW). The web angles are L3 × 3 × $\frac{3}{8}$ × 10 in. The beam reaction is 50 kips. All structural steel is A36.

Solution

A. *Design of shop weld to beam web*
1. Locate the center of gravity (find $\bar{x}$) of the weld by summation of moments of lengths of weld about the 10-in. side.

$$\bar{x} = \frac{\Sigma\, \ell x}{\Sigma\, x} = \frac{2(2.5)(1.25) + 10(0)}{10 + 2.5 + 2.5}$$

$$\bar{x} = 0.42 \text{ in.}$$

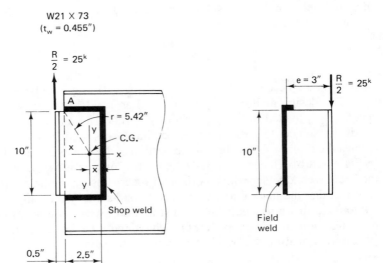

Figure 8-29 Weld configuration.

2. Determine the polar moment of inertia.

$$J = I_x + I_y$$

$$I_x = \left(\frac{1}{12}\right)(10)^3 + 2(2.5)(5)^2 = 208.3 \text{ in.}^3$$

$$I_y = (2)\left(\frac{1}{12}\right)(2.5)^3 + (2)(2.5)\left(\frac{2.5}{2} - 0.42\right)^2 + 10(0.42)^2$$

$$= 7.8 \text{ in.}^3$$

$$J = 208.3 + 7.8 = 216.1 \text{ in.}^3$$

3. The torsional moment is

$$M = Pe$$

$$= \frac{R}{2}(e)$$

$$= 25(3.00 - 0.42) = 64.5 \text{ in.-kips}$$

4. $r = \sqrt{(2.5 - 0.42)^2 + 5^2} = 5.42 \text{ in.}$

The force on the weld due to the torsional moment (at point A) is

$$P_m = \frac{Mr}{J} = \frac{64.5(5.42)}{216.1} = 1.62 \text{ kips/in.}$$

5. The horizontal component of P_m is

$$\frac{5}{5.42}(1.62) = 1.49 \text{ kips/in.}$$

The vertical component of P_m is

$$\frac{2.08}{5.42}(1.62) = 0.62 \text{ kips/in.}$$

6. The force P_v on the weld due to the axial effect of the eccentric load where $P = R/2$ is

$$P_v = \frac{P}{\ell} = \frac{25}{15} = 1.67 \text{ kips/in.}$$

7. Adding the forces vectorially and determining the resultant force F gives us

$$F^2 = 1.49^2 + (0.62 + 1.67)^2$$
$$F = 2.73 \text{ kips/in.}$$

8. The fillet weld leg size required (number of sixteenths) is

$$D = \frac{2.73}{0.925} = 2.95$$

Use a $\frac{3}{16}$-in. fillet weld.

9. Check to ensure that the shear capacity of the $\frac{3}{16}$-in. fillet weld (on each side of the beam web) does not exceed the shear capacity of the web. The shear capacity of the welds (per linear inch, with reference to Table 8-1 of this chapter) is

$$2(2.78) = 5.56 \text{ kips/in.}$$

The shear capacity of the web (A36 steel) is

$$F_v(t_w) = 0.40F_y t_w$$

$$= 0.40(36)(0.455) = 6.55 \text{ kips/in.}$$

$$5.56 \text{ kips/in.} < 6.55 \text{ kips/in.} \qquad \textbf{O.K.}$$

The $\frac{3}{16}$-in. fillet weld is satisfactory.

B. *Design of field weld to the supporting member (end return is neglected)*
   1. The horizontal shear due to the rotational effect, as shown in Fig. 8-28 (b), is

$$f_h = \frac{Re}{0.56\ell^2} = \frac{50(3)}{0.56(10)^2} = 2.68 \text{ kips/in.}$$

2. The vertical shear is

$$f_v = \frac{50}{2(10)} = 2.5 \text{ kips/in.}$$

3. The resultant shear is

$$f_r = \sqrt{f_h^2 + f_v^2}$$
$$= \sqrt{2.68^2 + 2.5^2} = 3.67 \text{ kips/in.}$$

4. The fillet weld size required (number of sixteenths) is

$$D = \frac{3.67}{0.925} = 3.97$$

Use a $\frac{1}{4}$-in. fillet weld.

8-9 WELDING SYMBOLS

The use of welding symbols as a means of communication has been standardized by the American Welding Society. To prepare drawings properly, the steel detailer must have a means of accurately conveying complete information about the welding to the shop and erection personnel so that this information conforms to the intent of the designer. Hence, to avoid misunderstanding and confusion, it is important that the same standard system of weld symbols be used by everyone involved. The reader is referred to the AISCM, Part 4, Welded Joints, Standard Symbols.

Since fillet welds and simple butt welds compose 95% of most structural steel welding, a few of the more common symbols for fillet welds and groove welds are shown in Fig. 8-30.

8-10 WELDING INSPECTION

In the inspection phase, one is concerned primarily with the soundness and quality of a welded joint or weldment. Inspection should begin prior to the actual welding and should continue during welding as well as after the welding is completed. All personnel engaged in inspection operations should be familiar with their company inspection methods as well as all governing codes or standards. The service conditions to which the weldment might be subjected must be known and carefully evaluated before an inspection method can be specified.

The inspection process is only as good as the quality of the inspectors. Employment of competent inspectors is only one aspect of assuring weld quality. In addition, good welding procedures and the use of qualified and certified welders will contribute to an acceptable weld. The weld testing methods generally used for structures may be

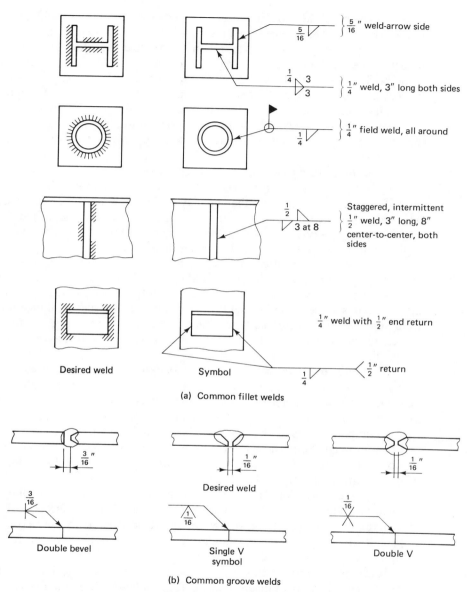

(a) Common fillet welds

(b) Common groove welds

Figure 8-30   Weld symbols.

categorized as nondestructive and include visual, magnetic particle, radiographic, liquid penetrant, and ultrasonic methods.

Visual inspection is probably the most widely used of all inspection methods. It is simple, inexpensive, and the only equipment commonly used is a magnifying glass. Although many factors are beyond the scope of visual examination, it must be regarded as one of the most important methods for determining weld quality. Visual inspection should begin before the first arc is struck. The materials should be examined to see if they meet specifications for quality, type, size, cleanliness, and freedom from defects. Foreign matter, such as grease, paint, oil, oxide film, and heavy scale, which could be detrimental to the weld, should be removed. The pieces to be joined should be checked for straightness, flatness, and dimensions. Warped, bent, improperly cut, or damaged pieces should be repaired or rejected. Alignment, fit-up of parts, and joint preparation should be checked. Inspection prior to welding also includes verification that the correct process and procedures are to be employed and that the electrode type and size are as specified.

Inspection during welding may detect errors and defects that could easily be remedied. It prevents minor defects from piling up into major defects and leading to ultimate rejection. When more than one layer of filler metal is to be deposited, it may be necessary to inspect each layer before a subsequent layer is deposited. The greater the degree of supervision and inspection during welding, the greater the probability of the joint being satisfactory and efficient in service.

Visual inspection after the weldment has been completed is also useful in evaluating quality even if ultrasonic, radiographic, or other methods are to be employed. The following quality factors can usually be determined by visual means: dimensional accuracy of the weldment, conformity to specification requirements, weld appearance, and surface flaws, such as cracks and porosity. With only surface defects visible to the eye, additional nondestructive methods may be necessary and specified.

Magnetic particle inspection is a nondestructive method used to detect the presence of cracks and seams in magnetic materials. It is not applicable to non-magnetic materials. This method will detect surface discontinuities which are too fine to be seen with the naked eye, those which lie slightly below the surface, and when special equipment is used, the more deeply seated discontinuities. The basic principle involved in magnetic particle inspection is that when a magnetic field is established in a piece of ferromagnetic material that contains one or more discontinuities in the path of the magnetic flux, minute poles are set up at the discontinuities. These poles have a stronger attraction for the magnetic particles than the surrounding surface of the material. The particles form a pattern or indication on the surface which assumes the approximate shape of the discontinuity. It is a relatively low cost method of inspection and is considered outstanding for detecting surface cracks. It is also used to advantage on heavy weldments and assemblies.

Radiographic inspection is one of the most widely used techniques for showing the presence and nature of macroscopic defects and other discontinuities in the interior of welds. This test method is based on the ability of X-rays and gamma rays to penetrate metal and other opaque materials and produce an image on sensitized film or a fluorescent screen. It is a nondestructive test method and offers a permanent record when recorded on film. It is a relatively expensive type of inspection and due to the radiation hazard requires extensive safety precautions. Considerable skill is required in choosing angles of exposure and operating equipment and in interpreting the results.

Liquid penetrant inspection is a nondestructive method for locating surface cracks and pinholes that are not visible to the naked eye. It is a favored technique for locating leaks in welds, and it can be applied where magnetic particle inspection cannot be used, such as with nonferrous metals. Fluorescent or dye penetrating substances may be used for liquid penetrant inspection.

Fluorescent penetrant inspection makes use of a highly fluorescent liquid with unusual penetrating qualities. It is applied to the surface of the part to be inspected and is drawn into extremely small surface openings by capillary action. The excess liquid is then removed from the part, a "developer" is used to draw the penetrant to the surface, and the resulting indication is viewed by ultraviolet (black) light. The high contrast between the fluorescent material and the background makes possible the detection of minute traces of penetrant.

Dye penetrant inspection is similar to fluorescent penetrant inspection except that dyes visible under ordinary light are used. By eliminating the need for ultraviolet light, greater portabillity in equipment is achieved.

Ultrasonic inspection is a rapid and efficient nondestructive method of detecting, locating, and measuring both surface and subsurface defects in the weldment and/or base materials. Flaws that cannot be discovered by other methods, and even cracks small enough to be termed microseparations, may be detected. Ultrasonic testing makes use of an electrically timed wave of the same nature as a sound wave, but of a higher pitch or frequency. The frequencies used are far above those heard by the human ear, hence, the name ultrasonic. The sound waves or vibrations are propagated in the metal which is being inspected until a discontinuity or change of density is reached. At these points, some of the vibrational energy is reflected back and indicated on a cathode ray tube. The pattern on the face of the tube is thus a representation of the reflected signal and of the defect. The ultrasonic method requires special commercial equipment and in addition a high degree of skill is required in interpreting the cathode ray tube patterns.

REFERENCES

1. *Structural Welding Code,* AWS D1.1-77, American Welding Society, 2501 N.W. Seventh Street, Miami, FL 33125.
2. A. J. Julicher, "Welding Fundamentals for Structural Engineers," *Civil Engineering,* Vol. 51, No. 1, January 1981.
3. *Welding Handbook,* latest edition, American Welding Society, 2501 N.W. Seventh Street, Miami, FL 33125. A multi-volume series covering practically all aspects of welding.

PROBLEMS

8-1. Determine the allowable tensile load that may be applied to the connection shown. The steel is A36 and the electrode used was an E70 (SMAW). The weld is a $\frac{3}{8}$-in. fillet weld.

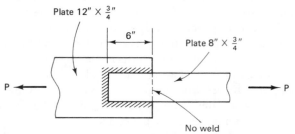

Problem 8-1

8-2. Determine the allowable tensile load that may be applied to the connection shown. The steel is A36 and the electrode used was an E70 (SMAW). The weld is a $\frac{1}{4}$-in. fillet weld.

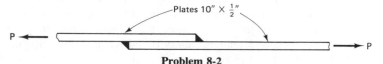

Problem 8-2

8-3. Determine the maximum tensile load that can be applied to the wide-flange tension member shown. The welds are $\frac{1}{4}$-in. fillet welds. The steel is A36 and the electrode used was an E70.

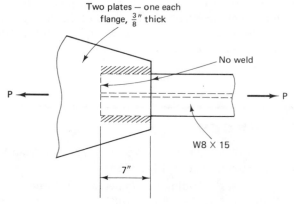

Problem 8-3

8-4. A fillet weld made between two steel pieces intersecting at right angles was determined to have one $\frac{1}{4}$-in. leg and one $\frac{9}{16}$-in. leg. The electrode was an E80 (SMAW). Determine the strength of this weld in kips/in.

8-5. Design longitudinal fillet welds to develop the full tensile capacity of the 8 in. $\times \frac{1}{2}$ in. plate shown. The steel is A36 and the electrode is an E70 (SMAW). Standard end returns are used.

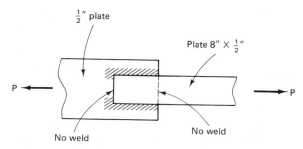

Problem 8-5

8-6. Design an end connection using longitudinal welds and an end transverse weld to develop the full tensile capacity of the angle shown. Use A36 steel and E70 electrodes (SMAW).

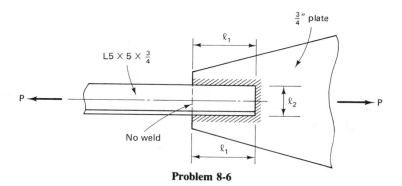

Problem 8-6

8-7. Design an end connection using only longitudinal welds to develop the full tensile capacity of the angle described. The member is subjected to repeated stress variations. Use A36 steel and an E70 electrode (SMAW). The angle is an L7 $\times$ 4 $\times \frac{3}{4}$ with the long leg connected to a $\frac{9}{16}$-in. gusset plate. Use the maximum allowable weld size.

8-8. Rework Problem 8-7 using a transverse weld at the end of the angle in addition to the longitudinal welds.

8-9. Design the connection of a C15 $\times$ 40 to a $\frac{3}{8}$-in. gusset plate as shown to develop the full tensile capacity of the channel. Welding is not permitted across the back of the channel. All structural steel is A36. Use an E70 electrode (SMAW). The maximum length of lap on the gusset plate is 6 in. due to space limitations.

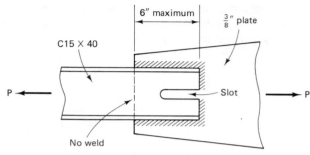

Problem 8-9

8-10. Compute the maximum allowable beam reaction for the end-plate shear connection shown. All structural steel is A36. The electrode is an E70 (SMAW).

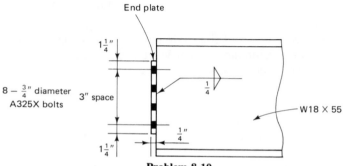

Problem 8-10

8-11. With reference to Problem 7-20, design the welded portion of the end-plate shear connection using an E70 electrode (SMAW).

8-12. Calculate the maximum resultant force (kips/in.) to be resisted by the welds shown.

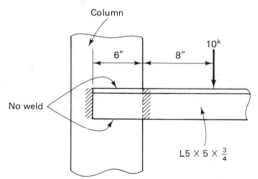

Problem 8-12

8-13. Rework Problem 8-12, but weld is "all around."

8-14. Determine the size of the fillet weld required to resist a 65-kip load on the bracket shown. The steel is A36. Use an E70 electrode (SMAW).

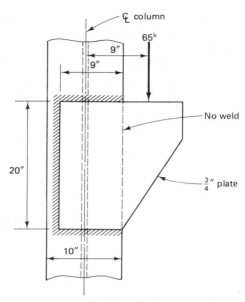

Problem 8-14

8-15. Rework Problem 8-14 using the AISCM ultimate strength method (using AISCM tables).

8-16. Design an unstiffened seated beam connection for a W16 × 77 beam to a W8 × 35 column web. All structural steel is A36. The beam reaction is 35 kips. Use an E70 electrode (SMAW). The long leg of the seat angle is to be shop-welded to the column web.

8-17. Compute the maximum allowable beam reaction for the welded framed beam connection shown. All structural steel is A36 and the electrode used was an E70 (SMAW).

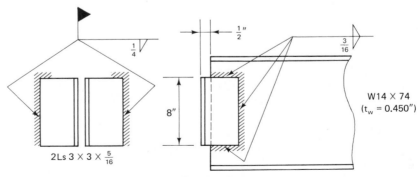

Problem 8-17

8-18. Design the shop and field welds for the framed connection for the W18 × 71 beam shown. Use an E70 electrode (SMAW). All structural steel is A36. The beam reaction is 40 kips and the web angles are L3 × 3 × $\frac{1}{4}$ × 10 in.

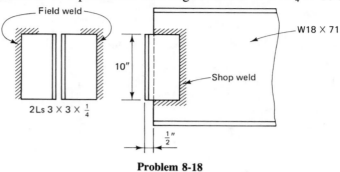

Problem 8-18

9 Open Web Steel Joists and Metal Deck

9-1 INTRODUCTION TO STEEL JOISTS

Steel joists are standardized prefabricated trusses generally used for the direct support of floor decks and/or roof decks in buildings. They are commonly used in combination with a corrugated steel metal deck and generally will provide an efficient and economical floor or roof system in lightly loaded buildings. A typical floor or roof system includes other suspended or supported materials (in addition to the joists and metal deck), such as suspended ceilings and roofing materials. Discussion of these other materials will not be included in this text. The reader is referred to texts on construction materials and appropriate manufacturers literature (see the references at the end of this chapter). There are three categories of steel joists available: (1) Open Web Steel Joists, H-Series; (2) Longspan Steel Joists, LH-Series; and (3) Deep Longspan Steel Joists, DLH-Series.

Open Web Steel Joists, H-Series refers to members that have the appearance of shallow trusses with parallel chords, and are completely standardized as to length, depth, and carrying capacities. They are suitable for use in both floor and roof applications. The standard depths for the H-Series joists are 8 to 30 in., varying by 2-in. increments. Clear span lengths of up to 60 ft are common for the H-Series joists.

In the United States, the design, fabrication, and erection of steel joists are generally accomplished in accordance with the requirements of the specifications published by the Steel Joist Institute (SJI).[1] The SJI publication is a valuable reference source and should be obtained by the reader who desires more in-depth information.

Typical roof and deck construction for a two-story office building. Open web steel joists for the roof are being welded to supporting joist girders.

The Institute is a nonprofit organization of steel joist manufacturers with the primary purpose of promoting the use of steel joists. The *Standard Specifications* first appeared in 1928 and has been modified over the years to reflect progress in research, manufacturing, materials and welding techniques.

Longspan Steel Joists, LH-Series and **Deep Longspan Steel Joists, DLH-Series** are similar in general appearance to the H-series joists, but are deeper and span greater distances. Longspan Series (LH) joists have been standardized in depths from 18 to 48 in. for clear spans to 96 ft and are generally used in both floor and roof applications. Deep Longspan Series (DLH) joists have been standardized in depths from 52 to 72 in. for clear spans up to 144 ft and are generally used for roof applications. The specifications applying to the LH-Series and the DLH-Series joists are found in Reference 1. Both series can be furnished with either parallel chords or with single- or double-pitched top chords to provide sufficient slope for roof drainage.

The design of all standardized joists has become the responsibility of the joist manufacturers. The selection of which joists to use, irrespective of series, involves the use of standard load tables as furnished by the SJI and, commonly, by the joist manufacturer as well. Hence, when the designer of a building decides on the use of steel joists as part of the floor or roof system, the designer does not *design* the joists, but rather *selects* the proper joists from the load tables based on span length, loading, and joist spacing. All joists are originally designed as simply supported *uniformly* loaded trusses supporting a floor and/or roof deck so that the top chords of the joists are adequately braced against lateral buckling. Where joists are used under conditions different from those for which they were originally designed, they must be investigated and modified as necessary since the load tables are no longer applicable.

The SJI standard load tables for H-Series and LH-Series joists are reproduced in Appendix A (Tables A-1 and A-2) in the back of this text. Note that the tables furnish the allowable *total* uniformly distributed load in pounds per linear foot for each joist designation. The *dead* load, including the weight of the joists, should be deducted to determine the *live*-load-carrying capacity of the joist. The lower number in a box in the H-Series table, normally printed in color in the SJI table, indicates the live load per linear foot of joist which will produce a deflection of span/360. Live loads which will produce a deflection of span/240 may be obtained by multiplying the lower number by 1.5. Example problems will demonstrate the use of the tables.

9-2 OPEN WEB STEEL JOISTS, H-SERIES

The design of the H-Series joist *chord* is based on a steel minimum yield strength of 50,000 psi. The design of the *web members* may be based on a steel minimum yield strength of 36,000 psi or 50,000 psi.

An example of the standard designation for H-Series joists is **22H7.** The depth of this joist is 22 in. H represents the series and the number 7 denotes the relative size of the chords of the joist. Chord sizes are designated by the numbers 3 through 11, the size increasing with increasing number. The chord and web member may vary in shape and makeup from manufacturer to manufacturer, but the design and the capacity of the joists must conform to the SJI specifications and to the standardized load tables. The H-Series standard load table is applicable where the joists are installed up to a maximum slope of $\frac{1}{2}$ in. per foot.

The use of the open web steel joist in any given application must be based on SJI requirements as furnished in its standard specifications. These requirements are summarized as follows:

1. The clear span of a joist must not exceed 24 times its depth.

2. The ends of joists must extend a distance of not less than $2\frac{1}{2}$ in. over steel supports. When bearing on concrete, the minimum bearing length is 4 in. If the joists bear on masonry, the bearing length may vary from 4 to 6 in. depending on the joist chord size.

3. In joist construction, bridging and bridging anchors are required for the primary purpose of furnishing lateral stability for the joists, particularly during the construction phase. The bridging spans between and perpendicular to the steel joists. As soon as joists are erected all bridging must be completely installed and the joists permanently fastened into place before the application of any construction loads (neglecting, of course, the weight of the erectors or steel workers). Even under the weight of an erector, the joists may exhibit some degree of lateral instability until the bridging is installed. Where more than three rows of bridging are required, caution must be exercised by the erector until all bridging is completely and properly installed. The bridging also serves a purpose of holding the steel joists in the position as shown on the plans. The minimum number of rows of bridging is a function of the joist chord size and clear span. A table is furnished in the standard specifications, which establishes the required number of rows of bridging. Spacing of bridging rows should be approximately equal. Also, Reference 2 contains pertinent information. Two permissible types of bridging may be observed in Fig. 9-1. *Horizontal bridging* [Fig. 9-1(a)] consists of two continuous horizontal steel members, one attached to the top chord and the other attached to the bottom chord by means of welding or mechanical fasteners. The attachment must be capable of resisting a horizontal force of not less than 700 lb. If the bridging member is a round bar the diameter must be at least $\frac{1}{2}$ in. The maximum slenderness ratio (ℓ/r) of the bridging member cannot exceed 300, where ℓ is the distance between bridging attachments and r is the least radius of gyration of the bridging member. *Diagonal bridging* [Fig. 9-1(b)] consists of cross-bracing with a maximum ℓ/r of 200, with ℓ and r as defined previously. Where the cross-bracing members connect at their intersection, ℓ is the distance between the intersection attachment and chord attachment. The ends of all bridging lines terminating at walls or beams must be properly anchored. A typical detail may be observed in Fig. 9-1(b).

4. Positive end anchorage for the joists must be provided in addition to the required end bearing length. Where joists rest on steel, the minimum positive connection must be two $\frac{1}{8}$-in. fillet welds, each being one inch in length, or a $\frac{1}{2}$-in.-diameter bolt. Where joists rest on masonry supports, an anchor bar $\frac{3}{8}$ in. in diameter and 8 in. long may be furnished and embedded in the mortar joints. This is commonly called

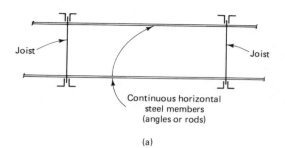

(a)

(b)

Figure 9-1 Typical bridging.

a **government anchor** [Fig. 9-2(a)]. Every joist in a roof and every third joist on floors must be furnished with end anchorage. Where roofs have no parapet walls, two $\frac{1}{2}$-in.-diameter anchor bolts, or equal, may be used in lieu of the steel bar. Where steel columns are not framed in at least two directions with structural steel members, the joists at column lines must be field-bolted at the columns to assure some measure of lateral stability during construction.

5. Extended ends are frequently used with H-Series joists [Fig. 9-2(b)]. These are designed as cantilever beams with their reactions carried back at least to the first interior panel point of the joist. Specific designs and load tables for extended ends are generally furnished by the various manufacturers and should be used. The standard extended end depth is $2\frac{1}{2}$ in., which conforms to the H-Series standard bearing end depth.

6. Ceiling extensions [Fig. 9-2(c)] in the form of an extended bottom chord element or a loose unit, whichever is standard with the joist manufacturer, are frequently used to support ceilings which are to be attached directly to the bottom of the joists. They are not furnished for the support of suspended ceilings.

7. When used in conjunction with a corrugated metal deck and concrete slab, the cast-in-place slab should not be less than 2 in. thick.

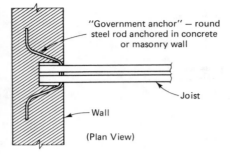

"Government anchor" — round
steel rod anchored in concrete
or masonry wall

Joist

Wall

(Plan View)

(a) End anchorage

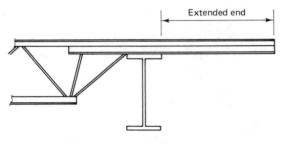

Extended end

(b) Extended top chord

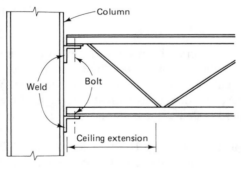

Column

Weld

Bolt

Ceiling extension

(c) Typical column connection

Figure 9-2   Typical joist details.

9-3 LONGSPAN STEEL JOISTS, LH-SERIES, AND DEEP LONGSPAN STEEL JOISTS, DLH-SERIES

The design of the LH- and the DLH-Series is based on a steel yield strength of at least 36,000 psi but not greater than 50,000 psi. The standard designation for LH-Series joists, for example, 28LH09, furnishes first the depth of the member, 28 in. The designation LH represents the series. The final number, 05 through 13 (for the 28-in.

deep members), denotes the relative size of the chords, the size increasing with the number. Other depth LH-Series joists may have the last two digits different than those of the 28-in. deep series. The deep longspan series designation is similar, but the designation DLH replaces LH.

The SJI *Standard Specifications* for these two series may be summarized as follows:

1. The clear span of a joist must not exceed 24 times its depth.

2. The ends of joists must extend a distance of not less than 4 in. over steel supports. When bearing on concrete the bearing length may vary from 6 to 9 in., and when bearing on masonry the bearing length may vary from 6 to 12 in., depending on the joist chord size for each condition.

3. Bridging must be of the cross-bracing type (diagonal) with the same slenderness ratio requirements as for the H-Series. The connections of bracing members to chords of joists must be capable of resisting horizontal forces as furnished in Table 104.5.1 of the SJI *Standard Specifications*. In the same table, the maximum spacing of lines of bridging is furnished. Bridging installation requirements are the same as for the H-Series.

4. Positive end anchorage for the joists must be provided in addition to the required end bearing length. Where joists rest on steel, the minimum positive connection must be two $\frac{1}{4}$-in. fillet welds, each 2 in. in length or with two $\frac{3}{4}$-in.-diameter bolts. Where joists rest on masonry or concrete supports an anchor bar $\frac{3}{4}$ in. in diameter and at least 12 in. long (or equivalent) may be furnished. All joists must be furnished with end anchorage. In roofs where masonry parapet walls are less than 2 ft high, two $\frac{3}{4}$-in. anchor bolts or equal must be used in lieu of the steel bar. Where steel columns are not framed in at least two directions with structural steel members, the joists at column lines must be field-bolted at the columns to provide lateral stability during construction.

5. Extended ends and ceiling extensions are also frequently used with the LH- and DLH-Series with the specific details conforming to the dimensions and requirements for each series. The various manufacturers should be contacted for necessary design data.

6. When used in conjunction with a corrugated metal deck and concrete slab, the cast-in-place slab should not be less than 2 in. thick.

Example 9-1

Select open web steel joists H-Series for a floor system of a typical interior bay of a commercial building, as shown in Fig. 9-3. The joists are to span in the direction indicated. The span length is 26 ft. The allowable live-load deflection is span/360. The floor loadings are 40 psf superimposed dead load (DL) and 100 psf live load (LL). Use a joist spacing of 2 ft–0 in. on center.

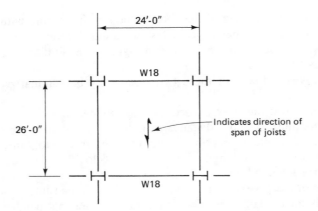

Figure 9-3 Framing plan—typical interior bay.

Solution   Calculate the load per linear foot on a joist:

$$DL = 40(2) = 80 \text{ lb/ft}$$

$$LL = 100(2) = \underline{200 \text{ lb/ft}}$$

$$Total = 280 \text{ lb/ft}$$

The minimum joist depth, as per the SJI *Standard Specifications,* is span/24:

$$\frac{26(12)}{24} = 13 \text{ in.}$$

Refer to H-Series standard load table and select the following possibilities:

Joist	Weight (lb/ft)	Tabulated load-carrying capacity (lb/ft)
14H6	8.6	303/156
16H6	8.6	339/199
18H5	8.0	321/216
20H5	8.4	360/263

The 18H5 is most economical (lightest, at 8.0 lb/ft); therefore, this joist will be checked first.

1. The superimposed load capacity = 321.0 − 8.0 = 313 lb/ft
2. The actual total superimposed load = 280.0 lb/ft < 313.0 lb/ft   **O.K.**
3. The LL that will produce a deflection of span/360 = 216 lb/ft
4. The actual LL = 200 lb/ft < 216 lb/ft   **O.K.**

Use 18H5 at 2 ft–0 in. on center.

Example 9-2

Select open web steel joists for a roof system of a typical interior bay of a commercial building as shown in Fig. 9-4. The joist span length is 50 ft. The allowable live load deflection is span/240. The roof loadings are 30 psf superimposed dead load (DL) and 45 psf snow load (LL). Use a joist spacing of 5 ft–0 in. on center.

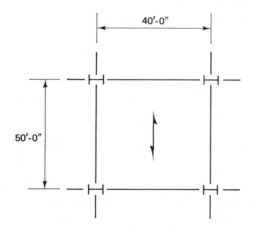

Figure 9-4   Framing plan—typical interior bay.

Solution   Calculate the load per linear foot on the joist.

$$DL = 30(5) = 150 \text{ lb/ft}$$

$$\text{snow load} = 45(5) = \underline{225 \text{ lb/ft}}$$

$$\text{total} = 375 \text{ lb/ft}$$

Minimum joist depth as per the SJI *Standard Specifications* is span/24:

$$\frac{1}{24}(50)(12) = 25 \text{ in.}$$

Refer to LH-Series standard load tables and select the following possibilities:

Joist	Weight (lb/ft)	Tabulated load-carrying capacity (lb/ft)
28LH09	24.0	463/243
32LH08	21.0	397/242

The 32LH08 is most economical (lightest); therefore, this member will be checked first.

1. The superimposed load capacity = $397 - 21 = 376$ lb/ft
2. The actual total superimposed load = 375 lb/ft $<$ 376 lb/ft   **O.K.**
3. The LL that will produce a deflection of span/240 is

$$242(1.5) = 363 \text{ lb/ft}$$

4. The actual LL = 225 lb/ft $<$ 363 lb/ft   **O.K.**

Use 32LH08 at 5 ft–0 in. on center.

9-4 FLOOR VIBRATIONS

Despite the fact that the structural design of the steel joists is accomplished in accordance with design specifications, a floor system may be susceptible to undesirable vibrations. This is a phenomenon that is separate and different from strength and has to do mainly with the psychological and physiological response of humans to motion. Large open floor areas without floor to ceiling partitions may be subject to such undesirable vibrations.

The AISCS Commentary recommends a minimum depth-to-span ratio of 1/20 for a steel beam supporting large open floor areas free of partitions. In addition, the SJI requires a minimum depth-to-span ratio of 1/24 for steel joists, although a generally accepted practice for steel joist roofs and floors is to use a minimum depth-to-span ratio of 1/20. Even if these recommendations and requirements are satisfied, a vibration analysis should be made, particularly when a floor system is composed of steel joists which support a thin concrete slab placed on steel metal deck. References 3 and 4 contain relatively brief and sufficiently accurate methods which can be used to determine (1) if disturbing vibrations will be present in a floor system, and (2) possible design solutions for the problem.

9-5 CORRUGATED STEEL DECK

Steel deck is commonly used in conjunction with steel joists in floor and roof systems as shown in Fig. 9-5. Most decking used in buildings today is designed, manufactured, and erected in accordance with the Steel Deck Institute specifications and code of recommended standard practice.[5] The *Specifications for the Design of Cold-Formed Steel Structural Members* of the American Iron and Steel Institute[6] also apply.

Steel decks are cold-formed steel products with longitudinal ribs of various configurations distinctive to individual manufacturers. Originally, steel decks were used only for roof construction. However, as a result of testing and research, applications now include floor decks as well. All steel decks are cold-formed from various thickness sheet steel. Whereas it has been common practice (and still is) to classify

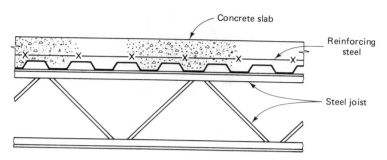

Figure 9-5 Corrugated steel deck on open web steel joists.

the steel deck by a gage designation, the Steel Deck Institute has replaced the gage value with a design thickness as the unit of measure in references to material thickness.

When used as a roof deck, the decking acts as a structural supporting member and must support the applied roof system such as rigid insulation and built-up roofing or insulating concrete and roofing material plus roof snow load. When used as the structural supporting member, the steel deck must be galvanized. Roof deck may be furnished with narrow ribs, intermediate ribs, or wide ribs and of various thicknesses depending on the manufacturer. All manufacturers have load tables indicating load-carrying capacity for the different designations and varying span lengths.

When used as a floor deck, the steel deck may act in two different ways.

1. As a form only, for a structural concrete slab until the concrete reaches its design strength. In this case the design load will consist of the weight of the wet concrete plus some arbitrary construction live load (which is generally assumed to be 20 psf). In this application the steel deck is generally furnished uncoated.

2. The steel deck may also act compositely with the concrete slab. The steel deck, in effect, provides the positive moment reinforcing for the concrete slab and may eliminate the need for any additional reinforcing. Composite floor decks are designed to interlock positively with the overlying concrete, resulting in unit action. The interlocking process is achieved by mechanical means, deck profile, and surface bond, or a combination of these. When used in this manner, the side of the deck that is to be in contact with the concrete must be uncoated. The underside should be galvanized. Welded wire fabric should be provided in all composite floor deck slabs, primarily for purposes of crack control rather than negative reinforcing.

All steel deck, whether roof or floor deck, when in place and properly attached is usually assumed to provide lateral restraint for the compression flange (assuming simple spans) of the supporting members. The steel deck offers satisfactory in-plane stiffness in a structure, assuming proper attachments to the supports. This is particu-

This metal deck has been welded to supporting open web steel joists. Square welding washers have been used to reinforce the thin decking in the weld area.

larly critical where steel deck is supported by open web steel joists since the joists have a minimum of lateral stability with their compression flange unsupported laterally. Connections to the supporting member flange is usually accomplished by self-tapping screws or by plug welding from the top through the metal deck. For thin metal decks (lighter than 22 gage, thickness 0.0283 in.) the use of welding washers is recommended. The spacing of the connection should not exceed 12 in. on centers, or as otherwise recommended, and the deck should be connected to all supporting members.

Prior to specifying any steel deck the reader is encouraged to consult the manufacturers' literature. Design load tables should be read carefully and used with

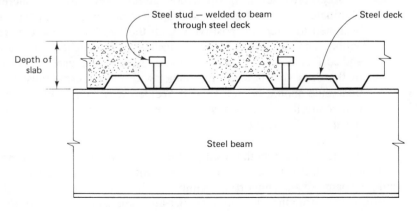

Figure 9-6   Composite system.

caution since there is little consistency from one manufacturer to another with respect to table format.

A relatively recent development has been the introduction of composite beams where "shear studs" are welded through the metal deck on to the top flange of the supporting beam as shown in Fig. 9-6. Composite action[7] depends on steel beam–concrete interaction and tests have demonstrated that the ribs of the steel deck do not interfere with this interaction. However, limitations as to stud diameter and length may exist along with other design criteria for the various steel decks.

REFERENCES

1. Steel Joist Institute, *Standard Specifications Load Tables and Weight Tables,* published annually by The Steel Joist Institute, 1205 48th Ave. North, Myrtle Beach, South Carolina 29577.

2. Steel Joist Institute, *Spacing of Bridging for Open Web Steel Joists,* Technical Digest No. 2, The Steel Joist Institute, 1205 48th Ave. North, Myrtle Beach, South Carolina 29577.

3. Kenneth H. Lenzen, "Vibration of Steel Joist–Concrete Floors," *Engineering Journal, American Institute of Steel Construction,* July 1966.

4. Steel Joist Institute, *Vibration of Steel Joist-Concrete Slab Floors,* Technical Digest No. 5, The Steel Joist Institute, 1205 48th Ave. North, Myrtle Beach, South Carolina 29577.

5. Steel Deck Institute, *Design Manual for Floor Decks and Roof Decks,* 1980, The Steel Deck Institute, P.O. Box 3812, St. Louis, MO 63122.

6. American Iron and Steel Institute, *Specifications for the Design of Cold-Formed Steel Structural Members,* American Iron and Steel Institute, 1000 16th Street, N. W., Washington, DC 20036.

7. L. Spiegel, and G. F. Limbrunner, *Reinforced Concrete Design* (Englewood Cliffs, NJ: Prentice-Hall, Inc., 1980), pp. 334–343.

8. R. C. Smith, *Materials of Construction,* 3rd ed. (New York: Gregg Div., McGraw-Hill, 1979).

9. W. J. Patton, *Construction Materials* (Englewood Cliffs, NJ: Prentice-Hall, Inc., 1976).

PROBLEMS

9-1. (a) What is the designation of the shallowest H-Series joist?
  (b) What is the designation of the deepest, heaviest H-Series joist?
  (c) What is the approximate weight of a 24H8?
  (d) What is the total safe load for a 22H8 on a 38-ft span?

    (e) For a 24H6 on a span of 34 ft, what live load will cause a deflection of span/360?

9-2. A 36LH09 joist spans 66 ft and supports a roof with a superimposed dead load of 20 psf excluding its own weight. Joists are spaced 4 ft on center. Determine the allowable live load (psf) if the allowable live load deflection = span/240.

9-3. Select steel joists for a floor system. The span length is to be 34 ft. The floor loading consists of 100 psf live load and 40 psf superimposed dead load. The allowable live-load deflection is span/360. Use a joist spacing of 2 ft–8 in. on center.

9-4. If 24H7 joists were available for the floor in Problem 9-3, determine the maximum spacing for these joists in order that they be acceptable to carry the load.

9-5. Select open web steel joists (H-Series) for a floor system with a span length of 21 ft. The floor loading is live load 60 psf and superimposed dead load 30 psf. The allowable live load deflection is span/240. Use a joist spacing of 2 ft–0 in.

9-6. Select open web steel joists (H- or LH-Series) for a roof system with a span length of 60 ft. The roof loading is snow load 45 psf and superimposed dead load 20 psf. The allowable live load deflection (snow load) is span/360. Use a joist spacing of 5 ft–0 in.

9-7. Select open web steel joists for a floor system with a span length of 30 ft. Floor loading is live load 100 psf and superimposed dead load 40 psf. Allowable live load deflection is span/360. Use a joist spacing of 2 ft–6 in.

10 Plate Girders

10-1 INTRODUCTION TO WELDED PLATE GIRDERS

Plate girders are built-up bending members which are designed and fabricated to fulfill requirements which exceed those of usual rolled sections. The most common form of plate girder currently being designed consists of two flange plates welded to a relatively thin web plate. The width and thickness of the flange plates may be changed somewhere along a span length, although the economy of such a change is questionable. The *thickness* of the web plate is generally constant. The *depth* of the web plate may be constant or it may be increased in areas of higher moment such as at supports of continuous or overhanging beams. Variable-depth plate girders are normally used only for long-span structures. For reasonable proportions and economical design the depth-to-span ratio of a girder should range anywhere from $\frac{1}{8}$ to $\frac{1}{12}$. The basic elements of a typical welded plate girder are shown in Fig. 10-1.

Bearing stiffeners are used at reactions and concentrated load points to transfer the concentrated loads to the full depth of the web. Transverse intermediate stiffeners are utilized at various spacings along the span length and serve to increase the web buckling strength, thereby increasing web resistance to shear and moment combinations. For deeper web plates, particularly in areas of high moments, longitudinal web stiffeners may also be required. Overall economy is sometimes realized by using a web plate of such thickness that stiffeners are not required.

The AISCS, Section 1.10.1, states that, in general, plate girders should be proportioned by the moment-of-inertia method. This approach requires the selection of a suitable trial cross section which would then be checked by the moment-of-inertia method. To obtain preliminary sizes of the flange and web plates, a flange-area method is generally utilized. This is discussed in Section 10-2 of this text.

Plate girder highway grade separation bridge near Rochester, New York. Girders are simple spans and carry a curved upper roadway. Note variable thickness bottom flange and intermediate transverse stiffeners. (Courtesy of the New York State Department of Transportation.)

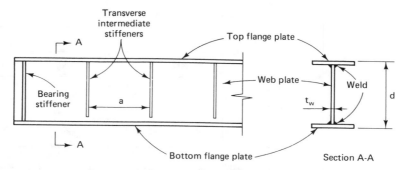

Figure 10-1   Welded plate girder.

10-2 AISC DESIGN CRITERIA FOR PRELIMINARY SELECTION OF PLATE GIRDER WEBS AND FLANGES

As mentioned previously, the total girder depth would generally range from $\frac{1}{8}$ to $\frac{1}{12}$ of the span length, depending on load and span requirements. Therefore, the web depth may be estimated to be from 2 to 4 in. less than the assumed girder total depth. The web thickness may then be selected based on permissible depth–thickness ratios as established in the AISCS. These ratios are based on buckling considerations. The web must have sufficient thickness to resist buckling tendencies which are created by girder curvature under load. As a girder deflects, a vertical compression is induced in the web due to the components of the flange stresses, the result of which constitutes a squeezing action. The buckling strength of the web must be capable of resisting this squeezing action. This is the basis for the AISCS criteria (Section 1.10.2) that the ratio of the clear distance between flanges to the web thickness must not exceed

$$\frac{h}{t_w} \leq \frac{14,000}{\sqrt{F_y(F_y + 16.5)}}$$

where h = clear distance between the flanges
t_w = web thickness
F_y = yield stress of the compression flange (ksi)

It is allowed for this ratio to be exceeded if transverse intermediate stiffeners are provided with a spacing not in excess of 1.5 times the total girder depth. The maximum permissible h/t_w ratio then becomes

$$\text{maximum } \frac{h}{t_w} = \frac{2000}{\sqrt{F_y}}$$

Resulting values for the preceding two expressions, as functions of various F_y values, are shown in Table 10-1.

TABLE 10-1 MAXIMUM h/t_w RATIOS

h/t_w	F_y			
	36 ksi	42 ksi	46 ksi	50 ksi
$\dfrac{14{,}000}{\sqrt{F_y(F_y + 16.5)}}$	322	282	261	243
$\dfrac{2000}{\sqrt{F_y}}$	333	309	295	283

In addition, web buckling considerations may require a reduction of the allowable bending stress in the compression flange. According to the AISCS, Section 1.10.6, when the web depth–thickness ratio exceeds $760/\sqrt{F_b}$, the maximum bending stress in the compression flange must be reduced to a value that may be computed from AISCS formula 1.10-5.

Plate girder webs which depend on **tension field action,** to be defined shortly, as provided in AISCS formula 1.10.2, must be so proportioned that the bending tensile stress not exceed $0.60F_y$ or

$$\left(0.825 - 0.375\,\frac{f_v}{F_v}\right)F_y \qquad \text{AISCS formula 1.10-7}$$

where f_v = actual web shear stress (total shear divided by web area) (ksi)
F_v = allowable web shear stress according to AISCS formula 1.10-2 (ksi)

This expression, in effect, constitutes an allowable bending stress reduction due to the interaction of concurrent bending and shear stress (AISCS, Section 1.10.7).

After preliminary web dimensions are selected, the required flange area may be determined using an approximate approach as follows. With reference to Fig. 10-2,

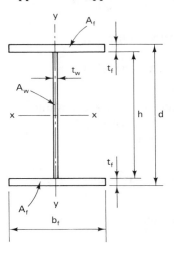

Figure 10-2 Girder nomenclature.

the moment of inertia of the total section with respect to axis x-x is

$$I_x = I_{x(\text{web})} + I_{x(\text{flanges})}$$

Neglecting the moment of inertia of the flange areas about their own centroidal axes and assuming that $h \simeq (d - t_f)$, an approximate gross moment of inertia may be expressed as

$$I_x = \frac{t_w h^3}{12} + 2A_f \left(\frac{h}{2}\right)^2$$

Expressing this in terms of the section modulus (S) and also assuming that $h \simeq d$:

$$S_x = \frac{t_w h^3/12}{h/2} + \frac{2A_f(h/2)^2}{h/2}$$

$$= \frac{t_w h^2}{6} + A_f h$$

However, the required $S_x = M/F_b$; therefore,

$$\frac{M}{F_b} = \frac{t_w h^2}{6} + A_f h$$

and

$$\text{required } A_f = \frac{M}{F_b h} - \frac{t_w h}{6}$$

where A_f = area of one girder flange
$\quad\ h$ = depth of the girder web
$\quad\ F_b$ = allowable bending stress for the compression flange
$\quad\ M$ = maximum bending moment with respect to x-x axis

The first portion of this expression $M/F_b h$ represents the required flange area necessary to resist the bending moment M assuming no contribution by the girder web. However, since the web does furnish some bending moment resistance, the second term $(t_w h/6)$ is included.

Based on the computed required flange area, actual proportions of the flange can be determined taking into account additional AISCS criteria.

To prevent a localized buckling of the compression flange, the AISCS, Section 1.9.1.2, states that the ratio of one-half flange width to flange thickness may not exceed $95.0/\sqrt{F_y}$ if the flange is to be considered fully effective. Values are tabulated in the AISCS, Appendix A, Table 6. For A36 steel, the maximum flange plate width would then be

$$15.8t_f + 15.8t_f = 31.6t_f$$

As mentioned previously, a reduction of the allowable bending stress in the compression flange due to web buckling will be necessary if the web depth-to-thickness ratio exceeds $760/\sqrt{F_b}$. When this occurs, the allowable flange stress may not exceed

$$F'_b \leq F_b\left[1.0 - 0.0005\frac{A_w}{A_f}\left(\frac{h}{t_w} - \frac{760}{\sqrt{F_b}}\right)\right] \qquad \text{AISCS formula 1.10-5}$$

where F_b = applicable bending stress as established by the AISCS, Section 1.5.1.4
  A_w = area of web
  A_f = area of compression flange
  F'_b = allowable bending stress in compression flange of plate girders as reduced because of large web depth-to-thickness ratio

After completing the preliminary selection of the girder web and flanges, the actual moment of inertia and section modulus must be calculated. The actual bending stress should then be calculated and compared with the allowable bending stress. Due consideration must be given to a laterally unsupported compression flange.

The flange plates whose sizes are determined based on the *maximum* bending moment may extend the full length of the girder. However, this is not necessary and they may be reduced in size when the applied moment has decreased appreciably. Changes in flange plates are best achieved by changing plate thickness and/or width with the ends of the two flange plates joined by a full-penetration groove butt weld. Any such reduction in plate size should be made only if the saving in the cost of the flange material more than offsets the added expense of making the butt welds at the transition locations.

The determination of the theoretical transition points for the flange plates is similar to the determination of the theoretical cutoff points for the cover plates of cover-plated beams. This is discussed in Chapter 3 of this text.

10-3 AISC DESIGN CRITERIA FOR TRANSVERSE INTERMEDIATE STIFFENERS

Transverse intermediate stiffeners primarily serve the purpose of stiffening the deep thin girder webs against buckling. However, the AISCS permits the girder web to go into the post buckling range since research has shown that after a *stiffened* thin web panel buckles in shear, it can still continue to resist increasing load. When this occurs, the buckled web is subject to a diagonal tension and the intermediate stiffeners to a compressive force. This behavior is termed **tension field action** and the design of the stiffeners must consider the added compressive force.

No intermediate stiffeners are required, and tension field action is not considered, if the ratio h/t_w for the web is less than 260 (as well as being less than the limit stipulated in Table 10-1) and the maximum web shear stress f_v is less than that permitted by AISCS formula 1.10-1, where

$$\text{maximum } f_v = \frac{V_{max}}{ht_w}$$

and the allowable shear stress is

$$F_v = \frac{F_y}{2.89}(C_v) \leq 0.40F_y \qquad \text{AISCS formula } 1.10\text{-}1$$

where

$$C_v = \frac{45,000k}{F_y(h/t_w)^2} \qquad \text{when } C_v < 0.8$$

$$= \frac{190}{h/t_w}\sqrt{\frac{k}{F_y}} \qquad \text{when } C_v > 0.8$$

$$k = 4.0 + \frac{5.34}{(a/h)^2} \qquad \text{when } a/h < 1.0$$

$$= 5.34 + \frac{4.00}{(a/h)^2} \qquad \text{when } a/h > 1.0$$

where t_w = web thickness
a = clear distance between intermediate stiffeners
h = depth of web

The allowable shear stress F_v based on AISCS formula 1.10-1 may also be obtained from the AISCS, Appendix A, Tables 10-36 and 10-50, for 36 ksi yield stress steel and 50 ksi yield stress steel, respectively. These values are based on tension field action *not* occurring. *With* tension field action included, the allowable shear stress F_v may be obtained from AISCS formula 1.10-2, or the AISCS, Appendix A, Tables 11-36 and 11-50 for 36 ksi yield stress steel and 50 ksi yield stress steel, respectively.

The spacing of intermediate stiffeners, where stiffeners are required, must be such that the actual web shear stress must not exceed the value of F_v given by AISCS formula 1.10-1 or 1.10-2 as applicable. The ratio (sometimes designated aspect ratio) a/h must not exceed

$$\left(\frac{260}{h/t}\right)^2$$

with a maximum spacing of three times the girder web depth h.

When intermediate stiffeners are required, the design procedure is to locate the first stiffener from the end of the girder and then determine the spacing for the remaining stiffeners. The size of the stiffener is then determined. Generally, for welded plate girders the stiffeners are plates welded to the web on each side.

Whenever stiffeners are required, they must satisfy minimum moment of inertia requirements, whether tension field action is counted upon or not. To provide adequate lateral support for the web, the AISCS, Section 1.10.5.4, requires that all stiffeners have a moment of inertia at least equal to $(h/50)^4$. The stiffeners must also satisfy a minimum cross-sectional area requirement as provided by AISCS formula 1.10-3. The gross area (in.^{2}) of intermediate stiffeners, spaced as required for formula 1.10-2, must not be less than

$$A_{st} = \frac{1 - C_v}{2} \left[\frac{a}{h} - \frac{(a/h)^2}{\sqrt{1 + (a/h)^2}} \right] YDht_w \qquad \text{AISCS formula 1.10-3}$$

where C_v, a, h, and t_w are as previously defined

        Y = ratio of yield stress of web steel to yield stress of stiffener steel

        D = 1.0 for stiffeners furnished in pairs

            = 1.8 for single angle stiffeners

            = 2.4 for single plate stiffeners

When stiffeners are furnished in pairs, the area determined is *total* area.

    The required A_{st} may also be obtained in most cases from the AISCS, Appendix A, Table 11-36, using the italicized tabulated values. This gross area requirement may be reduced by the ratio f_v/F_v when $f_v < F_v$ in a panel (AISCS, Section 1.10.5.4).

    In addition, the AISCS, Section 1.9.1.2, states that the ratio of width to thickness for plate girder stiffeners must not exceed $95/\sqrt{F_y}$.

    Generally, intermediate stiffeners are stopped short of the girder tension flange. A minimum length of stiffener, based on AISCS requirements for the attaching weld (AISCS, Section 1.10.5.4), may be taken as

minimum length = web depth − 6(web thickness) − web to flange weld size

10-4 AISC DESIGN CRITERIA FOR BEARING STIFFENERS

Bearing stiffeners are generally placed in pairs at unframed ends on the webs of plate girders and where required at points of concentrated loads. In addition to transferring reactions or concentrated loads to the full depth of the web, bearing stiffeners will prevent localized web crippling as well as a more general vertical buckling of the web. If a plate girder is connected to columns at its ends by plates and/or angles, end bearing stiffeners are usually unnecessary.

    Bearing stiffeners should have close contact against the flanges and should extend approximately to the edges of the flanges as shown in Fig. 10-3. The stiffeners

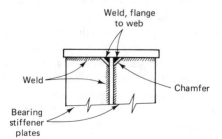

Figure 10-3   Bearing stiffeners.

are to be designed as columns assuming the column section to comprise the pair of stiffeners and a centrally located strip of the web whose width is equal to not more than 25 times its thickness when the stiffeners are located at the end of the web. The effective column length shall be taken as not less than three-fourths the length of the stiffeners in computing the slenderness ration ℓ/r (AISCS, Section 1.10.5.1). The stiffeners must also be checked for local bearing pressure. Only that portion of the stiffener outside the flange to web welds shall be considered effective in bearing and the bearing stress shall not exceed the allowable value of $0.90F_y$ (AISCS, Section 1.5.1.5.1).

Where bearing stiffeners do not exist, web crippling should be checked as a result of the uniformly distributed load bearing directly on the girder flange. Where the compression flange is continuously restrained against rotation, the allowable web crippling (compressive) stress in ksi may be computed from

$$\left[5.5 + \frac{4}{(a/h)^2}\right]\frac{10,000}{(h/t_w)^2} \qquad \text{AISCS formula 1.10-10}$$

When the flange is not restrained, the allowable web crippling stress (ksi) may be computed from

$$\left[2 + \frac{4}{(a/h)^2}\right]\frac{10,000}{(h/t_w)^2} \qquad \text{AISCS formula 1.10-11}$$

The actual compressive stress on the compression edge of the web may be computed as follows:

1. *For concentrated loads:* Divide the load (in kips) by the product of t_w and the lesser of either the girder depth d or the length of the panel in which the load is placed.

2. *For distributed loads:* Divide the load (in kips per linear inch) by web thickness.

The actual web crippling (compressive) stress cannot exceed the allowable value computed from the AISCS formulas 1.10-10 or 1.10-11 (AISCS, Section 1.10.10.2).

10-5 AISC DESIGN CRITERIA FOR CONNECTION OF GIRDER ELEMENTS

1. Connection of intermediate stiffeners to web. AISCS, Section 1.10.5.4, estimates the total shear (f_{vs}) in kips per linear inch which must be transferred between the intermediate stiffeners and web due to tension field action. This expression furnishes a minimum value

$$f_{vs} = h\sqrt{\left(\frac{F_y}{340}\right)^3} \qquad \text{AISCS formula 1.10-4}$$

However, if the actual web shear based on

$$f_v = \frac{V}{ht_w}$$

is less than the allowable shear based on AISCS formula 1.10-2, the shear to be transferred (f_{vs}) may be reduced in direct proportion.

Generally, this connection is made with intermittent fillet welds where the clear distance between welds cannot exceed 16 times the web thickness or 10 in.

2. Connection of bearing stiffener to web. Since bearing stiffeners are load-carrying elements, the weld connection is generally a continuous fillet weld on both sides of each stiffener plate. The weld is designed to transmit the total reaction or concentrated load into the web.

3. Connection of flange plate to web. These connections are designed to resist the total horizontal shear resulting from the bending forces on the girder. In addition, the welds must be proportioned to transmit to the web any loads applied directly to the flange, unless provision is made to transmit such loads by direct bearing (such as through bearing stiffeners).

This weld may be designed as an intermittent fillet weld; however, it is the contention of the authors that for the same reasons that the weld of the bearing stiffeners to the web should be continuous, the weld of the flange to the web should also be continuous.

The total horizontal shear force v_h (kips per linear inch) may be obtained from the expression (see any strength-of-materials text)

$$v_h = \frac{VQ_f}{I}$$

where Q_f = statical moment of the flange area with respect to the girder neutral axis (in.^{3})
V = maximum shear (kips)
I = moment of inertia of total girder section (in.^{4})

The load applied directly to the flange (where no bearing stiffeners exist) may be considered a vertical shear force per inch and added vectorially to the horizontal shear to determine the resultant shear between the web and the flange.

Using $v_v = w/12$ (where w is distributed load in kips per foot or pounds per foot) as the shear per linear inch applied directly to the flange,

$$v_r = \sqrt{v_h^2 + v_v^2}$$

The connection must be capable of transferring the shear force v_r.

Example 10-1

Design a uniform depth welded plate girder for the span and loading as shown in Fig. 10-4. The compression flange is laterally supported at the supports and

Example 10-1   Plate Girder Design   **313**

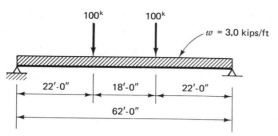

Figure 10-4 Load diagram.

at points of concentrated load. Use A36 steel and the AISCS. Welding is to be accomplished with E70 electrodes (SMAW).

Solution   The maximum moment due to the applied loads is

$$M = Pa + \frac{wL^2}{8} = 100(22) + \frac{3(62)^2}{8} = 3642 \text{ ft-kips}$$

This exceeds the M_R for the largest commonly available wide-flange shape (W36 × 300). Therefore, a built-up member or plate girder is a logical solution. However, it should again be noted that larger rolled wide-flange shapes are available with M_R values (A36 steel) ranging in excess of 6000 ft-kips. These may be economical in particular applications. See Section 1-4 of this text. Nonetheless, in this example problem we will demonstrate the design of a plate girder.

A. *Selection of web (preliminary)*
1. Assume a total girder depth of $\ell/10$:

$$\frac{62(12)}{10} = 74.4 \text{ in.}   \text{(say 75 in.)}$$

Assume a web depth of 72 in.
2. The maximum h/t_w ratio for no reduction in allowable flange stress (AISCS, Section 1.10.6) is

$$\text{maximum } \frac{h}{t_w} = \frac{760}{\sqrt{F_b}} = \frac{760}{\sqrt{22}} = 162$$

Therefore, the minimum web thickness t_w for no stress reduction is

$$\text{minimum } t_w = \frac{h}{162} = \frac{72}{162} = 0.444 \text{ in.}$$

3. For web buckling considerations, the maximum h/t_w ratio (AISCS, Section 1.10.2) from Table 10-1 is 322 if stiffeners of specified make-up are not provided. The corresponding minimum thickness of web t_w is

$$\text{minimum } t_w = \frac{h}{322} = \frac{72}{322} = 0.224 \text{ in.}$$

For exterior girders, general practice is not to use less than $\frac{5}{16}$ in. thickness for webs. Therefore, try a $\frac{5}{16}$ in. × 72 in. web plate.

$$A_w = 22.5 \text{ in.}^2$$

$$\frac{h}{t_w} = \frac{72}{\frac{5}{16}} = 230.4$$

Since $230.4 > 162$, the allowable bending stress in the compression flange will have to be reduced.

B. *Selection of girder flanges (preliminary).* Since the required flange area is a function of the maximum bending moment, a preliminary estimate of the girder weight could be made at this time and shear and moment diagrams drawn.

The area of the web is 22.5 in.². Therefore, the weight of the web is

$$22.5 \text{ in.}^2 \times \frac{490 \text{ lb/ft}^3}{144 \text{ in.}^2/\text{ft}^2} = 22.5 \text{ in.}^2 \times 3.4 \frac{\text{lb}}{\text{in.}^2} \text{ per ft} = 76.5 \text{ lb/ft}$$

Assume the weight of the flanges and stiffeners to be 150 lb/ft. The total assumed weight of the girder is then

$$76.5 + 150 = 226.5 \text{ lb/ft} \qquad \text{(say 225 lb/ft)}$$

Load, shear, and moment diagrams (Fig. 10-5) may now be drawn.

Since the actual depth-to-thickness ratio of the web is 230.4, which is in excess of the maximum ratio for no bending stress reduction (162), assume that $F_b = 20$ ksi. Then the required flange area is

$$\text{required } A_f = \frac{M}{F_b h} - \frac{t_w h}{6}$$

$$= \frac{3750(12)}{20(72)} - \frac{\frac{5}{16}(72)}{6}$$

$$= 31.25 - 3.75$$

$$= 27.5 \text{ in.}^2$$

Try a plate $1\frac{1}{2}$ in. × 18 in. for each flange.

$$A_f = 27 \text{ in.}^2 \simeq 27.5 \text{ in.}^2$$

Checking the width–thickness ratio according to the AISCS, Section 1.9.1.2, gives us

$$\frac{b_f}{2t_f} = \frac{18}{2(1.5)} = 6.0 < \frac{95.0}{\sqrt{36}} = 15.8$$

Therefore, the width–thickness ratio of the flange plate is O.K.

C. Compute the *properties of the trial girder* with reference to Fig. 10-6, and check actual and allowable stresses.

Example 10-1   Plate Girder Design   **315**

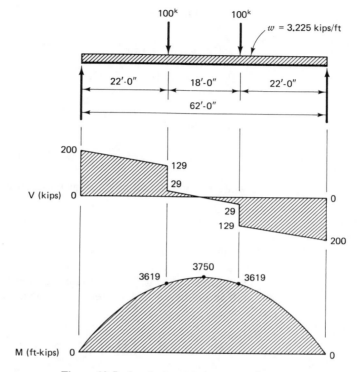

Figure 10-5 Load, shear, and moment diagrams.

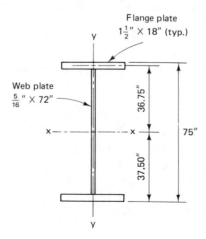

Figure 10-6 Cross-section of trial plate girder.

1. The moment of inertia is

$$\text{web} = \frac{1}{12}\left(\frac{5}{16}\right)(72)^3 \qquad = 9{,}720 \text{ in.}^4$$

$$\text{flanges} = 2(18)(1.5)(36.75)^2 = \underline{72{,}930} \text{ in.}^4$$

$$\text{total } I_x = 82{,}650 \text{ in.}^4$$

2. The section modulus is

$$S_x = \frac{I_x}{d/2} = \frac{82{,}650}{37.5} = 2204 \text{ in.}^3$$

3. The maximum actual bending stress at midspan is

$$f_b = \frac{M}{S_x} = \frac{3750(12)}{2204} = 20.4 \text{ ksi}$$

4. Calculate the allowable bending stress in the 18 ft–0 in. panel at the center of the girder span (the unbraced length is 18 ft–0 in.).

   The radius of gyration of the compression flange plus one-sixth of the web about the y-y axis is (neglecting the moment of inertia of the web)

$$I_y = \left(\frac{1}{12}\right)(1.5)(18)^3 = 729 \text{in.}^4$$

$$A_f + \left(\frac{1}{6}\right)A_w = 27.0 + \frac{1}{6}(22.5) = 30.75 \text{ in.}^2$$

$$r_T = \sqrt{\frac{I_y}{A_f + A_w/6}} = \sqrt{\frac{729}{30.75}} = 4.87 \text{ in.}$$

With reference to the AISCS, Section 1.5.1.4.5,

$$\sqrt{\frac{102 \times 10^3 C_b}{F_y}} = 53\sqrt{C_b}$$

But since $M_{\max} > M$ at each end of the panel, $C_b = 1$. Therefore, $53\sqrt{C_b} = 53$.

$$\frac{\ell}{r_T} = \frac{18(12)}{4.87} = 44.4 < 53$$

Therefore, $F_b = 0.60F_y = 22.0$ ksi.

   The reduced allowable bending stress in the 18 ft–0 in. panel due to the excessive h/t_w ratio is calculated from

$$F'_b = F_b\left[1.0 - 0.0005 \frac{A_w}{A_f}\left(\frac{h}{t_w} - \frac{760}{\sqrt{F_b}}\right)\right]$$

Example 10-1   Plate Girder Design   **317**

$$= 22\left[1.0 - 0.0005\left(\frac{22.5}{27.0}\right)\left(230.4 - \frac{760}{\sqrt{22}}\right)\right]$$

$$F'_b = 21.37 \text{ ksi}$$

$$f_b = 20.4 \text{ ksi} < F'_b = 21.37 \text{ ksi} \qquad\qquad \textbf{O.K.}$$

5. Calculate the allowable bending stress in the 22 ft–0 in. panel due to an unbraced length of 22 ft–0 in. Again, with reference to the AISCS, Section 1.5.1.4.5:

$$C_b = 1.75 + 1.05\left(\frac{M_1}{M_2}\right) + 0.3\left(\frac{M_1}{M_2}\right)^2$$

   Since the moment at one end of the panel is 0 (M_1), then $C_b = 1.75$. Therefore,

$$53\sqrt{C_b} = 53\sqrt{1.75} = 70.1$$

$$\frac{\ell}{r_T} = \frac{22(12)}{4.87} = 54.2 < 70.1$$

   Therefore, $F_b = 0.6F_y = 22.0$ ksi. Also, as computed for the 18 ft–0 in. panel:

$$F'_b = 21.37 \text{ ksi}$$

6. The actual maximum bending stress in the 22 ft–0 in. panel is

$$f_b = \frac{M}{S_x} = \frac{3619(12)}{2204} = 19.7 \text{ ksi}$$

$$f_b = 19.7 \text{ ksi} < F'_b = 21.37 \text{ ksi} \qquad\qquad \textbf{O.K.}$$

   Therefore, use the following for the plate girder:

$$\text{Web plate: } \frac{5}{16} \text{ in.} \times 72 \text{ in.}$$

$$\text{Two flange plates: } 1\frac{1}{2} \text{ in.} \times 18 \text{ in.}$$

D. *Bearing stiffeners.* Bearing stiffeners will be designed at the end reactions and under the point loads.
1. At the end of the girder, design for a reaction of 200 kips. The maximum width–thickness ratio of the stiffener is 15.8 (AISCS, Section 1.9.1.2). Since bearing stiffeners should extend to edge of flange plates (approximately), try two 8-in. plates.

$$\text{required } t = \frac{\text{width}}{15.8} = \frac{8}{15.8} = 0.51 \text{ in.}$$

Therefore, try two $\frac{9}{16} \times 8$ in. plates. Check the compressive stress in the plates, assuming column action (AISCS, Section 1.10.5.1). With reference to Fig. 10-7, the moment of inertia of the plates about an axis in the web of the girder is

$$I = \frac{1}{12}\left(\frac{9}{16}\right)(16.31)^3 = 203 \text{ in.}^4$$

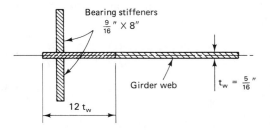

Bearing stiffeners
$\frac{9}{16}$ " $\times$ 8"

Girder web

$t_w = \frac{5}{16}$ "

12 t_w

Figure 10-7 Effective column section.

The effective area is

$$2(8)\left(\frac{9}{16}\right) + 12\left(\frac{5}{16}\right)\left(\frac{5}{16}\right) = 10.17 \text{ in.}^2$$

The radius of gyration is

$$r = \sqrt{\frac{I}{A}} = \sqrt{\frac{203}{10.17}} = 4.47 \text{ in.}$$

The effective length $K\ell$ is taken as $\frac{3}{4}(72) = 54$ in. and the slenderness ratio is

$$\frac{K\ell}{r} = \frac{54}{4.47} = 12.1$$

The allowable compressive stress F_a (AISCS, Appendix A, Table 3-36) is 21.05 ksi. The actual compressive stress is

$$f_a = \frac{P}{A_{\text{eff}}} = \frac{200}{10.17}$$

$$= 19.67 \text{ ksi} < 21.05 \text{ ksi} \qquad \textbf{O.K.}$$

Therefore, use two plates $\frac{9}{16} \times 8 \times 6$ ft–0 in. as end bearing stiffeners with a close fit against the flange receiving the reaction.

2. Select bearing stiffeners to be placed under the interior concentrated loads (100 kips each). The minimum stiffener size is $9/16 \times 8$ (same as for the end bearing stiffeners), the applied load is less, and the effective column area is greater ($25t_w$ compared with $12t_w$) than for the end bearing stiffeners. Therefore, use the same-size bearing stiffeners under the interior concentrated loads.

Example 10-1   Plate Girder Design   **319**

3. Check the bearing stress for the bearing stiffeners selected thus far. From the AISCS, Section 1.5.1.5.1, the allowable bearing stress is

$$F_p = 0.90F_y = 0.90(36) = 32.4 \text{ ksi}$$

Assume a $\frac{3}{8}$-in. weld for the flange-to-web connection. Therefore, effective area for bearing of the stiffeners on the flange is calculated as

$$\left(8 - \frac{3}{8}\right)\left(\frac{9}{16}\right)(2) = 8.58 \text{ in.}^2$$

The actual bearing stress is

$$f_p = \frac{200}{8.58} = 23.3 \text{ ksi} < 32.4 \text{ ksi} \qquad \textbf{O.K.}$$

4. Check web crippling resulting from a distributed load of 3.225 kips/ft at center of 18 ft–0 in. panel (AISCS, Section 1.10.10.2.). The allowable compressive stress (assuming that the compression flange is continuously restrained against rotation) is calculated as

$$\left[5.5 + \frac{4}{(a/h)^2}\right]\frac{10,000}{(h/t_w)^2} \qquad \text{AISCS formula 1.10-10}$$

Assuming that there will be no intermediate stiffeners in the center 18 ft–0 in. panel,

$$\frac{a}{h} = \frac{18(12)}{72} = 3.0$$

Substituting gives us

$$\left[5.5 + \frac{4}{(3.0)^2}\right]\frac{10,000}{230.4^2} = 1.12 \text{ ksi}$$

The actual compressive stress on the web is

$$f_a = \frac{3.225}{12(5/16)} = 0.86 \text{ ksi}$$

$$1.12 \text{ ksi} > 0.86 \text{ ksi} \qquad \textbf{O.K.}$$

E. Design *intermediate stiffeners* as required.
1. Check to see if intermediate stiffeners are required within the 22 ft–0 in. panel length. Several criteria must be checked:

$$\frac{h}{t_w} = \frac{72}{5/16} = 230.4 < 260 \qquad \textbf{O.K.}$$

$$f_v = \frac{V}{ht_w} = \frac{200}{72(5/16)} = 8.89 \text{ ksi} < 0.40F_y \qquad \textbf{O.K.}$$

The relationship $f_v \leq F_v$ must exist if stiffeners are not to be used. F_v is determined from AISCS formula 1.10-1 or by entering Table 10-36, AISCS, Appendix A, with

$$\frac{h}{t_w} = 230.4$$

and

$$\frac{a}{h} = \frac{22(12)}{72} = 3.67$$

from which F_v is determined to be 1.5 ksi by interpolation. Therefore,

$$1.5 \text{ ksi} = F_v < f_v = 8.89 \text{ ksi}$$

and intermediate stiffeners *are* required within the 22 ft–0 in. panel length.

2. Locate the *first* intermediate stiffener from the girder end bearing stiffener. The object of the stiffener placement is to increase F_v to the point where $F_v = f_v$. The required maximum stiffener spacing for the end panel is found by utilizing AISCS formula 1.10-1:

$$F_v = \frac{F_y}{2.89}(C_v) \leq 0.40F_y$$

from which

$$\text{required } C_v = \frac{2.89(8.89)}{36} = 0.71$$

When $C_v < 0.8$,

$$C_v = \frac{45{,}000k}{F_y(h/t_w)^2}$$

from which

$$\text{required } k = \frac{C_v F_y(h/t_w)^2}{45{,}000}$$

$$= \frac{0.71(36)(230.4)^2}{45{,}000}$$

$$k = 30.15$$

Assuming that $a/h < 1.0$,

$$k = 4.00 + \frac{5.34}{(a/h)^2}$$

Example 10-1   Plate Girder Design   **321**

and solving for the required a/h:

$$\frac{a}{h} = \sqrt{\frac{5.34}{30.15 - 4.00}}$$

$$= 0.45$$

From which

$$\text{maximum } a = 0.45h = 0.45(72) = 32.4 \text{ in.}$$

$$[\text{use 2 ft–6 in. (30 in.)}]$$

3. Locate other intermediate stiffeners within the 22 ft–0 in. panel length (up to the bearing stiffeners under the concentrated loads). As previously done, determine whether stiffeners are required. The shear stress at the first intermediate stiffener is calculated as (refer to Fig. 10-5)

$$V = 200 - 3.225(2.5) = 191.9 \text{ kips}$$

$$f_v = \frac{V}{ht_w} = \frac{191.9}{72(\frac{5}{16})} = 8.53 \text{ ksi}$$

The distance between the first intermediate stiffener and the concentrated load is

$$a = 22(12) - 30 = 234 \text{ in.}$$

Therefore,

$$\frac{a}{h} = \frac{234}{72} = 3.25$$

and since $a/h > 1.0$,

$$k = 5.34 + \frac{4.00}{(a/h)^2}$$

$$= 5.34 + \frac{4.00}{3.25^2} = 5.72$$

$$C_v = \frac{45,000k}{F_y(h/t_w)^2} = \frac{45,000(5.72)}{36(230.4)^2} = 0.135$$

From Table 11-36, AISCS, Appendix A,

$$F_v = 1.5 \text{ ksi} < 8.53 \text{ ksi} \qquad \textbf{N.G.}$$

Therefore, intermediate stiffeners must be provided between the first intermediate stiffeners and the concentrated load.

The reader should take note at this point that Table 11-36 of the AISCS, Appendix A (which reflects AISCS formula 1.10-2), was utilized for this spacing determination instead of Table 10-36 (which reflects AISCS formula 1.10-1). Table 11-36 takes into consideration web postbuckling strength which, in effect, provides better economy in plate girder design. The spacing from the girder end bearing stiffener to the first intermediate stiffener must be determined without the benefit of tension field action. Therefore, Table 10-36 (and AISCS formula 1.10-1) was used for that computation (AISCS, Section 1.10.5.3).

The spacings required for the intermediate stiffeners in the 234-in.-long section of the end panel may be determined. The maximum a distance, with reference to the AISCS, Section 1.10.5.3, is

$$\text{maximum value of } \frac{a}{h} = \left(\frac{260}{h/t_w}\right)^2$$

and since $h/t_w = 230.4$,

$$\text{maximum } a = 72\left(\frac{260}{230.4}\right)^2 = 91.7 \text{ in.}$$

Try intermediate stiffeners at the one-sixth points:

$$a = \frac{234}{6} = 39 \text{ in.}$$

and

$$\frac{a}{h} = \frac{39}{72} = 0.54$$

From Table 11-36, AISCS, Appendix A,

$$F_v \simeq 11.0 \text{ ksi} > 8.53 \text{ ksi} \qquad \textbf{O.K.}$$

Since this is relatively conservative, try intermediate stiffeners at the quarter points.

$$a = \frac{234}{4} = 58.5 \text{ in.}$$

and

$$\frac{a}{h} = \frac{58.5}{72} = 0.81$$

From Table 11-36,

$$F_v \simeq 9.5 \text{ ksi} > 8.53 \text{ ksi} \qquad \textbf{O.K.}$$

Therefore, place these intermediate stiffeners at a spacing of 4 ft–$10\frac{1}{2}$ in.

Example 10-1   Plate Girder Design   **323**

4. Locate intermediate stiffeners between the two concentrated loads.

$$V = 29 \text{ kips}$$

$$f_v = \frac{V}{ht_w} = \frac{29}{72\left(\frac{5}{16}\right)} = 1.29 \text{ ksi}$$

Assuming no intermediate stiffeners,

$$a = 18(12) = 216 \text{ in.}$$

and

$$\frac{a}{h} = \frac{216}{72} = 3.0$$

From Table 11-36 (with $h/t_w = 230.4$), $F_v = 1.5 \text{ ksi}(+)$,

$$f_v < F_v$$

therefore, no intermediate stiffeners are required.

F. Check any requirement for *allowable bending stress reduction* due to concurrent bending and shear stress in the web. Perform this check at the concentrated load:

$$f_v = \frac{V}{ht_w} = \frac{129}{72\left(\frac{5}{16}\right)} = 5.73 \text{ ksi}$$

The allowable bending tensile stress in the web is

$$F_b = \left(0.825 - 0.375\frac{f_v}{F_v}\right)F_y < 0.60F_y$$

$$F_b = \left[0.825 - 0.375\left(\frac{5.73}{9.5}\right)\right]36 = 21.56 \text{ ksi}$$

The actual bending stress at the concentrated loads is

$$f_b = \frac{Mc}{I} = \frac{3619(12)\,(36.0)}{82,650} = 18.92 \text{ ksi} < 21.56 \text{ ksi} \qquad \textbf{O.K.}$$

G. Select the *size of the intermediate stiffeners.*
   1. The minimum required gross area of stiffener may be determined using AISCS formula 1.10-3:

$$A_{st} = \frac{1 - C_v}{2}\left[\frac{a}{h} - \frac{(a/h)^2}{\sqrt{1 + (a/h)^2}}\right]YDht_w$$

   or may be obtained from AISCS, Appendix A, Table 11-36, where minimum A_{st}, as a percent of the web area, is tabulated.

For $h/t_w = 230.4$ and $a/h = 58.5/72 = 0.81$,

minimum area $= A_{st} =$ approximately 10.6% $\times$ (web area)

$$A_{st} = 0.106(72)\left(\frac{5}{16}\right) = 2.39 \text{ in.}^2$$

Also, there is a permissible reduction of A_{st} based on the ratio f_v/F_v (AISCS, Section 1.10.5.4). At the first intermediate stiffener,

$$\text{actual required } A_{st} = \frac{f_v}{F_v} A_{st}$$

$$= \frac{8.53}{9.5}(2.39) = 2.15 \text{ in.}^2$$

Try two plates $\frac{5}{16} \times 4$:

$$A_{st} = 2\left(\frac{5}{16}\right)(4) = 2.5 \text{ in.}^2 > 2.15 \text{ in.}^2 \qquad \textbf{O.K.}$$

2. Check the width–thickness ratio (AISCS, Section 1.9.1.2)

$$\frac{4}{\frac{5}{16}} = 12.8 < \frac{95}{\sqrt{F_y}} = 15.8 \qquad \textbf{O.K.}$$

3. Check the moment of inertia (AISCS, Section 1.10.5.4) of the inter-mediate stiffeners:

$$\text{required } I = \left(\frac{h}{50}\right)^4 = \left(\frac{72}{50}\right)^4 = 4.3 \text{ in.}^4$$

$$\text{furnished } I = 0.313(8.31)^3\left(\frac{1}{12}\right) = 14.8 \text{ in.}^4$$

$$14.8 \text{ in.}^4 > 4.3 \text{ in.}^4 \qquad \textbf{O.K.}$$

4. For these stiffeners, the minimum length required (AISCS, Section 1.10.5.4) is

$$h - 6(\text{web thickness}) - \text{assumed weld size}$$

$$= h - 6\left(\frac{5}{16}\right) - \frac{1}{4}$$

$$= 72 - 1.88 - 0.25 = 69.87 \text{ in.}$$

Use for intermediate stiffeners, two plates $\frac{5}{16} \times 4 \times$ 5 ft-10 in. fillet-welded to the compression flange and web.

H. The final step is to *design the welded connections* between the various parts of the plate girder.

Example 10-1   Plate Girder Design   **325**

1. The weld between the web and the intermediate stiffeners (AISCS, Section 1.10.5.4) must transfer a total shear of

$$f_{vs} = h\sqrt{\left(\frac{F_y}{360}\right)^3}$$

$$= 72\sqrt{\left(\frac{36}{340}\right)^3} = 2.48 \text{ kips/in.}$$

for a pair of stiffeners (two stiffener welds). The minimum weld size for a $\frac{5}{16}$-in. plate is $\frac{3}{16}$ in. (AISCS, Table 1.17.2A).

Therefore, use a $\frac{1}{4}$-in. weld which has a strength of 3.70 kips/in. (Table 8-1 of this text). The minimum length of fillet weld (from the AISCS, Section 1.17.4) is

$$4(\text{leg size}) = 4\left(\frac{1}{4}\right) = 1 \text{ in.}$$

However, the minimum length of intermittent fillet welding is $1\frac{1}{2}$ in. (AISCS, Section 1.17.5). Therefore, use a $\frac{1}{4}$-in. weld 2 in. long.

The required weld spacing is

$$\frac{\text{weld strength}}{\text{total shear transfer}} = \frac{2(2)(3.70)}{2.48} = 5.97 \text{ in.}$$

Try intermittent welds spaced $5\frac{1}{2}$ in. on center. The clear distance between welds cannot exceed 16 times the web thickness, nor more than 10 in.

$$\text{clear distance} = 5\frac{1}{2} - 2$$

$$= 3\frac{1}{2} \text{ in.}$$

$$3\frac{1}{2} \text{ in.} < 16\left(\frac{5}{16} \text{ in.}\right) < 10 \text{ in.} \qquad \textbf{O.K.}$$

Therefore, for the intermediate stiffeners, use $\frac{1}{4} \times 2$ in. welds $5\frac{1}{2}$ in. on center.

2. The welds between the web and the bearing stiffeners should be continuous on both sides of the stiffener plate since the bearing stiffeners are load-carrying elements. The minimum weld size is $\frac{1}{4}$ in. (AISCS, Table 1.17.2A), which has a strength of weld of 3.70 kips/in. The total weld strength is

$$2(3.70)(2)(72) = 1066 \text{ kips} > 200 \text{ kips} \qquad \textbf{O.K.}$$

Therefore, use a $\frac{1}{4}$-in. continuous fillet weld on both sides for all bearing stiffeners.

3. The weld between the flange and the web must transfer a horizontal shear of

$$v_h = \frac{VQ_f}{I}$$

$$Q_f = (27.0)(36.75) = 992.3 \text{ in.}^3$$

$$V = 200 \text{ kips}$$

$$I = 82,650 \text{ in.}^4$$

$$v_h = \frac{200(992.3)}{82,650} = 2.40 \text{ kips/in.}$$

The vertical shear to be transmitted is due to the uniformly distributed load on the top flange of 3.225 kips/ft.

$$v_v = \frac{3.225}{12} = 0.27 \text{ kip/in.}$$

The resultant shear is calculated as

$$v_r = \sqrt{v_h^2 + v_v^2}$$
$$= \sqrt{2.40^2 + 0.27^2} = 2.42 \text{ kips/in.}$$

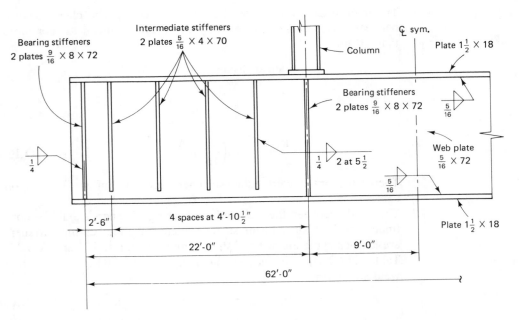

Figure 10-8 Design sketch.

Example 10-1   Plate Girder Design   **327**

The minimum weld size for the $1\frac{1}{2}$-in.-thick flange plate is $\frac{5}{16}$ in. (AISCS, Table 1.17.2A). For the $\frac{5}{16}$-in. weld,

$$\text{weld strength} = 4.63 \text{ kips/in.} > 2.42 \text{ kips/in.} \qquad \textbf{O.K.}$$

Therefore, use a $\frac{5}{16}$-in. continuous weld.

$$\text{total weld strength} = 2(4.63) = 9.26 \text{ kips/in.} > 2.42 \text{ kips/in.} \quad \textbf{O.K.}$$

As may be observed, the minimum weld required by AISCS governs the weld size rather than stress considerations. This is generally the case for this connection.

The entire design is summarized in the design sketch shown in Fig. 10-8.

11 Continuous Construction and Plastic Design

11-1 *INTRODUCTION*

Continuous construction may be defined as a structural system in which individual members are integrally attached together so that they behave as single members. For example, if three simply supported beams were placed end to end with their ends connected by moment-resisting connections, the resulting beam would be considered a **continuous beam** spanning four supports as shown in Fig. 11-1. A continuous beam may be considered to be any beam that spans more than two supports. This type of member is frequently used in modern structures and generally offers economy with respect to the beam itself when compared with a series of simple beams over the same spans. The effect of the beam continuity is to reduce the maximum bending moment from that of the simple beam, thereby reducing the required beam size. Continuity in a structure also generally serves to increase stiffness and decrease deflections. However, other factors associated with continuous beams may offset the beam savings making the question of total structural economy difficult to establish.

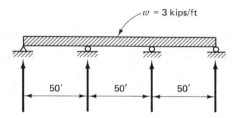

Figure 11-1   Three-span continuous beam.

Continuous frames, which may include one-story rigid frames as well as multistory frames, are structures whose individual members are rigidly connected to each other with moment-resisting connections as shown in Fig. 11-2, thereby preventing relative rotation under load.

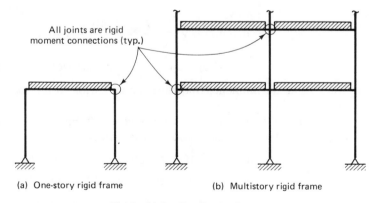

(a) One-story rigid frame   (b) Multistory rigid frame

Figure 11-2   Continuous frames.

Continuous beams and frames are categorized as statically indeterminate structures which means that the moments, shears, and external reactions cannot be found by the condition of static equilibrium alone. It is not the intent of this text to review or introduce the many analytical techniques that may be used for the determination of the moments, shears, and reactions in indeterminate structures. Several appropriate references are included at the end of this chapter. However, since the structural design of continuous beams is similar to the design of simple beams and since multistory frame design is beyond the scope of this text, the coverage herein is limited to continuous beams.

Example 11-1

Compare the maximum bending moments and shears for the beams shown in Figs. 11-3 and 11-4. Use the AISCM, Part 2.

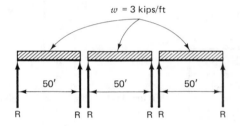

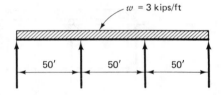

Figure 11-3   Simple beams (three spans).   **Figure 11-4**   Three-span continuous beam.

Solution   For the simple beam of Fig. 11-3:

$$\text{maximum } M = \frac{wL^2}{8} = \frac{3(50)^2}{8} = 938 \text{ ft-kips}$$

$$\text{maximum } V = \frac{wL}{2} = \frac{3(50)}{2} = 75 \text{ kips}$$

Reaction at the end of each span $(R) = 75$ kips

The maximum moment is positive (compression in the top) and occurs at the middle of the 50-ft spans.

For the three-span continuous beam of Fig. 11-4, utilizing shear and moment coefficients (discussed in Section 11-2),

$$\text{maximum } M = 0.100wL^2 = 0.100(3)(50)^2 = 750 \text{ ft-kips}$$

$$\text{maximum } V = 0.600wL = 0.600(3)(50) = 90 \text{ kips}$$

$$\text{maximum reaction} = 1.10wL = 1.10(3)(50) = 165 \text{ kips}$$

The maximum moment is negative and occurs at the interior supports. The maximum shear occurs at the end-span side of the interior supports.

The least maximum bending moment occurs with the three-span continuous beam, hence economy will result with respect to beam size.

11-2 ELASTIC DESIGN OF CONTINUOUS BEAMS

In an effort to simplify continuity and expedite the planning and design phases of a structure, the AISCM, Part 2, furnishes moment and shear coefficients for two-, three-, and four-span continuous beams. These coefficients are based on equal span lengths and are applicable for steel beams with a constant moment of inertia and subject to uniformly distributed loading conditions as shown in the diagrams.

For conditions such as unequal span lengths, varying moment of inertia, and combinations of types of loading numerous analytical solutions are available. In addition, numerous commercial computer programs are available for continuous beams, which greatly facilitate analysis.

After determination of the continuous beam positive and negative bending moments, the AISCS, Section 1.5.1.4, permits a modification of these values. According to Section 1.5.1.4, a continuous structural steel *compact* member may be designed on

Variable depth, multispan, continuous, welded plate girder bridge over the Hudson River between Troy and Watervliet, New York. (Courtesy of the New York State Department of Transportation.)

the basis of nine-tenths of the maximum negative moments produced by gravity loading provided that the maximum positive moments are increased by one-tenth of the average negative moments at the adjacent supports. The moment adjustments only apply to gravity loads and not to lateral loads (such as wind). In addition, it does not apply to A514 steel members, hybrid girders, or to moments produced by loads on cantilevers.

Example 11-2

Design a W-shape continuous beam for the conditions shown in Fig. 11-5. Use A36 steel and elastic design approach. Assume continuous lateral support for the compression flange.

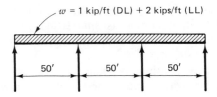

w = 1 kip/ft (DL) + 2 kips/ft (LL)

50'   50'   50'

Figure 11-5   Continuous beam.

Solution

1. The dead-load negative moment (all spans loaded) at the interior supports is

$$M = 0.100wL^2$$
$$= 0.100(1.0)\,(50)^2 = 250 \text{ ft-kips}$$

The dead-load positive moment in the end spans is

$$M = 0.080wL^2$$
$$= 0.080(1.0)\,(50)^2 = 200 \text{ ft-kips}$$

This occurs at a distance of $0.40L$ from the end support:

$$0.40(50) = 20 \text{ ft}$$

2. The live-load moments are determined with careful regard to the loading patterns which produce maximum conditions.

   a. Maximum positive moment in end spans—load the end spans only:

$$M = 0.1013wL^2$$
$$= 0.1013(2)(50)^2 = 507 \text{ ft-kips}$$

This occurs at a distance of $0.450L$ from the end support:

$$0.450(50) = 22.5 \text{ ft}$$

b. Maximum negative moment at one interior support—load two adjacent spans with one end span unloaded:

$$M = 0.1167wL^2$$

$$= 0.1167(2)\,(50)^2 = 584 \text{ ft-kips}$$

c. Negative moment at interior supports with only the end spans loaded:

$$M = 0.050wL^2$$

$$= 0.050(2)\,(50)^2 = 250 \text{ ft-kips}$$

3. Maximum negative moment for design (AISCS, Section 1.5.1.4.1):

$$M = 0.9(250 + 584) = 751 \text{ ft-kips}$$

The maximum positive moment for design (assume maximum dead-load and live-load moments occur at the same location) is determined by increasing the calculated maximum positive moment by one-tenth of the average of the negative moments at each end of the span.

$$M = (200 + 507) + 0.1\left(\frac{0 + 250 + 250}{2}\right)$$

$$= 707 + 25 = 732 \text{ ft-kips}$$

4. Negative moment > positive moment; therefore (assuming that $F_b = 24$ ksi),

$$\text{required } S_x = \frac{M}{F_b} = \frac{751(12)}{24} = 376 \text{ in.}^3$$

5. From the AISCM, Part 2, select a W33 × 130, $S_x = 406$ in.^{3}. The section is compact; therefore, the assumed F_b is satisfactory.

6. Check the shear:

$$\text{maximum } V_{\text{DL}} = 0.6wL = 0.6(1)\,(50) = 30 \text{ kips}$$

$$\text{maximum } V_{\text{LL}} = 0.617wL = 0.617(2)\,(50) = 62 \text{ kips}$$

$$\text{total maximum } V = 92 \text{ kips}$$

The maximum shear V occurs at the end span face of the interior support when the opposite end span is unloaded. From the Tables of Uniform Load Constants, AISCM, Part 2, the allowable shear for a W33 × 130 is 278 kips. Since 278 kips > 92 kips, the section is satisfactory. **Use a W33 × 130.**

Depending on the structure, a complete continuous beam design would normally involve other design considerations such as the determination of deflections or the need for bearing stiffeners at supports. In addition, bearing plates may be necessary as may be a beam splice.

The designer should also be aware that in areas of negative moment in continuous beams, the bottom flange is the compression flange. If only the top flange is laterally supported, the lateral support conditions for the bottom flange must be considered. F_b may be affected. Since L_b is defined as the distance between lateral support points on the *compression flange,* the unsupported length may be considered to end at a point of lateral support or at a point of inflection (moment $= 0$). Moment diagrams should be carefully studied when considering L_b for continuous beams.

11-3 INTRODUCTION TO PLASTIC DESIGN

As discussed in Chapter 1, most structural steel beams are designed in accordance with the allowable stress design method whereby an actual bending stress f_b induced by applied loads may not exceed an allowable bending stress F_b. The AISCS, Section 1.5.1.4, indicates that this allowable bending stress is always substantially less than the steel yield stress F_y and therefore, as depicted in Fig. 1-4, must lie on the initial straight-line portion of the steel stress-strain curve. Hence, in this method of design, bending stresses will be kept within the elastic range.

The **plastic design method** takes advantage of the substantial reserve strength of a steel beam that exists after the yield stress has been reached at some location. Extensive tests have shown that bending members can support loads in excess of those that would initially induce the yield stress. Therefore, the plastic theory utilizes the stress-strain relationships through the plastic range up to the start of strain hardening. The strain-hardening range could theoretically permit steel members to withstand additional stress. However, the corresponding strains and resulting structural deformations would be so large that the structure would no longer be usable. The assumption is made in plastic design, therefore, that strains do not reach the strain-hardening range.

The idealized stress-strain diagram of Fig. 1-4 is based on the assumption that the maximum stress through the plastic range does not exceed the yield stress F_y. The *stress* within the plastic range is assumed to be constant despite the fact that *strain* increases. As strains increase from the elastic range into the plastic range on a beam cross section, however, there is a distinct change in the shape of the resulting bending stress distribution diagram. The shape of the bending stress diagram which is assumed in allowable stress design is shown in Fig. 11-6(a). This general shape exists up to the time at which the maximum bending stress (at extreme outside fiber of the beam) becomes F_y. In this range, unit strain varies linearly from zero at the neutral axis to a maximum at the outer fibers and since unit stress is proportional to unit strain (in the elastic range only) the stress variation also varies linearly from zero at the neutral axis to a maximum at the outermost fibers. When the outermost fibers *first* reach the yield stress F_y, and the rest of the cross section is still stressed to less than F_y, the resisting

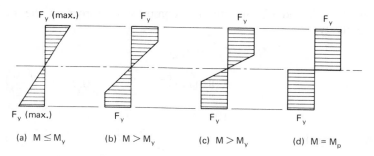

Figure 11-6 Bending stress distribution.

moment existing in the beam is

$$M_y = F_y S_x$$

and may be termed the **yield moment.** If the moment is increased beyond the yield moment, the outer fibers, which have been stressed to their yield stress, will continue to have the same stress, but at the same time additional strains will occur and no longer will unit stress be proportional to unit strain. Any required *additional* resisting moment will then be furnished by *the fibers nearer to the neutral axis* and the stress distribution will take the form shown in Fig. 11-6(b). This process will continue with more parts of the beam cross section stressed to the yield point as shown in Fig. 11-6(c) until a fully **plastic rectangular stress distribution** develops as shown in Fig. 11-6(d). At this point, the unit strain has become so large that practically the entire cross section has yielded and it is assumed that no additional moment can be resisted. The moment that exists at this point is called the **plastic moment** M_p. For the hypothetical cross section shown in Fig. 11-7, the magnitude of M_p is determined as

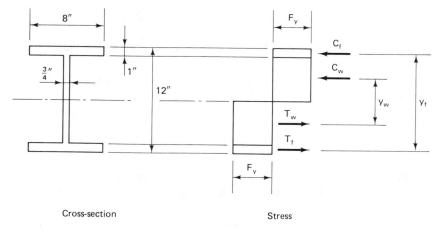

Cross-section   Stress

Figure 11-7 Diagram for determination of M_p.

follows (C_f and C_w are internal compressive forces in the flange and web, respectively):

$$M_p = C_f y_f + C_w y_w$$

$$= F_y(8)\,(1)\,(12 - 1) + F_y(0.75)\,(5)\left(\frac{10}{2}\right)$$

$$= 106.75 F_y$$

Assuming that $F_y = 36$ ksi,

$$M_p = 106.75(36) = 3843 \text{ in.-kips}$$

The resistance to bending at this point may also be expressed as

$$M_p = F_y Z$$

where M_p = plastic moment
   Z = plastic section modulus (in.^{3})
   F_y = yield stress

The plastic section modulus of the cross section is equal to the numerical sum of the moments of the areas of the cross section above and below the neutral axis, taken about the neutral axis. For the cross section of Fig. 11-7, Z was determined as 106.75 in.^{3} in the calculation for M_p. The plastic section modulus is tabulated for W and M shapes in the AISCM, Part 2, Plastic Design Selection Table.

The ratio M_p/M_y is called the **shape factor** and may be described as being a measure of the plastic moment strength in comparison to the yield moment strength. For the cross section of Fig. 11-7, the shape factor is calculated as follows:

$$M_y = F_y S_x = F_y \frac{I_x}{c}$$

I_x may be determined to be 548 in.^{4}, and again assuming that $F_y = 36$ ksi,

$$M_y = 36\left(\frac{548}{6}\right) = 3288 \text{ in.-kips}$$

from which

$$\text{shape factor} = \frac{M_p}{M_y} = \frac{3843}{3288} = 1.17$$

For most wide-flange shapes (the cross section of Fig. 11-7 is hypothetical) with bending occurring about the strong axis, the value of the shape factor lies between 1.10 and 1.23.

Assuming a progressive increase in a beam loading, the actual bending moment induced at some location would eventually reach the plastic moment strength M_p.

When this occurs, a **plastic hinge** is said to have formed and no additional moment can be resisted at that location. Although the effect of a plastic hinge may extend for some distance along the beam, it is assumed for analysis and design purposes to be localized in a single plane. When sufficient plastic hinges have formed so that no further loading may be supported, a **mechanism** is said to have been created. This may be defined as an arrangement of plastic hinges and/or real hinges which would permit collapse of a structural member.

All structural members are designed based on some factor of safety. In allowable stress design a bending member is designed to support working loads so that an allowable bending stress is not exceeded. Hence the factor of safety against yielding may be considered to be F_y/F_b. In plastic design, the working loads are increased by a **load factor** and the bending member is designed on the basis of the plastic or **collapse** strength. In essence, the factor of safety is the load factor and as a minimum must equal 1.70 times the given live load and dead load (AISCS, Part 2, Section 2.1).

11-4 PLASTIC DESIGN APPLICATION: SIMPLY SUPPORTED BEAMS

Plastic design is of little advantage for simply supported beams. However, it may be economical for statically indeterminate members such as fixed end (restrained) or continuous beams. A simple beam will fail if one plastic hinge develops since real hinges exist at each support.

Assuming a W shape subjected to a concentrated load at midspan, a plastic hinge will develop at the point of maximum moment (under the load in this case) as the loading is increased. The combination of the plastic hinge with the two real hinges at the supports creates a collapse mechanism as shown in Fig. 11-8 and failure is assumed to have occurred.

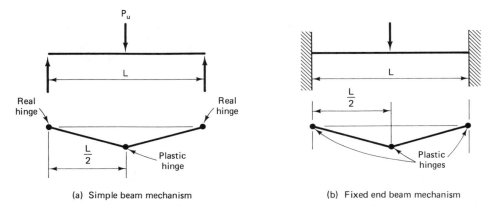

(a) Simple beam mechanism   (b) Fixed end beam mechanism

Figure 11-8   Collapse mechanisms.

Example 11-3

A laterally supported ($L_b = 0$) simple span beam with a 24 ft–0 in. span length must support a uniformly distributed working load of 3.0 kips/ft (which includes an assumed weight of beam). Use A36 steel and the AISCS, and disregard shear and deflection. Select the lightest W shape by

(a) The elastic design method.
(b) The plastic design method.

Solution

(a) *Elastic (allowable stress design) method:*

$$M = \frac{wL^2}{8} = \frac{3(24)^2}{8} = 216 \text{ ft-kips}$$

$$\text{required } S_x = \frac{M}{F_b} = \frac{216(12)}{24} = 108 \text{ in.}^3$$

From AISCM, Part 2, Allowable Stress Design Selection Table:

use W24 × 55 $S_x = 114$ in.^{3}

(b) *Plastic design method.*   Knowing that the moment will reach M_p at collapse and will occur at midspan, a beam must be selected which has an M_p value at least equal to the moment M_u created by the factored loads:

$$M_u = \frac{w_u L^2}{8} = \frac{1.7(3)\,(24)^2}{8} = 367.2 \text{ ft-kips}$$

$$M_u = M_p = F_y Z$$

Therefore,

$$\text{required } Z_x = \frac{M_u}{F_y} = \frac{367.2(12)}{36} = 122.4 \text{ in.}^3$$

From the AISCM, Part 2, Plastic Design Selection Table:

use W24 × 55 $Z_x = 134$ in.^{3}

Alternatively, the beam could have been selected on the basis of its tabulated M_p.

11-5 PLASTIC DESIGN APPLICATION: FIXED-END BEAMS

For a beam that is fixed at both ends and supports a uniform load (AISCM, Part 2, Beam Diagrams and Formulas, Case 15), expressions for positive and negative moments are shown in Fig. 11-9. These expressions are valid for the allowable stress

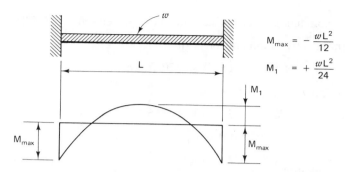

$$M_{max} = -\frac{wL^2}{12}$$

$$M_1 = +\frac{wL^2}{24}$$

Figure 11-9   Fixed-end beam.

design method. They may also be used for the case where the uniformly distributed load is increased to that load w_y which will induce a maximum bending stress equal to the yield stress.

As the load is further increased, plastic hinges will form at the fixed ends. These are the points of maximum moment. These points will allow rotation to take place without resisting any more of the applied moment. That is, a constant moment M_p will exist at the hinges. A further increase in load must then be resisted by sections of the beam that are less stressed. The load may be increased until the moment at some other point reaches the plastic moment M_p. In this case, this will occur at midspan and a third plastic hinge will be developed. The combination of the three plastic hinges, when formed, constitutes a collapse mechanism and the beam is no longer capable of supporting additional load.

Example 11-4

A laterally supported ($L_b = 0$) fixed-end beam with a 20 ft–0 in. span length must support a uniformly distributed working load of 5.0 kips./ft (which includes an assumed weight of beam). Use A36 steel and the AISCS, and disregard shear and deflection. Select the lightest W shape by

(a) The elastic design method.
(b) The plastic design method.

Solution

(a) *Elastic (allowable stress design) method:*

Maximum negative moment (at supports):

$$M = -\frac{wL^2}{12} = \frac{5.0(20)^2}{12} = 166.7 \text{ ft-kips}$$

Maximum positive moment (at midspan):

$$M = +\frac{wL^2}{24} = \frac{5.0(20)^2}{24} = 83.3 \text{ ft-kips}$$

Modifying the moments in accordance with AISCS, Section 1.5.1.4.1:

Maximum negative moment for design:

$$-M = 0.9(166.7) = 150 \text{ ft-kips}$$

Maximum positive moment for design:

$$+M = 83.3 + (0.1)(166.7)$$

$$= 100.0 \text{ ft-kips}$$

$$\text{required } S_x = \frac{M}{F_b} = \frac{150(12)}{24} = 75 \text{ in.}^3$$

Use W21 × 44

$$S_x = 81.6 \text{ in.}^3 > 75 \text{ in.}^3 \qquad\qquad \textbf{O.K.}$$

(b) *Plastic design method.* Knowing that at the instant of collapse the moment M_p will occur at supports as well as midspan (three plastic hinge mechanism), an expression for M_p will be developed utilizing a free-body approach as shown in Fig. 11-10.

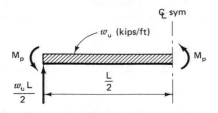

Figure 11-10   Free-body diagram.

ΣM about midspan:

$$\frac{w_u L}{2}\left(\frac{L}{2}\right) - w_u\left(\frac{L}{2}\right)\left(\frac{L}{4}\right) - M_p - M_p = 0$$

$$\frac{w_u L^2}{4} - \frac{w_u L^2}{8} = 2M_p$$

$$M_p = \frac{w_u L^2}{16}$$

Using a load factor of 1.7, the factored load

$$w_u = 1.7(5) = 8.5 \text{ kips/ft}$$

$$M_p = \frac{w_u L^2}{16} = \frac{8.5(20)^2}{16} = 212.5 \text{ ft-kips}$$

$$\text{required } Z_x = \frac{M_p}{F_y} = \frac{212.5(12)}{36} = 70.8 \text{ in.}^3$$

From the AISCM, Part 2, Plastic Design Selection Table:

$$\text{Use } \mathbf{W16 \times 40} \qquad Z_x = 72.9 \text{ in.}^3$$

$$72.9 \text{ in.}^3 > 70.8 \text{ in.}^3 \qquad\qquad \mathbf{O.K.}$$

11-6 PLASTIC DESIGN APPLICATION: CONTINUOUS BEAMS

A more practical and realistic application of plastic design is in the design of continuous beams. They are relatively common in structures compared to fixed-end beams, which are seldom encountered.

If the loads supported by a continuous beam are increased proportionately, the ultimate or maximum loading would be reached when the weakest span is reduced to a mechanism by the formation of plastic hinges progressively at points of maximum moment. For the case of three or more identically loaded equal spans having simple supports at the outer ends, mechanisms will form in the end spans at a magnitude of loading less than that required to form mechanisms in the interior spans. As a means of comparison between plastic design and elastic design, the continuous beam of Example 11-2 will be redesigned using the plastic design method.

Example 11-5

Redesign the beam of Example 11-2 using the AISCS, Part 2, Plastic Design method. The steel is A36.

Solution   Using a load factor of 1.7, the total factored load on any one of the spans is

$$1.7(3.0) = 5.1 \text{ kips/ft}$$

Considering the end span first, a plastic hinge will develop initially at the interior support since this is the location of the elastic maximum moment within that span. The ultimate or maximum load condition would be reached when another plastic hinge develops within the span. The end span would then have one real hinge and two plastic hinges, thereby forming a collapse mechanism as shown in Fig. 11-11. This second plastic hinge will occur as a result of some ultimate load and will form at a point $0.414L$ from the simply supported end. This location is applicable for the end spans of continuous beams of two or more equal spans having identical uniformly distributed loading. It is the location of maximum moment M_p within the end span and shear will be zero at this point.

The required plastic moment strength M_p that must exist to resist the ultimate load may be obtained using a free-body approach as shown in Fig.

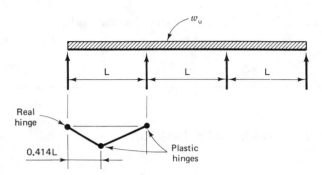

Real hinge

0.414L

Plastic hinges

Figure 11-11   Continuous beam mechanism.

11-12. Since shear is zero at the right end of the free-body diagram, the left reaction must be $0.414w_uL$. Then taking ΣM about plane P,

$$0.414w_uL(0.414L) - w_u(0.414L)\left(0.414\frac{L}{2}\right) - M_p = 0$$

$$M_p = 0.0858w_uL^2$$

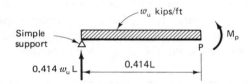

Simple support

w_u kips/ft

M_p

$0.414\,w_u\,L$ $0.414L$ P

Figure 11-12   Free-body diagram.

If the interior span is now considered, plastic hinges would be developed simultaneously at each end and after an increase in loading a third plastic hinge will form at midspan in a manner similar to the single-span fixed-end beam previously discussed. The required M_p to develop the plastic hinge at midspan would also be the same as that of the fixed-end single span shown previously as

$$M_p = \frac{w_uL^2}{16} = 0.0625\ w_uL^2$$

Since the interior and end spans are of the same length, the beam selection will be based on the largest required M_p.

Using

$$M_p = 0.0858w_uL^2$$

$$= 0.0858(5.1)\,(50)^2$$

$$= 1094\ \text{ft-kips}$$

the required plastic modulus is

$$Z_x = \frac{M_p}{F_y} = \frac{1094(12)}{36} = 364.7\ \text{in.}^3$$

$$\text{Use W30} \times \textbf{116} \qquad Z_x = 378 \text{ in.}^3$$

$$378 \text{ in.}^3 > 364.7 \text{ in.}^3 \qquad\qquad \textbf{O.K.}$$

For a complete design, shear and deflection should also be checked as well as the width–thickness ratios of the beam flange and web in accordance with the AISCS, Part 2, Section 2.7.

The plastic design method used in the previous problems is generally designated the **equilibrium method.** However, as beams and their spans and loadings become more complex, a more practical and simpler method is recommended. This alternative method is generally designated the **virtual work method** or **mechanism method.** This method is also recommended for statically indeterminate frames which constitute another type of structure where the plastic design method is applicable.

For a more extensive coverage of plastic design, the reader is referred to the given reference material.

REFERENCES

1. H. H. West, *Analysis of Structures* (New York: John Wiley & Sons, Inc., 1980).
2. Robert L. Ketter, et al., *Structural Analysis and Design* (New York: McGraw-Hill Book Company, 1979).
3. L. S. Beedle, *Plastic Design of Steel Frames* (New York: John Wiley & Sons, Inc., 1958).
4. *Plastic Design in Steel,* ASCE Manual of Engineering Practice, No. 41, 2nd ed., 1971.
5. *Plastic Design of Braced Multistory Steel Frames,* Publication M004, American Institute of Steel Construction, 1968.

12 Structural Steel Detailing: Beams

12-1 INTRODUCTION

The construction sequence of a steel-framed building, as discussed in Section 1-5 of this text, consists of three sequential phases which occur (generally) after the design sequence. These phases may be categorized as the detailing, fabrication, and erection phases.

In the detailing phase, information from the design drawings, which have been prepared by the architect and/or engineer, is used to develop detail drawings. These are generally called **shop drawings** and they must convey all the information necessary for shop fabrication of the multitude of structural members and their connections in a given structure. They will also convey information relative to field erection procedures and sequences.

The actual production of the shop drawings is the job of the **detailer**. The detailer must develop the ideas conveyed by the design drawings to the point where individual members and all the many required components of the structure may be fabricated. A great deal of practical knowledge is required. The drawings and schedules generated by the detailer will be instrumental in coordination of the work in the fabrication and erection phases. Good detailers can do much to promote economy in steel construction.

12-2 OBTAINING THE STEEL

Most structural steel fabricators are staffed and equipped for the detailing, fabrication, and erection portions of the project but must purchase the steel from the rolling mills. Economy does not permit most fabricators to keep in stock a very large inventory of

Structural steel in the fabricator's shop. The ends of the stacked wide flange beams on the left have been coped and punched in readiness for application of other detail material which is placed nearby.

structural steel. In an effort to expedite the structural steel portion of a project, the fabricator will prepare an advance bill of materials based purely on the design drawings information for the purpose of ordering the steel from the rolling mills. This order is placed before shop drawings are prepared in order to expedite getting some steel into the shop. In reality, the preparation of the shop drawings has started at the same time the advance bill of materials has been prepared; however, shop fabrication cannot be started until the ordered material has arrived at the fabricator's plant. When the order arrives, quantities and items are checked and subsequently stored, ready for fabrication. To facilitate fabrication and with efficient planning, it may be possible to order the larger and longer pieces cut to detail sizes at the mill. However, because the mills assess extra charges for cutting to detail sizes, it is generally more economical to order the structural shapes in long lengths from which several shorter lengths can be shop cut.

12-3 DRAWING PREPARATION

The initial step in the preparation of the shop drawings is to prepare a set of erection drawings which give all information required for the layout and installation of the structural steel. These drawings show each steel piece or subassembly of pieces with their assigned shipping or erection mark to identify and locate them in their correct position in the structure. The erection plans include an **anchor bolt plan.** This is one of the first drawings made since the anchor bolts must be set prior to the erecting of the steel. This drawing locates all the anchor bolts that will be embedded in the concrete foundations and which serve to anchor the steel frame to the supporting foundation. This aspect is discussed further in Chapter 13 of this text.

Sometimes reproductions of the architect's and/or engineer's design drawings are used as erection plans. Erection marks and instructions are added to the plans and the drawing is given a new number.

Erection drawings not only show the exact location of every piece with its erection mark, but also the sequence of erection when the size of the project is extensive. Large areas of framing are usually divided into separate sections called **installments.** These installments permit fabricated pieces to be delivered on a detailed schedule to predetermined locations at the project site without expensive rehandling costs. This advance planning helps to establish detailing, fabrication, shipping, and erection schedules.

The main objective of the detailing phase is to prepare the structural steel detail or shop drawings. These drawings, which are prepared by detailers, are subsequently used in the shop to fabricate each individual piece of steel required in the structure. From the information furnished in the contract documents (design plans and specifications), the detailer prepares complete and explicit details for each structural member. To avoid the repetition of labeling each sketch with the same information, shop notes are placed on the shop drawings indicating bolt size, size of open holes, type of material, paint, and other pertinent information required by the shop. In addition, since the design drawings generally show only a few connections and those are seldom completely developed, it is up to the detailer to develop them fully, including modifications to facilitate fabrication and erection. The detailer must also design other typical connections as well as any special connections not shown on the design drawings. Hence the detailer must be familiar with the AISCM as well as standard design and detailing practices. After the structural members are detailed and shop drawings completed, it is general practice to have more experienced personnel, called **checkers,** review and verify all sketches and dimensions on the drawings. To help the shop assemble the material required to fabricate various shipping pieces detailed on the shop drawings, a shop bill of material is prepared as part of each shop drawing. Most fabricators provide standard shop drawing sheets with a preprinted shop bill form on the sheet. To provide space on the drawing, separate shop bill forms are sometimes used.

Before fabrication can begin, the finished shop drawings must be approved by the architect/engineer or some other owner-designated representative. This applies to all shop details and erection plans since all the shop drawings contain additional information not specifically shown on the design drawings.

12-4 BEAM DETAILS

The required details for fabrication of a beam are shown on a shop drawing. Generally, each beam in a roof or floor framing system makes a convenient shipping and erection unit. All features that have a bearing on the erection of the beam must be investigated. Beam connection holes must match the location of similar holes in the supporting members. Proper erection clearances must be provided and possible interferences must be eliminated so that the beam can be swung or lowered into position for connection to its supporting members.

In detailing a beam, the detailer must first design the end connections to transmit the beam load to its supporting members. The necessary information for the connection design is generally furnished on the design drawings. This should include information on the type of construction, design loads, shears, reactions, moments, and axial forces where applicable.

A partial framing plan taken from a design drawing is shown in Fig. 12-1. It represents a portion of a typical interior bay of a steel-framed building floor system. Unless otherwise noted, one may assume that members shown on such a framing plan are to be

1. Placed parallel or at right angles to one another with their webs in a vertical plane
2. Located at some specific elevation and set level end to end

Beam designations (A3, B4, etc.) are generally not furnished on a design drawing, but are furnished here for purposes of explanation. As a means of introduction to the detailing process, Fig. 12-2 depicts the detailed beam A3. The various

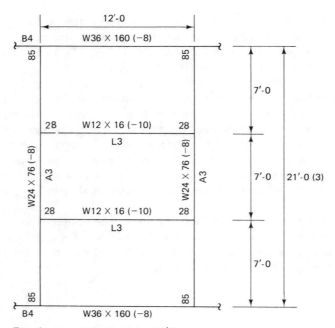

Top of concrete slab elevation = 100'-0
Top of steel below top of slab noted thus: (−8)

Figure 12-1 Partial framing plan; first floor.

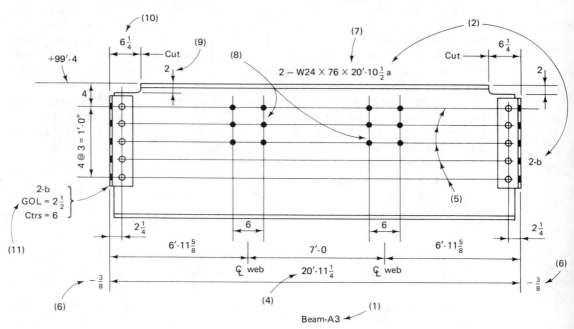

Figure 12-2 Beam details.

dimensions, together with other information numbered in the sketch, are referenced sequentially in the following discussion.

1. **Beam A3** is the beam designation and is sometimes called the **shipping and erection mark.** The derivation of this designation may be according to some method—for instance, a capital letter followed by the number of the drawing on which the beam is detailed. Various marking systems are used throughout the industry and the reader is referred to the AISC publication *Structural Steel Detailing*[2] for further discussion.

2. The **assembly piece marks** are a and b. These are used where the assembly to be shipped is composed of several pieces. In this case, there are five pieces per shipping unit. If a piece is plain (nothing attached to it), it may not have an assembly mark.

3. The 21'-0 dimension for A3 (Fig. 12-1) represents the beam's **theoretical span length,** center to center of supporting members. Note that the inch symbol (″) is omitted from the dimensions. This is common practice (and we will follow it for the duration of our detailing discussion).

4. The 20'-11$\frac{1}{4}$ dimension represents the **beam assembly unit length** back to back of connection angles. It is established as the theoretical span length minus one-half the web thickness of the supporting member at each end. Another $\frac{1}{16}$ in. is subtracted for each end of the beam (see Fig. 12-3). The beam assembly unit length is then calculated from

$$\text{theoretical span length} - \frac{t_w}{2} - \frac{t_w}{2} - 2 \times \left(\frac{1}{16}\right)$$

The length should then be rounded to the nearest $\frac{1}{16}$ in. Calculating the beam assembly unit length, we have

$$(21'\text{-}0) - \left(\frac{5}{16} + \frac{1}{16}\right)2 = 20'\text{-}11\frac{1}{4}$$

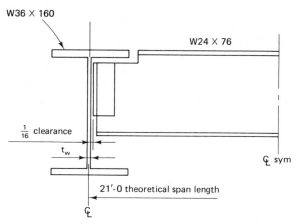

W36 × 160

W24 × 76

$\frac{1}{16}$ clearance

t_w

$\underset{\text{sym}}{\mathcal{C}}$

21'-0 theoretical span length

$\mathcal{C}$

Figure 12-3   Working sketch—each end of beam.

5. These lines are the gage lines on the beam web. Preferably, all bolt holes should fall on these gage lines. In establishing the gage lines, it is necessary to consider the beam's own end connections as well as all the members that frame into the beam web.

6. The **setback distance** is $-\frac{3}{8}$. It is the distance from the centerline of the supporting beam to the back of the connection angle:

$$\text{setback} = \frac{t_w}{2} + \frac{1}{16}$$

In this case,

$$\text{setback} = \frac{5}{16} + \frac{1}{16} = \frac{3}{8}$$

7. The **ordered, or billed, length** of the beam is $20'\text{-}10\frac{1}{2}$. It is sometimes shown only in the bill of materials. In this case, it is shown together with the assembly piece mark. It should be computed and specified to the nearest $\frac{1}{2}$ in., so that the beam will stop about $\frac{1}{2}$ in. short of the backs of the connection angles. This allows for inaccurate cutting of the beam length and eliminates possible recutting or trimming. The actual difference between the ordered beam length and length back to back of connection angles will be between $\frac{3}{8}$ and $1\frac{5}{8}$ in. based on AISCM cutting tolerances of $\frac{3}{8}$ in. over and under (AISCM, Part 1).

8. Represents the hole locations for the connections for beams L3. The vertical location of the holes must be coordinated with the connections on the supported member L3. Note that the centerline of the L3 beam web is referenced from the back of the connection angles on beam A3. The dimension is determined by subtracting $\frac{1}{16}$ in. clearance and one-half the web thickness of B4 from the centerline-to-centerline dimension of $7'\text{-}0$.

$$(7'\text{-}0) - \frac{1}{16} - \frac{5}{16} = 6'\text{-}11\frac{5}{8}$$

9. Represents the depth of cope. Coping is required where tops of beams and supporting girders are at the same elevation or will interfere with each other. The cope depth is generally established so that the horizontal cut is at the level of the toe of fillet of the supporting member. The depth of cope dimension is usually rounded up to the next $\frac{1}{4}$ in.

10. Represents the length of cope. It is dimensioned from the back of the connection angles and should allow for $\frac{1}{2}$ to $\frac{3}{4}$ in. clearance at the edge of the flange. The dimension is usually rounded up to the next $\frac{1}{4}$ in. With reference to Fig. 12-4, in this case

$$\text{length of cope} = \frac{b_f}{2} - \frac{t_w}{2} + \frac{1}{2}$$

$$= \frac{12}{2} - \frac{5}{16} + \frac{1}{2} = 6\frac{3}{16}$$

Use $6\frac{1}{4}$ in.

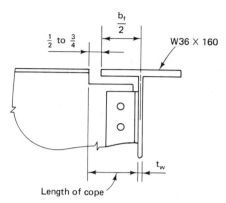

Figure 12-4 Working sketch.

11. GOL $= 2\frac{1}{2}$ refers to the gage distance on the outstanding legs of the connection angles. Ctrs $= 6$ represents the transverse spacing of the gage lines for the two outstanding legs of the angles. It is generally called the *spread,* as shown in Fig. 12-5. This value must be computed taking into consideration the assembling clearances for threaded fasteners as furnished in the AISCM, Part 4. A detailed example is furnished in Example 12-1. If adequate clearances cannot be furnished, the bolts through the outstanding legs of the connection angles must be staggered with respect to those through the beam web.

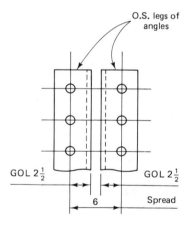

Figure 12-5 Working sketch.

Example 12-1

The partial framing plan shown in Fig. 12-6 is that of a typical interior bay of a roof system in a one-story steel-framed building. All structural steel is A36. The bolts are $\frac{3}{4}$-in.-diameter A325N in standard holes (where welding is used, use E70 electrodes, shielded metal arc welding).

Detail beams B1, B2, and G1 in accordance with the latest AISCS.

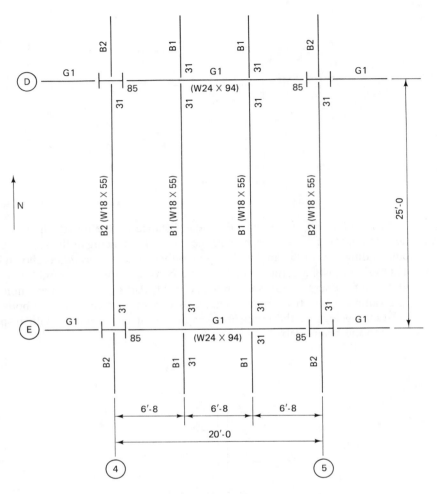

Figure 12-6 Partial framing plan.

Solution   This is a lengthy detailing problem. It will be accomplished step by step in accordance with the following outline:

I. Beam B1
    A. Connection to B1 web
    B. Connection to G1 web
    C. Framing angle leg sizes
    D. Required dimensions

  II. Beam B2
       A. Select seated beam connection
       B. Required dimensions
  III. Girder G1
       A. Connection to G1 web
       B. Connection to column flange
       C. Required dimensions

All table references are to the AISCM, Part 4, unless otherwise noted.

  I. **Beam B1**   Since the detailing process includes the design of the end connections, the initial step will be to select the most economical end connections for each end of the beam. Beam B1 is a W18 × 55 which frames into girder G1, which is a W24 × 94. High-strength bolted framed connections will be used. The end reaction of 31 kips is given on the framing plan. In some instances the end reactions are not furnished. When this occurs and when it is obvious that the beam loading is only a uniformly distributed load, the end connection must be designed to support one-half the uniform load capacity of the beam as furnished in the AISCM, Part 2, Uniform Load Constants Tables.

  Checking the W18 × 55 (A36 steel), the uniform load constant W_c is given as 1570 ft-kips. The maximum end reaction is

  $$\text{end reaction (kips)} = \frac{W_c}{2L} = \frac{1570}{2(25)} = 31.4 \text{ kips}$$

  Since 31 kips is the furnished reaction, it will be used for design.
  A. Consider the part of the connection through the W18 × 55 web.
      1. To provide stability during erection, it is recommended (AISCM, Part 2, Table II, discussion) that the minimum length of connection angle be at least one-half the T dimension as furnished in the AISCM, Part I. For a W18 × 55, T is $15\frac{1}{2}$ in. Therefore, the minimum angle length L is

      $$\text{minimum } L = \frac{T}{2} = \frac{15.5}{2} = 7.75 \text{ in.}$$

      2. Based on bolt shear, from Table II-A select a connection with three rows ($n = 3$) of $\frac{3}{4}$-in.-diameter A325N bolts. The angles for this connection are to be $\frac{5}{16}$ in. thick and the capacity based on bolt shear is 55.7 kips. This exceeds the 31-kip reaction. The length of the angles is $8\frac{1}{2}$ in., which is greater than the required minimum of 7.75 in.
      3. Check the net shear in the connection angles. This may be accomplished using Table II-C. However, this is critical only for those values in Table II-A which are footnoted below the table. The value 55.7 kips is *not* footnoted; therefore, net

shear in the $\frac{5}{16}$-in.-thick angles is *not* critical. (Table II-C reveals a capacity in net shear of 82.2 kips.)

4. The bearing on both the beam (B1) web and connection angles must be checked. Figure 12-7 shows a tentative layout of the connection with regard to bolt spacing, vertical edge distance, and horizontal edge distance. Tops of flanges are at the same elevation and dimension Q_1 is determined so that the horizontal cut of the cope is at or below the toe of the fillet of the girder G1 ($k = 1\frac{5}{8}$ in. for the W24 × 94). This dimension is rounded up to the next $\frac{1}{4}$ in. Therefore, use Q_1 of $1\frac{3}{4}$ in.

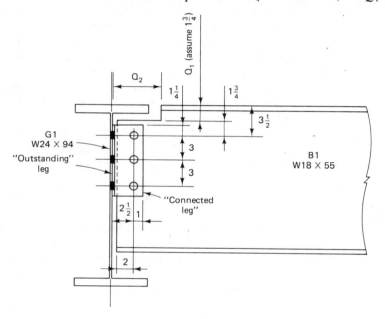

Figure 12-7   Working sketch.

The capacity in bearing, based on the beam web ($t_w = 0.390$ in.), is:

Considering that $\ell_v = 1\frac{3}{4}$ in. (vertical edge distance) controls edge distance, from Table I-F:

$$0.390(3)(50.8) = 59.4 \text{ kips}$$

Considering 3-in. bolt spacing from Table I-E:

$$0.390(3)(65.3) = 76.4 \text{ kips}$$

The capacity based on the connection angles (total thickness for two angles = 0.625 in.) is:

Considering that $\ell_h = 1$ in. (horizontal edge distance) controls edge distance, from Table I-F:

$$0.625(3)(29.0) = 54.4 \text{ kips}$$

The least bearing capacity of 54.4 kips exceeds the reaction of 31 kips. Therefore, this aspect of the web connection is satisfactory.

5. Check the capacity of the connection based on web tear-out (block shear). The working sketch (Fig. 12-7) indicates that B1 must be coped at each end.

From Table I-G, with $\ell_v = 1\frac{3}{4}$ in. and $\ell_h = 2$ in., the tabulated coefficients are

$$C_1 = 1.53$$

$$C_2 = 0.99$$

$$\text{connection capacity} = (C_1 + C_2)F_u t_w$$

$$= (1.53 + 0.99)(58)(0.390)$$

$$= 57.0 \text{ kips}$$

$$57.0 \text{ kips} > 31 \text{ kips} \qquad \textbf{O.K.}$$

The connection to the web of B1 as sketched in Fig. 12-7 is adequate.

B. Consider the part of the connection through the web of the supporting member G1, which is a W24 × 94. The six high-strength bolts must support a 31-kip reaction from each side of G1. Therefore, the total reaction to G1 is 62 kips.

1. Check the capacity of the connection in double shear. From Table II-A the capacity of three $\frac{3}{4}$-in.-diameter A325N bolts in double shear is 55.7 kips. For six bolts, the capacity is

$$55.7(2) = 111.4 \text{ kips}$$

$$111.4 \text{ kips} > 62 \text{ kips} \qquad \textbf{O.K.}$$

2. Check the capacity of the connection in bearing on the web of girder G1, based on bolt spacing. (Edge distances are not applicable for this check.) The web thickness of G1 is 0.515 in. The connection capacity from Table I-B is

$$65.3(0.515)(6) = 202 \text{ kips}$$

$$202 \text{ kips} > 62 \text{ kips} \qquad \textbf{O.K.}$$

3. Check the capacity of the connection in bolt bearing on the $\frac{5}{16}$-in. angle thickness. Considering the bolt spacing of 3 in.

and the vertical edge distance of $1\frac{1}{4}$ in., Table I-E shows that edge distance controls. There are four connection angles with three bolts through each one:

$$12(11.3) = 135.6 \text{ kips}$$

$$135.6 \text{ kips} > 62 \text{ kips} \qquad \textbf{O.K.}$$

The six-bolt connection to the web of G1 is adequate. (The angle thickness is $\frac{5}{16}$ in. and the length of angle is $8\frac{1}{2}$ in.)

C.   Establish the leg sizes of the connection angles.

1.   With reference to Fig. 12-7, the length of the connected leg (so-called because it is the leg of the angle connected to the beam when the beam assembly is shipped) may be determined by summing the clearance distance and the edge distances on the beam web and the angle:

$$\frac{1}{2} + 2 + 1 = 3\frac{1}{2} \text{ in.}$$

Use a $3\frac{1}{2}$-in. connected leg with a gage as shown.

2.   The length required for the outstanding leg of the connection angles is a function of the required assembling clearances.

The angles will be shop-bolted to the web of B1 and field-bolted to the web of G1. Therefore, adequate wrench tightening clearance must be provided for this field installation.

Pertinent data for these calculations are found in the AISCM, Part 4, Assembling Clearances for Threaded Fasteners. With reference to Fig. 12-8, the following dimensions may be established:

$$C_1 = \text{clearance for tightening} = 1\frac{1}{4} \text{ in.}$$

$$H_2 = \text{shank extension} = 1\frac{3}{8} \text{ in. (includes flat}$$

$$\text{washer, nut, and projection)}$$

$$t = \text{angle thickness} = \frac{5}{16} \text{ in.}$$

The sum of the foregoing represents the minimum gage distance for the angle outstanding leg:

$$C_1 + H_2 + t = 1\frac{1}{4} + 1\frac{3}{8} + \frac{5}{16} = 2\frac{15}{16} \text{ in.}$$

The minimum edge distance required to the rolled edge

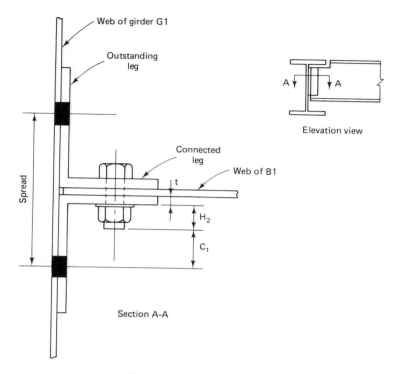

Figure 12-8   Working sketch.

of the angle (AISCS, Table 1.16.5.1) is 1 in. Therefore, the minimum outstanding leg size is

$$2\frac{15}{16} + 1 = 3\frac{15}{16} \text{ in.}$$

Use a 4-in. outstanding leg. Therefore, the connection angles will be

$$2 \text{ L}4 \times 3\frac{1}{2} \times \frac{5}{16} \times 8\frac{1}{2} \text{ in.}$$

with a gage distance of 3 in. for the outstanding leg.

The spread must be, as a minimum,

$$2(3) + \frac{3}{8} = 6\frac{3}{8} \text{ in.}$$

Use a $6\frac{3}{8}$-in. spread.

D. In addition to the end connections, various dimensions must be computed to complete the detailing of B1.

1. Q_1 has already been determined to be $1\frac{3}{4}$ in. The length of cope Q_2 is measured from the back of connection angles and is computed based on the flange width and web thickness of the supporting member G1. A minimum clearance of $\frac{1}{2}$ in. is commonly provided.

$$\text{required } Q_2 = \frac{b_f}{2} - \frac{t_w}{2} + \frac{1}{2}$$

$$= \frac{9.125}{2} - \frac{1}{4} + \frac{1}{2}$$

$$= 4.81 \text{ in.}$$

Q_2 should be rounded up to the next $\frac{1}{4}$ in. Therefore, use Q_2 of 5 in.

The vertical and horizontal cuts (Q_1 and Q_2) should be connected and shaped notch free to a radius of $\frac{1}{2}$ in. (see AISCM, Part 4, Fabricating Practices).

2. The setback distance required is the distance from the center-line of the supporting beam to the back of the connection angle:

$$= \frac{t_w}{2} + \frac{1}{16}$$

$$= \frac{1}{4} + \frac{1}{16} = \frac{5}{16} \text{ in.}$$

3. The beam assembly unit length is the distance back to back of connection angles and is calculated from

$$\text{theoretical span length} - \left(\frac{t_w}{2} + \frac{1}{16}\right)(2)$$

$$= (25'\text{-}0) - \left(\frac{5}{16}\right)(2)$$

$$= 24'\text{-}11\frac{3}{8}$$

4. The ordered, or billed, length of the beam should be established so that the ends stop approximately $\frac{1}{2}$ in. short of the backs of the connection angles. This dimension is rounded to the nearest $\frac{1}{2}$ in.

$$\left(24'\text{-}11\frac{3}{8}\right) - \frac{1}{2} - \frac{1}{2} = 24'\text{-}10\frac{3}{8}$$

The ordered length is $24'\text{-}10\frac{1}{2}$. The complete detail of beam B1 is shown in Fig. 12-9.

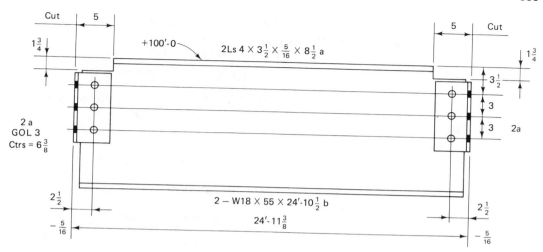

Figure 12-9   Beam B1.

II.   **Detail Beam B2**   With reference to Fig. 12-6, it may be observed that beam B2 is supported at each end by columns and must fit in between the column flanges. Beam B2 is a W18 × 55 and the column is a W10 × 60. An unstiffened seated beam connection will be used at each end with the beam seat shop bolted to the column and field bolted to beam B2. The beam reaction is 31 kips. Use $\frac{3}{4}$-in.-diameter A325N bolts in standard holes.

A.   Select the bolted seated beam connection.

1.   Select the angle thickness based on an angle length of 6 in. determined as follows. The angle length cannot exceed the T dimension of the supporting column (W10 × 60) since the angle will be bolted to the column web. The T dimension is $7\frac{5}{8}$ in. Therefore, use an angle length of 6 in.

   Enter Table V-A, AISCM, Part 4. With a beam (B2) web thickness of $\frac{3}{8}$ in., the required angle thickness is 1 in. The outstanding leg capacity (based on an outstanding leg of 4 in.) is 37.6 kips.

$$37.6 \text{ kips} > 31.0 \text{ kips} \qquad \textbf{O.K.}$$

2.   Enter Table V-C to select the type of seat angle. With $\frac{3}{4}$-in.-diameter A325N high-strength bolts, a Type B connection must be used. Fastener shear capacity is 37.1 kips.

$$37.1 \text{ kips} > 31.0 \text{ kips} \qquad \textbf{O.K.}$$

3.   Enter Table V-D to select the seat angle. With a Type B connection an 8 × 4 angle is available in 1 in. thickness.

4.   Check the capacity of the connection in bearing on the col-

umn web. The Type B connection has four bolts and the
column web thickness is 0.420 in.

Based on 3-in. bolt spacing (Table I-E), the capacity is

$$0.420(65.3)(4) = 109.7 \text{ kips}$$

Since beam B2 is framing into the column web from
both sides, the reaction to the column is

$$31(2) = 62 \text{ kips}$$

$$109.7 \text{ kips} > 62 \text{ kips} \qquad \textbf{O.K.}$$

5. Check the capacity of the connection in bearing on an angle
    thickness of 1 in.:

    Based on 3-in. bolt spacing (Table I-E) the capacity is

    $$65.3(4) = 261 \text{ kips}$$

    Based on a 2-in. vertical edge distance (Table I-F), the
    capacity is

    $$58(1)(4) = 232 \text{ kips}$$

    $$232 \text{ kips} > 31 \text{ kips} \qquad \textbf{O.K.}$$

    Therefore, use a seat angle L8 × 4 × 1 × 6 with four
    $\frac{3}{4}$-in.-diameter A325N high-strength bolts in the vertical leg
    (Type B pattern) at each end of beam B2.

    In addition, select a top angle discussed in Section 7-7
    of this text). Since this is a roof framing system with the top
    of the column at approximately the same elevation as the tops
    of the beams and girders, the top angle cannot be located on
    the top flange but must be placed at the optional location as
    indicated in the AISCM, Part 4. Use an angle L4 × 3 ×
    $\frac{1}{4}$ × $5\frac{1}{2}$ connected to the beam web and column web with a
    total of four $\frac{3}{4}$-in.-diameter A325 high-strength bolts.

B. In addition to the selection of the seat angle, various dimensions
    must be computed to complete the detailing of B2. This is best
    accomplished with the use of working sketches as shown in Fig.
    12-10.

    1. The setback distance required is the distance from the center-
        line of the column to the end of the beam. Approximately $\frac{1}{2}$
        in. erection clearance should be used between the end of the
        beam and the face of the column web:

        $$\frac{t_w}{2} + \frac{1}{2} = \frac{1}{4} + \frac{1}{2} = \frac{3}{4} \text{ in.}$$

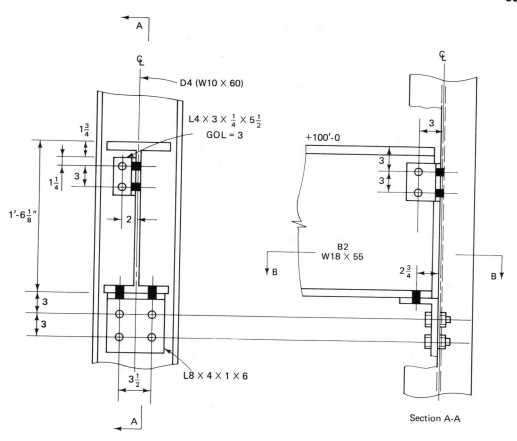

Section A-A

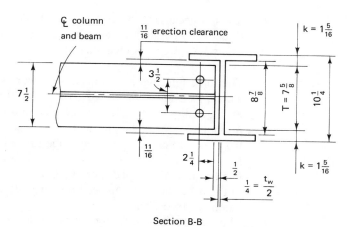

Section B-B

Figure 12-10 Working sketch.

2. The beam overall length or ordered length is the theoretical span length minus the setback:

$$(25'\text{-}0) - \frac{3}{4} - \frac{3}{4} = 24'\text{-}10\frac{1}{2}$$

3. As shown in the AISCM, Part I, Standard Mill Practice, the cutting tolerance for this beam is $\frac{3}{8}$ in. over and $\frac{3}{8}$ in. under.

    Depending on the final length of the beam, adjustments by the shop may be necessary for the location of holes and edge distances. However, the distance between the holes at the ends of the bottom flange is fixed at 24'-6, as shown in Fig. 12-11.

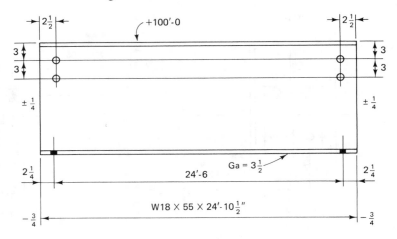

Figure 12-11 Beam B2.

4. It is not necessary to cut the top and bottom beam flanges to fit between column flanges since an $\frac{11}{16}$-in. erection clearance exists. An erection clearance of at least $\frac{1}{2}$ in. must be furnished.

5. The top angle, which connects the beam web to the column web is located below the toe of the fillet of beam B2 (defined by the k dimension). If a vertical edge distance of $1\frac{1}{4}$ in. is provided for the top angle and the angle is located $1\frac{3}{4}$ in. down from the top of B2, the upper gage line will be conveniently located 3 in. from the top of B2. This angle will be bolted hand-tight and shipped with the column. It will be removed in the field and then replaced properly after the placement of beam B2 (see Chapter 13).

6. The distance between gage lines on the column web may be

set at $3\frac{1}{2}$ in., which conforms to the angle length of 6 in. as shown in the AISCM, Part 4, Seated Beam Connections.

7. The distance from the end of beam B2 to the web holes for the top angle is the angle gage distance of 3 in. minus the erection clearance of $\frac{1}{2}$ in. from the end of the beam to the face of the column web. The $2\frac{1}{2}$-in. edge distance is shown in Fig. 12-11. Note that the detail of beam B2 in Fig. 12-11 does not include the seat angle or the top angle. Since they will be shipped with the column, they are detailed with the column (Chapter 13).

III. **Detail Girder G1**   G1 is a W24 × 94 supporting the end reactions of four B1 beams and in turn is supported by W10 × 60 columns (which will be designated D4). The girder end connections will be framed connections shop-welded to the girder web and field-bolted to the column flange. Recall that the beam-to-girder connections were detailed with (and will be shipped with) beams B1. High-strength bolts (A325N) $\frac{3}{4}$ in. in diameter in standard holes and E70 electrodes will be used. The end reaction as furnished on the framing plan (Fig. 12-6) is 85 kips.

A. Consider the part of the connection to the W24 × 94 web.

1. The framing angle length will again be taken as at least one-half of the T dimension, which for the W24 × 94 is 21 in. Therefore, use an angle length of at least $10\frac{1}{2}$ in. Enter Table III, AISCM, Part 4, under weld A and find the closest allowable load greater than 85 kips.

Select a weld size of $\frac{1}{4}$ in., with a capacity of 102 kips and a required length of angle of $11\frac{1}{2}$ in. This requires a $\frac{5}{16}$-in.-thick angle to meet the weld requirement of the AISCS, Section 1.17.3. The 0.515-in. web thickness of the W24 × 94 is greater than the minimum web thickness tabulated in Table III. Therefore, no capacity reduction is required.

2. Note in Table IIA, AISCM, Part 4, that the angle length provides for four bolts of $\frac{3}{4}$-in.-diameter A325N high-strength bolts, with a capacity of 74.2 kips. This represents the capacity of the eight bolts through the outstanding legs of the connection angles (in single shear) to the column flange. This is less then the beam reaction of 85 kips; therefore, the angle length must be increased.

Using five bolts with an angle length of $14\frac{1}{2}$ in. and an angle thickness of $\frac{5}{16}$ in. from Table IIA, the allowable bolt shear is 92.8 kips:

$$92.8 \text{ kips} > 85 \text{ kips} \qquad\qquad \textbf{O.K.}$$

3. Referring back to Table III, with an angle length of $14\frac{1}{2}$ in. and a fillet weld of $\frac{3}{16}$ in., weld A capacity is 93.5 kips:

$$93.5 \text{ kips} > 85 \text{ kips} \qquad \textbf{O.K.}$$

The data determined thus far is shown in Fig. 12-12. Data determined in part B are also reflected in Fig. 12-12.

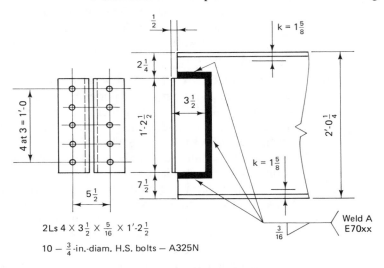

$2\text{Ls } 4 \times 3\frac{1}{2} \times \frac{5}{16} \times 1\text{'-}2\frac{1}{2}$

$10 - \frac{3}{4}$-in.-diam. H.S. bolts — A325N

Figure 12-12 Working sketch.

B. Consider the part of the connection through the column flange. The ten $\frac{3}{4}$-in.-diameter bolts must support an 85-kip reaction. The shear capacity of the bolts has been checked in part A.

1. Check the adequacy in the bearing on the column flange and on the outstanding legs of the angles. For the column, $t_f = 0.680$ in. The angles are $\frac{5}{16}$ in. (0.313 in.) thick. Considering the bolt spacing of 3 in., the angles are more critical (they are thinner). Considering the vertical edge distance of $1\frac{1}{4}$ in. for the angles (vertical edge distance consideration is not applicable to the column flange and horizontal edge distance will not be critical; refer to Fig. 13-5). Table I-E reveals that the angle edge distance is more critical. The capacity in bearing is calculated as

$$10(11.3) = 113 \text{ kips}$$

$$113 \text{ kips} > 85 \text{ kips} \qquad \textbf{O.K.}$$

The 10-bolt connection to the column flange is adequate.

To accommodate usual gages, the angle leg widths are generally taken as $4 \times 3\frac{1}{2}$ with the 4-in. leg outstanding.

Therefore, the connection angles for this connection are

$$2L4 \times 3\frac{1}{2} \times \frac{5}{16} \times 1'\text{-}2\frac{1}{2}$$

C. In addition to the end connections various dimensions must be computed to complete the detailing of G1.
   1. The setback distance required is the distance from the centerline of the supporting column to the back of the connection angle.

$$\frac{d}{2} + \frac{1}{16} = \frac{10\frac{1}{4}}{2} + \frac{1}{16} = 5\frac{3}{16} \text{ in.}$$

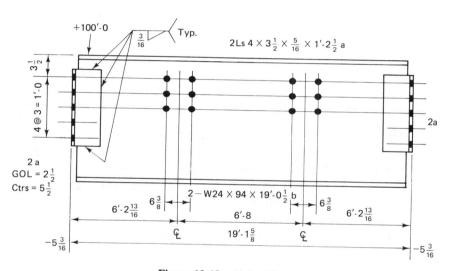

Figure 12-13   Girder G1.

2. The beam assembly unit length is the distance back to back of angles and is calculated as

$$\text{theoretical span length} - \left(\frac{d}{2} + \frac{1}{16}\right)2$$

$$= (20'\text{-}0) - \left(\frac{10\frac{1}{4}}{2} + \frac{1}{16}\right)2$$

$$= 19'\text{-}1\frac{5}{8}$$

3. The ordered or billed length of the beam is established so that the ends are $\frac{1}{2}$ in. short of the backs of the connection angles.

The ordered length is rounded to the nearest $\frac{1}{2}$ in.

$$\left(19'\text{-}1\frac{5}{8}\right) - \frac{1}{2} - \frac{1}{2} = 19'\text{-}0\frac{5}{8}$$

Use an ordered length of $19'\text{-}0\frac{1}{2}$.

4. The usual gage for the 4-in. outstanding leg of the framing angle is $2\frac{1}{2}$ in. The maximum spread is calculated as

$$2 \times 2\frac{1}{2} + t_w = 5 + \frac{1}{2} = 5\frac{1}{2} \text{ in.}$$

5. The dimension from the backs of the connection angles to the centerline of the connections for the B1 beams is determined by subtracting the setback distance from the centerline of column to the centerline of B1 dimension:

$$(6'\text{-}8) - 5\frac{3}{16} = 6'\text{-}2\frac{13}{16}$$

REFERENCES

1. *Manual of Steel Construction,* 8th ed., American Institute of Steel Construction, 400 North Michigan Avenue, Chicago, IL 60611.
2. *Structural Steel Detailing,* 2nd ed., American Institute of Steel Construction, 400 North Michigan Avenue, Chicago IL 60611.
3. *Structural Shop Drafting,* Vols. 1, 2, and 3, American Institute of Steel Construction, 400 North Michigan Avenue, Chicago, IL 60611, 1950, 1953, 1955.

PROBLEMS

12-1. A W16 × 36 beam is to connect to the flange of a W12 × 45 column. The beam reaction is 40 kips. All structural steel is A36. Select a suitable framed connection using $\frac{3}{4}$-in.-diameter A325F high-strength bolts. Include a working sketch of the connection and complete data for the angles to be used.

12-2. A W24 × 68 beam is to connect to the web of a W33 × 118 girder. The elevation of the top of the beam is to be 3 in. below the top of the girder. The beam reaction is 90 kips. Select a suitable framed connection using $\frac{3}{4}$-in.-diameter A325X high-strength bolts. Include a working sketch of the connection and complete data for the angles to be used.

12-3. Rework Problem 12-2, except that the beam reaction is 75 kips and tops of the beam and girder are to be at the same elevation.

12-4. Select a suitable framed beam connection for the beam to girder connection shown. All structural steel is A36. Use $\frac{7}{8}$-in.-diameter A325N high-strength bolts. Include a working sketch and complete data for the angles to be used.

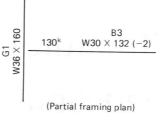

(Partial framing plan)

Problem 12-4

12-5. Select framed connections for beam B4 shown. Completely detail the beam. All structural steel is A36. Beam reactions are not given on the design drawings. Use $\frac{7}{8}$-in.-diameter A325N high-strength bolts for the connections.

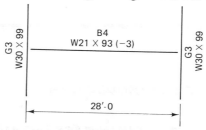

Problem 12-5

12-6. Rework Problem 12-5 but use E70 electrodes and weld the framing angles to the web of B4. Use $\frac{7}{8}$-in.-diameter A325N bolts for the girder web connection.

12-7. Select unstiffened seated beam connections for the beam shown. Completely detail the beam. All structural steel is A36. Use $\frac{3}{4}$-in.-diameter A325N high-strength bolts for the connections. Note that the columns are not similar.

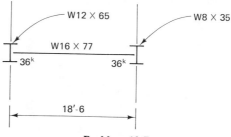

Problem 12-7

13 Structural Steel Detailing: Columns

13-1 INTRODUCTION

Columns and their associated base plates are the first structural steel members to be erected and they must be among the first to be fabricated. Therefore, shop drawings of the columns are generally prepared before the shop drawings of the other structural members.

The information needed for detailing columns is furnished on the structural drawings of the contract documents. The structural drawings (often termed **engineering drawings** or **design drawings**) will depict the floor and roof framing plans. These drawings also locate the column centerlines as well as orient the column with respect to direction of flanges and webs. Special enlarged details clarify special framing conditions such as off-center beams for spandrel framing or for framing around stairwells.

The structural drawings will also normally contain a column schedule which provides the fabricator with information on the size and length of the columns as well as splice location and column base plate information. Some form of grid system is used throughout the entire set of contract documents (structural, architectural, mechanical, etc.) to establish a consistent means of identifying all columns. The grid system usually reflects, in some way, the general shape of the building. The simplest is the rectangular grid, although grid shapes may be radial for circular buildings, repetitive triangles for sprawling buildings, or irregular patterns to meet some other shape. For the common rectangular grid, one normally uses a simple numerical sequence, beginning with number 1, in one direction and a letter sequence, beginning with A, in the other direction. Thus a column at the intersection of lines E and 5 would be uniquely identified by designation E5.

A sample column schedule is shown in Fig. 13-1. As may be observed, the column schedule furnishes column loads, various building elevations, and all column base plate information. Various forms of column schedules are used by the many design offices; however, they all convey the same important information.

13-2 COLUMN BASE DETAILS

Typical column bases and column base plate design are discussed in Chapter 5 of this text. Typical column base details for any given building are generally furnished on the structural drawings. Two such typical details are shown in Fig. 13-2. All column base details require a base plate and anchor bolts as well as other detail items as necessary. The anchor bolts fix the column base to the foundation. The construction of the foundation is usually the responsibility of the general contractor, not the steel fabricator/erector. The anchor bolts are generally set in place by the masonry contractor, but detailed and furnished by the fabricator. Since foundation construction will always precede the steel erection it is necessary for the fabricator to prepare an anchor bolt plan as soon as possible. This plan, which may be similar in appearance to the foundation plan, gives complete information for field placement. It furnishes anchor

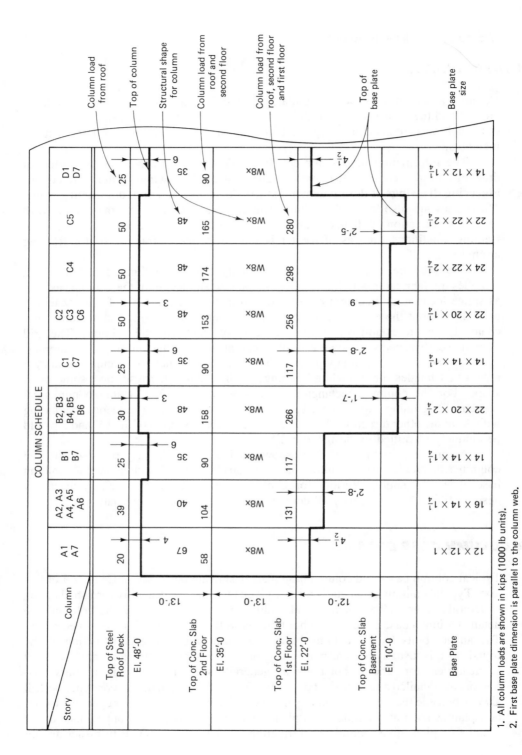

Figure 13-1 Typical column schedule.

1. All column loads are shown in kips (1000 lb units).
2. First base plate dimension is parallel to the column web.

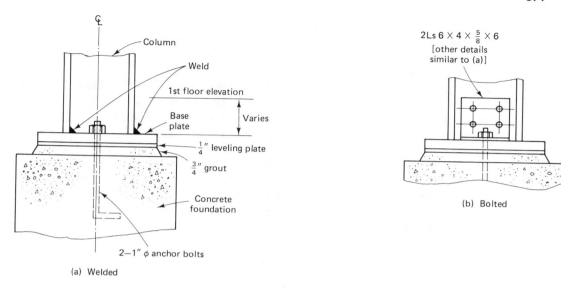

Figure 13-2 Typical column base details (as shown on design drawings).

bolt sizes, lengths, exact locations, erection marks, elevations at top of base plates, grout thickness, and the length of anchor bolts above the top of concrete. In addition, base plate erection marks and location as well as top of concrete elevations are furnished. Not only does the masonry contractor use the anchor bolt plan to set the bolts, but the steel erector will also use the plan to set the base plates. On occasion, loose base plates may be small enough to be set manually and may be placed by the masonry contractor. Heavy base plates (which may weigh several tons) are generally set by the erector with the use of a crane.

Small base plates are often attached to the columns in the shop. When this is the case, $\frac{1}{4}$-in.-thick steel leveling plates are normally installed on the foundation to provide a smooth bearing area. The leveling plates are easy to handle and are conveniently set level to the prescribed elevation prior to the erection of the columns. Leveling plates are generally furnished by the fabricator and placed by the masonry contractor.

Large base plates are set to elevation and leveled using shims of various thicknesses or by leveling screws with nuts welded to the edges of the base plate. The top of the rough masonry foundation is usually set approximately 1 in. below the bottom of the base plate to provide for adjustment and subsequent grouting. After a plate has been carefully leveled up to the proper elevation (through the use of the shims or leveling screws), cement grout is worked under the plate to build up the foundation and provide contact bearing over the full area of the underside of the plate. The design may have specified one or more large size holes near the center of the plate through which the grout may be poured. The object is to ensure an even distribution across the entire undersurface. Such holes need not be drilled but may be burned with an acetylene torch.

Anchor bolt holes in base plates and in any of the other fittings at the base of column are generally made $\frac{5}{16}$ to 1 in. larger than the diameter of the bolts to allow for inaccuracies in the setting of the bolts. The anchor bolts serve additional purposes during erection of a frame: they aid in centering the column and also serve to prevent displacement or collapse of columns due to accidental collisions. In addition, horizontal loads may tend to induce a vertical uplift on any given column of a completed structure, thereby requiring the anchor bolts to transmit these forces to the foundation. Therefore, all columns and their base plates must be fixed to the supporting foundations.

In the past, it was required that the bottom end of the column be milled to provide a true overall contact bearing. Some saws used in present-day shops, however, produce end surfaces which do not require the milling operation. The finishing of the column base plate must conform to the AISCS, Section 1.21.

Figure 13-3 is a portion of a drawing which shows an anchor bolt plan. Note that

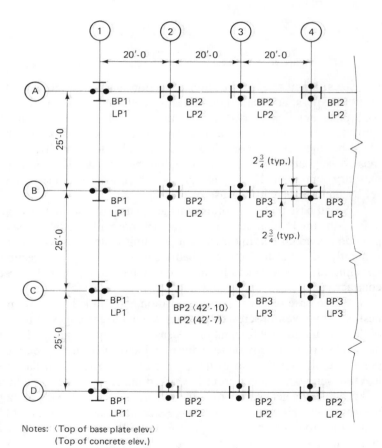

Notes: ⟨Top of base plate elev.⟩
⟨Top of concrete elev.⟩

Figure 13-3 Typical anchor bolt plan.

the plan shows bolt locations, base plate mark numbers, base plate locations, leveling plate mark numbers, leveling plate locations, elevations for tops of base plates, and elevations for tops of concrete supports.

13-3 COLUMN DETAILS

In multistory structures, the elevations at which the column sizes change provide a convenient means of dividing the framing vertically into **tiers** for ease in handling and erection. This requires the use of field splices to hold the column sections together and transfer the column loads. Generally, columns are spliced at two-story intervals. The splices are located far enough above the floor lines so that the connections will not interfere with beam framing details.

Structural steel in the fabricator's shop. These wide-flange columns have been prepared for bolted splices. The lower two members have shim plates tack-welded in place. Note that the direction that the flange is to face (West or East) has been marked on the columns.

The transfer of load in multistory building columns is generally achieved entirely by direct bearing through finished contact surfaces. The splice material, which usually consists of plates and fasteners, serves principally to hold all parts securely in place. Suggested details for column splices are furnished in the AISCM, Part 4. One- or two-story buildings are usually designed with a constant column section for the full height of the column. Therefore, no splices are needed. Pinholes are occasionally furnished in the splice plates for erection purposes, as shown in the AISCM, Part 4.

13-4 SHOP DRAWINGS OF COLUMNS

It is traditional in the preparation of shop drawings of columns to either detail the columns in the vertical position with the base of the column at the bottom of the sheet or in the horizontal position with the column base to the left. If the column is simple with little detail material, it is detailed upright. A complex column with complicated beam bracing and, perhaps, truss connections, would be detailed in the horizontal position.

It is fairly common that columns have some form of connection to both flanges and to both sides of the web. As a result, it is standard practice to assign a letter to each of the four faces of the column. This identification by letter is helpful to the shop worker in laying out the work and reduces the probability of shop errors. Usually, the detailer will select the **flange face** which contains the most detail (fittings and fabrication) and label that face "A." Then, looking down on top of the column, the lettering will continue alphabetically in a counterclockwise direction around the section.

As shown in Fig. 13-4, for a W shape, faces A and C are always flange faces and faces B and D are always web faces. The lettering could also be arranged so that the web with the most detail material is assigned the letter B. It is seldom necessary to show a separate view of face D (a web face) for W shapes, as any fittings on face D which differ from those on face B can be shown by dashed (invisible) lines. The detail materials to be placed on face B will be shown by visible lines and may be noted N.S. (near side). Face D detail material may be noted F.S. (far side). Faces that require no detail fittings or fabrication of any kind need not be shown. However, a note such as "Face C Plain" is advisable. Where detail material and fabrication on face C is identical with face A, a note such as "Face C Same as Face A" may be added.

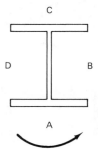

Figure 13-4　Column marking (W shape)

Where a transverse section through a column is needed to describe detail material, it should be taken from the face B view and shown looking down toward the column base. Columns which are alike except for minor differences may be detailed on the same sketch and the differences defined with notes. Combining details for too many columns on one sketch can result in a complicated drawing and cause shop errors. If the columns cannot be detailed with simple notes describing the exceptions, different sketches should be used. Shop errors are expensive to correct in the field.

Columns, like beams, must be given shipping marks. The mark shown on the drawing is painted on one of the flanges and near the base of the column. This is done in the shop when the detail material is assembled. This mark identifies the column during shop fabrication, shipping, and field erection.

It is generally necessary to provide a compass or direction mark on one of the column flanges so that the column can be oriented correctly in the field. A note such as "Face A North" instructs the shop to paint "North" on face A. In the field the erector

Structural steel in the fabricator's shop. This wide-flange column is being prepared in the shop by welding on various brackets and seats. Accurate placement of such detail material is essential for rapid field erection.

will then turn the column so that the word "North" is facing north when the column is erected and in place.

Example 13-1

Refer to Example 12-1 and Fig. 12-6 in Chapter 12 of this text. Using all the previously established design and detail data, detail column E4 in accordance with the latest AISCS. The column is a W10 × 60. The top of base plate elevation for this column is $81'-10\frac{1}{2}$. The top of steel roof deck is at elevation $100'-3$.

Solution   It is shown in Fig. 12-6 that girder (beam) G1 frames into each flange of column E4 and that beam B2 frames into each side of the web of column E4. The connections have been selected in Example 12-1. Typical column base details are shown in Fig. 13-2. The design drawings would indicate the desired type of column base detail. The base detail shown in Fig. 13-5 is of the common type shown in Fig. 13-2(b). Explanations of the various dimensions in Fig. 13-5 follow.

1. The top of steel roof deck elevation of $100'-3$ and top of base plate (bottom of column) elevation of $81'-10\frac{1}{2}$ would be furnished by the design drawings. The top of steel elevation of $100'-0$ is shown 3 in. below the top of steel roof deck. The top of columns is set $1\frac{1}{2}$ in. below top of steel.

2. In the detailing of columns, overall dimensions and floor and roof elevations are placed prominently farthest away from the views.

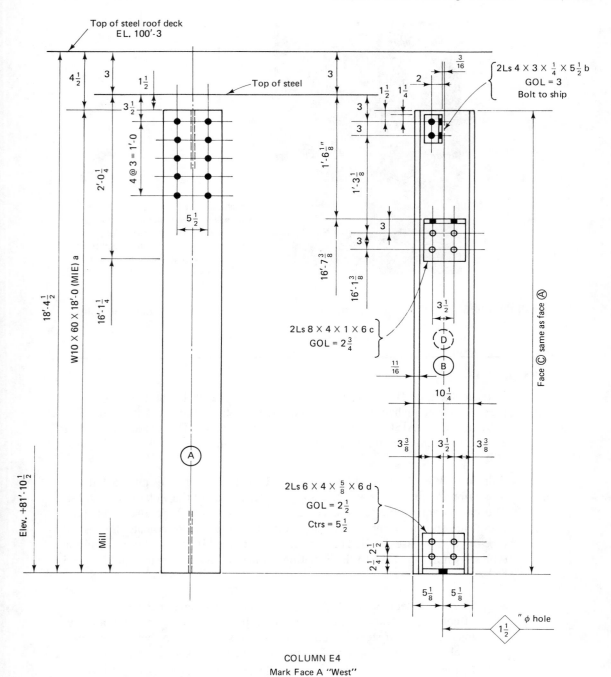

COLUMN E4
Mark Face A "West"

Figure 13-5 Typical column detail.

3. Detail dimensions showing hole spacing and detail location are placed closest to the related view.

4. Extension dimensions, used by the shop to establish and check the location of open holes or the tops of seat angles, are measured from the finished bottom of the column shaft.

5. All dimensions relating to a connection and holes should be tied to a floor or roof level at which the connected beam exists.

6. Depths of beams framing into the column are given so that clearances and connection angles can be checked.

7. The number and location of open holes on face A were established in Chapter 12 and shown in Fig. 12-13 (girder G1).

8. The seated connection material and details on face B were selected in Chapter 12 and shown in Fig. 12-10 (beam B2).

9. Girder G1 may be swung into place from either side to meet the flange holes on face A.

10. Beam B2 must be placed from above by moving the beam downward between the flanges of column E4. Therefore, top angle b must be shop-bolted hand-tight to the column for shipment, then removed and replaced in the field after placement of B2.

11. Based on the north arrow shown in Fig. 12-6, and the orientation of the columns, the shop will mark "West" on face A.

12. The bottom of the column (shaft) must be milled (M1E: "Mill One End") for full and level bearing on the base plate.

13. The total assembly unit will include one 18'-0 length of a W 10 × 60 with assembly mark "a" plus two angles marked "b," two angles marked "c," and two angles marked "d." All pieces will be assembled as shown in Fig. 13-5 and the assembly marked E4.

Loose base plates, since they constitute separate shipping pieces, are not detailed with the columns but on separate drawings and are given individual shipping marks. Typical shop details for such loose base plates together with leveling plates are usually shown by a simple plan view. An example is shown in Fig. 13-6. The base plate would be designed according to the principles discussed in Chapter 5 of this text. The connecting angles are from the AISCM, Part 4, Suggested Details for Column Base Plates. These details can be drawn to any convenient scale. To assist the erector in placing base plates, column center lines are scribed in both directions across the top surfaces of the plates.

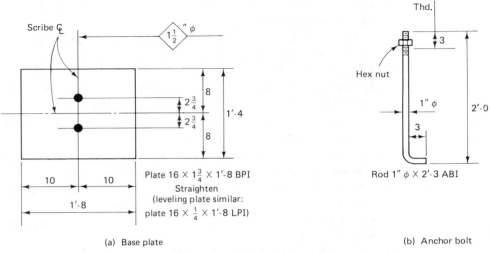

$1\frac{1}{2}$" φ

Scribe ℄

$2\frac{3}{4}$

$2\frac{3}{4}$

8

8

1'-4

10   10

1'-8

Plate 16 × $1\frac{3}{4}$ × 1'-8 BPI
Straighten
(leveling plate similar:
plate 16 × $\frac{1}{4}$ × 1'-8 LPI)

Thd.

3

Hex nut

1" φ

2'-0

3

Rod 1" φ × 2'-3 ABI

(a) Base plate

(b) Anchor bolt

Figure 13-6   Typical details.

PROBLEMS

13-1. Detail the column shown assuming that only beams B1 are framing into the column flanges. The top of base plate elevation is $63'\text{-}8\frac{1}{2}$ and the column base detail is that shown in Fig. 13-2(b) using 2L6 × 4 × 1/2 and two 1-in.-diameter anchor bolts. The building top of steel elevation is 85'-9 with the

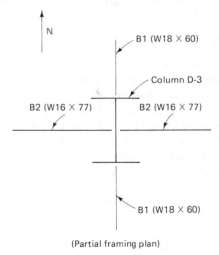

N

B1 (W18 × 60)

Column D-3

B2 (W16 × 77)   B2 (W16 × 77)

B1 (W18 × 60)

(Partial framing plan)

Problem 13-1

top of the steel roof deck at elevation 86'-0. The end connection for beams B1 consists of a 2L4 $\times$ 3 $\frac{1}{2}$ $\times$ $\frac{5}{16}$ $\times$ 8 $\frac{1}{2}$ in. long framed connection with three $\frac{7}{8}$-in.-diameter A325N bolts in standard holes in each leg of each angle. All structural steel is A36.

13-2. Detail the column shown for Prob. 13-1 assuming that beams B1 and B2 are framing into the column. All data from Problem 13-1 are applicable. The end connection for beams B2 is an unstiffened seated beam connection consisting of an L8 $\times$ 4 $\times$ $\frac{3}{4}$ $\times$ 6 in. seat angle with six $\frac{7}{8}$-in.-diameter A325N bolts in standard holes.

Appendix: Standard Load Tables for Open Web Steel Joists

TABLE A-1 STANDARD LOAD TABLE: OPEN WEB STEEL JOISTS, H-SERIES

Joist Designation	8H3	10H3	10H4	12H3	12H4	12H5	12H6	14H3	14H4	14H5	14H6	14H7	16H4	16H5	16H6	16H7	16H8
Nominal *Depth (in.)	8	10	10	12	12	12	12	14	14	14	14	14	16	16	16	16	16
Resist. Moment (in.-lbs.)	91,000	116,000	148,000	140,000	180,000	222,000	260,000	165,000	212,000	259,000	307,000	369,000	221,000	289,000	344,000	413,000	478,000
Max. End React. (lbs.)	2400	2500	2800	2800	3200	3600	3900	3200	3500	3800	4200	4600	3800	4300	4600	4900	5200
†Approx. Wt. (lbs./ft.)	5.0	5.0	6.1	5.2	6.2	7.1	8.2	5.5	6.5	7.4	8.6	10.0	6.6	7.8	8.6	10.3	11.4
Clear Span in Feet — 8 or less	600	500	560	467	533	600	650	457	500	543	600	657	475	538	575	613	650
9	533	500	560	467	533	600	650	457	500	543	600	657	475	538	575	613	650
10	480 / 460	500	560	467	533	600	650	457	500	543	600	657	475	538	575	613	650
11	436 / 345	455	509	467	533	600	650	457	500	543	600	657	475	538	575	613	650
12	400 / 266	417	467	467	533	600	650	457	500	543	600	657	475	538	575	613	650
13	359 / 209	385 / 337	431 / 417	431	492	554	600	457	500	543	600	657	475	538	575	613	650
14	310 / 167	357 / 270	400 / 334	400 / 393	457	514	557	457	500	543	600	657	475	538	575	613	650
15	270 / 136	333 / 219	373 / 271	373 / 320	427 / 418	480	520	427	467	507	560	613	475	538	575	613	650
16	232 / 112	302 / 181	350 / 223	350 / 264	400 / 345	450 / 404	488 / 480	400 / 366	438	475	525	575	475	538	575	613	650
17		268 / 151	329 / 186	323 / 220	376 / 287	424 / 337	459 / 400	376 / 305	412 / 398	447	494	541	447	506	541	576	612
18		239 / 127	305 / 157	288 / 185	356 / 242	400 / 284	433 / 337	340 / 257	389 / 336	422 / 393	467	511	422 / 413	478	511	544	578
19		214 / 108	273 / 133	259 / 157	332 / 206	379 / 241	411 / 286	305 / 218	368 / 285	400 / 334	442 / 399	484 / 470	400 / 351	453 / 432	484	516	547
20		193 / 92	247 / 114	233 / 135	300 / 177	360 / 207	390 / 246	275 / 187	350 / 245	380 / 287	420 / 342	460 / 403	368 / 301	430 / 370	460 / 437	490	520
21				212 / 117	272 / 152	336 / 179	371 / 212	249 / 162	320 / 212	362 / 248	400 / 295	438 / 348	334 / 260	410 / 320	438 / 377	467 / 454	495
22				193 / 101	248 / 133	306 / 155	355 / 185	227 / 141	292 / 184	345 / 215	382 / 257	418 / 302	304 / 226	391 / 278	418 / 328	445 / 395	473 / 454
23				176 / 89	227 / 116	280 / 136	328 / 162	208 / 123	267 / 161	326 / 189	365 / 225	400 / 265	279 / 198	364 / 243	400 / 287	426 / 346	452 / 398
24				162 / 78	208 / 102	257 / 120	301 / 142	191 / 108	245 / 142	300 / 166	350 / 198	383 / 233	256 / 174	334 / 214	383 / 253	408 / 314	433 / 351
25								176 / 96	226 / 125	276 / 147	327 / 175	368 / 206	236 / 154	308 / 193	367 / 224	392 / 269	416 / 310
26								163 / 85	209 / 111	255 / 131	303 / 156	354 / 183	218 / 137	285 / 169	339 / 199	377 / 239	400 / 275
27								151 / 76	194 / 99	237 / 117	281 / 139	337 / 164	202 / 122	264 / 151	315 / 177	363 / 214	385 / 246
28								140 / 68	180 / 89	220 / 104	261 / 125	314 / 147	188 / 110	246 / 135	293 / 159	350 / 197	371 / 220
29													175 / 99	229 / 121	273 / 143	327 / 172	359 / 198
30													164 / 89	214 / 110	255 / 129	306 / 156	347 / 179
31													153 / 81	200 / 99	239 / 117	287 / 141	332 / 162
32													144 / 74	188 / 90	224 / 107	269 / 128	311 / 148

Notes: Based on maximum allowable tensile stress of 30,000 psi. Loads above the heavy lines are governed by shear. The Steel Joist Institute publishes both Specifications and Load Tables; each of these contains standards which are to be used in conjunction with one another.

*Indicates nominal depth of steel joist only.

†Approximate weight per lineal foot for steel joist only. Accessories not included.

Source: Reproduced by permission of the Steel Joist Institute

TABLE A-1 STANDARD LOAD TABLE: OPEN WEB STEEL JOISTS, H-SERIES *(cont.)*

Joist Designation	18H5	18H6	18H7	18H8	18H9	18H10	18H11	20H5	20H6	20H7	20H8	20H9	20H10	20H11	22H6	22H7	22H8	22H9	22H10	22H11
Nominal *Depth (in.)	18	18	18	18	18	18	18	20	20	20	20	20	20	20	22	22	22	22	22	22
Resist. Moment (in.-lbs.)	325,000	383,000	466,000	540,000	627,000	705,000	814,000	365,000	406,000	499,000	602,000	701,000	789,000	912,000	422,000	526,000	653,000	776,000	873,000	1,009,000
Max. End React. (lbs.)	4500	4800	5200	5400	5900	6600	7600	4800	5100	5400	5600	6400	7000	7900	5400	5600	5800	6700	7200	8100
†Approx. Wt. (lbs./ft.)	8.0	9.2	10.4	11.6	12.6	14.0	15.8	8.4	9.6	10.7	12.2	13.2	14.6	16.4	9.7	10.7	12.0	13.8	15.2	16.9
Clear Span in Feet 18 or less	500	533	578	600	621	629	633	480	510	540	560	640	636	632	491	509	527	609	626	648
19	474	505	547	568	621	629	633	480	510	540	560	640	636	632	491	509	527	609	626	648
20	450	480	520	540	590	629	633	480	510	540	560	640	636	632	491	509	527	609	626	648
21	429 / 409	457	495	514	562	629	633	457	486	514	533	610	636	632	491	509	527	609	626	648
22	409 / 356	436 / 420	473	491	536	600	633	436	464	491	509	582	636	632	491	509	527	609	626	648
23	391 / 312	417 / 368	452 / 441	470	513	574	633	417 / 380	443 / 434	470	487	557	609	632	470	487	504	583	626	648
24	375 / 274	400 / 324	433 / 388	450 / 444	492 / 484	550 / 546	633	400 / 335	425 / 382	450	467	533	583	632	450 / 446	467	483	558	600	648
25	347 / 243	384 / 286	416 / 343	432 / 393	472 / 428	528 / 483	608 / 548	384 / 296	408 / 338	432 / 411	448	512	560	632	432 / 395	448	464	536	576	648
26	321 / 216	369 / 255	400 / 305	415 / 349	454 / 380	508 / 429	585 / 487	360 / 263	392 / 300	415 / 365	431	492 / 476	538	608	415 / 351	431 / 426	446	515	554	623
27	297 / 193	350 / 227	385 / 272	400 / 312	437 / 340	489 / 383	563 / 435	334 / 235	371 / 268	400 / 326	415 / 392	474 / 425	519 / 480	585 / 545	386 / 313	415 / 380	430	496	533	600
28	276 / 173	326 / 204	371 / 244	386 / 280	421 / 305	471 / 344	543 / 390	310 / 211	345 / 240	386 / 292	400 / 352	457 / 381	500 / 431	564 / 488	359 / 281	400 / 341	414	479 / 468	514	579
29	258 / 155	304 / 184	359 / 220	372 / 252	407 / 274	455 / 309	524 / 351	289 / 190	322 / 216	372 / 263	386 / 317	441 / 343	483 / 388	545 / 440	335 / 253	386 / 307	400 / 379	462 / 421	497 / 473	559 / 539
30	241 / 140	284 / 166	345 / 199	360 / 227	393 / 248	440 / 280	507 / 317	270 / 171	301 / 195	360 / 238	373 / 286	427 / 310	467 / 350	527 / 397	313 / 228	373 / 277	387 / 343	447 / 381	480 / 428	540 / 487
31	225 / 127	266 / 150	323 / 180	348 / 206	381 / 224	426 / 253	490 / 287	253 / 155	282 / 177	346 / 215	361 / 259	413 / 281	452 / 317	510 / 360	293 / 207	361 / 311	374 / 345	432 / 345	465 / 387	523 / 441
32	212 / 116	249 / 137	303 / 164	338 / 187	369 / 204	413 / 230	475 / 261	238 / 141	264 / 161	325 / 196	350 / 236	400 / 255	438 / 288	494 / 327	275 / 188	342 / 228	363 / 282	419 / 314	450 / 352	506 / 401
33	199 / 106	234 / 125	285 / 149	327 / 171	358 / 186	400 / 210	461 / 238	223 / 129	249 / 147	305 / 178	339 / 215	388 / 233	424 / 263	479 / 298	258 / 172	322 / 208	352 / 257	406 / 286	436 / 321	491 / 366
34	187 / 96	221 / 114	269 / 136	311 / 156	347 / 170	388 / 192	447 / 218	210 / 118	234 / 134	288 / 163	329 / 196	376 / 213	412 / 240	465 / 273	243 / 157	303 / 190	341 / 235	394 / 261	424 / 294	476 / 335
35	177 / 88	208 / 104	254 / 125	294 / 143	337 / 156	377 / 176	434 / 200	199 / 108	221 / 123	272 / 150	320 / 180	366 / 195	400 / 220	451 / 250	230 / 144	286 / 175	331 / 216	383 / 240	411 / 269	463 / 307
36	167 / 81	197 / 96	240 / 115	278 / 132	323 / 143	363 / 162	419 / 183	188 / 99	209 / 113	257 / 137	310 / 166	356 / 179	389 / 203	439 / 230	217 / 132	271 / 160	322 / 198	372 / 220	400 / 247	450 / 282
37								178 / 91	198 / 104	243 / 127	293 / 152	341 / 165	378 / 187	427 / 212	206 / 122	256 / 148	314 / 183	362 / 203	389 / 228	438 / 260
38								169 / 84	187 / 96	230 / 117	278 / 141	324 / 153	364 / 172	416 / 195	195 / 112	243 / 136	301 / 169	353 / 187	379 / 210	426 / 240
39								160 / 78	178 / 89	219 / 108	264 / 130	307 / 141	346 / 159	400 / 181	185 / 104	231 / 126	286 / 156	340 / 173	369 / 195	415 / 222
40								152 / 72	169 / 82	208 / 100	251 / 121	292 / 131	329 / 148	380 / 168	176 / 96	219 / 117	272 / 145	323 / 161	360 / 180	405 / 205
41															167 / 89	209 / 109	259 / 134	308 / 149	346 / 167	395 / 191
42															159 / 83	199 / 101	247 / 125	293 / 139	330 / 156	381 / 177
43															152 / 78	190 / 94	235 / 116	280 / 129	315 / 145	364 / 165
44															145 / 72	181 / 88	225 / 109	267 / 121	301 / 136	347 / 154

TABLE A-1 STANDARD LOAD TABLE: OPEN WEB STEEL JOISTS, H-SERIES (*cont.*)

Joist Designation	24H6	24H7	24H8	24H9	24H10	24H11	26H8	26H9	26H10	26H11	28H8	28H9	28H10	28H11	30H8	30H9	30H10	30H11
Nominal *Depth (in.)	24	24	24	24	24	24	26	26	26	26	28	28	28	28	30	30	30	30
Resist. Moment (in.-lbs.)	462,000	576,000	716,000	851,000	957,000	1,106,000	784,000	925,000	1,040,000	1,203,000	846,000	1,000,000	1,124,000	1,300,000	909,000	1,075,000	1,207,000	1,397,000
Max. End React. (lbs.)	5600	5800	6000	7000	7500	8200	6700	7200	7600	8300	6700	7200	7700	8400	6800	7500	8100	8700
†Approx. Wt. (lbs./ft.)	10.3	11.5	12.7	14.0	15.5	17.5	12.8	14.8	16.2	17.9	13.5	15.2	16.8	18.3	14.2	15.4	17.3	18.8
Clear Span in Feet 24 or less	467	483	500	583	625	631	515	554	585	638	479	514	550	600	453	500	540	580
25	448	464	480	560	600	631	515	554	585	638	479	514	550	600	453	500	540	580
26	431	446	462	538	577	631	515	554	585	638	479	514	550	600	453	500	540	580
27	415/375	430	444	519	556	607	496	533	563	615	479	514	550	600	453	500	540	580
28	393/336	414/406	429	500	536	586	479	514	543	593	479	514	550	600	453	500	540	580
29	366/303	400/365	414	483	517	566	462	497	524	572	462	497	531	579	453	500	540	580
30	342/273	387/330	400	467/457	500	547	447	480	507	553	447	480	513	560	453	500	540	580
31	320/248	374/299	387/373	452/414	484/465	529	432/418	465	490	535	432	465	497	542	439	484	523	561
32	301/225	363/272	375/339	438/376	469/423	513/482	419/380	450/445	475	519	419	450	481	525	425	469	506	544
33	283/205	352/248	364/309	424/343	455/386	497/440	406/346	436/405	461/456	503	406/404	436	467	509	412	455	491	527
34	266/188	332/227	353/283	412/314	441/353	482/402	394/317	424/371	447/417	488/476	394/370	424	453	494	400	441	476	512
35	251/172	313/208	343/259	400/288	429/323	469/369	383/290	411/340	434/383	474/437	383/339	411/396	440	480	389	429	463	497
36	238/158	296/191	333/238	389/264	417/297	456/339	372/267	400/312	422/352	461/401	372/311	400/364	428/410	467	378/359	417	450	483
37	225/146	280/176	324/219	378/243	405/274	443/312	362/246	389/288	411/324	449/370	362/287	389/336	416/378	454/432	368/330	405/387	438/436	470
38	213/135	266/162	316/202	368/225	395/253	432/288	353/227	379/266	400/299	437/341	353/265	379/310	405/349	442/399	358/305	395/357	426/402	458
39	202/124	252/150	308/187	359/208	385/234	421/266	344/210	369/246	390/276	426/316	344/245	369/287	395/322	431/369	349/282	385/331	415/372	446/426
40	193/115	240/139	298/174	350/193	375/217	410/247	327/194	360/228	380/256	415/292	335/227	360/266	385/299	420/342	340/262	375/306	405/345	435/395
41	183/107	228/129	284/161	337/179	366/201	400/229	311/181	351/211	371/238	405/272	327/211	351/247	376/278	410/318	332/243	366/285	395/320	424/367
42	175/100	218/120	271/150	322/166	357/187	390/213	296/168	343/197	362/221	395/253	319/196	343/229	367/258	400/295	324/226	357/265	386/298	414/341
43	167/93	208/112	258/140	307/155	345/174	381/199	283/156	334/183	353/206	386/235	305/183	335/214	358/241	391/275	316/211	349/247	377/278	405/318
44	159/87	198/105	247/130	293/145	330/163	373/186	270/146	319/171	345/193	377/220	291/171	327/200	350/225	382/257	309/196	341/230	368/259	395/297
45	152/81	190/98	236/122	280/135	315/152	364/173	258/137	305/160	338/180	369/205	279/159	320/187	342/210	373/240	299/184	333/215	360/242	387/278
46	146/76	181/92	226/114	268/127	302/142	348/162	247/128	291/150	328/168	361/192	267/149	313/175	335/197	365/225	286/172	326/202	352/227	378/260
47	139/71	174/86	216/107	257/119	289/133	334/152	237/120	279/140	314/158	353/180	255/140	302/164	328/184	357/211	274/161	319/189	345/213	370/244
48	134/67	167/81	207/100	246/111	277/125	320/143	227/112	268/132	301/148	346/169	245/131	289/154	321/173	350/198	263/151	311/177	338/200	363/229
49							218/106	257/124	289/139	334/159	235/124	278/144	312/163	343/186	252/142	298/167	331/188	355/215
50							209/100	247/117	277/131	321/150	226/116	267/136	300/153	336/175	242/134	287/157	322/177	348/202
51							201/94	237/110	267/124	308/141	217/110	256/128	288/144	329/165	233/126	276/148	309/166	341/191
52							193/88	228/104	256/117	297/133	209/103	247/121	277/136	321/156	224/119	265/139	298/157	335/180
53											201/98	237/114	267/128	309/147	216/112	255/132	286/148	328/170
54											193/92	229/108	257/121	297/139	208/106	246/125	276/140	319/161
55											186/87	220/102	248/115	287/132	200/101	237/118	266/133	308/152
56											180/83	213/97	239/109	276/125	193/95	229/112	257/126	297/144
57															187/90	221/106	248/119	287/137
58															180/86	213/101	239/113	277/130
59															174/81	206/95	231/108	268/123
60															168/77	199/91	224/102	259/117

TABLE A-2 STANDARD LOAD TABLE: LONGSPAN STEEL JOISTS, LH-SERIES

Joist Designation	Approx. Wt. in Lbs. per Linear Ft.	Nominal Depth in Inches	SAFE LOAD** in Lbs. Between	CLEAR OPENING OR NET SPAN IN FEET																
			21–24	25	26	27	28	29	30	31	32	33	34	35	36				
18LH02	13	18	12000	468	442	418	391	367	345	324	306	289	273	259	245				
				313	284	259	234	212	193	175	160	147	135	124	114				
18LH03	14	18	13300	521	493	467	438	409	382	359	337	317	299	283	267				
				348	317	289	262	236	213	194	177	161	148	136	124				
18LH04	16	18	15500	604	571	535	500	469	440	413	388	365	344	325	308				
				403	367	329	296	266	242	219	200	182	167	153	141				
18LH05	17	18	17500	684	648	614	581	543	508	476	448	421	397	375	355				
				454	414	378	345	311	282	256	233	212	195	179	164				
18LH06	19	18	20700	809	749	696	648	605	566	531	499	470	443	418	396				
				526	469	419	377	340	307	280	254	232	212	195	180				
18LH07	21	18	21500	840	809	780	726	678	635	595	559	526	496	469	444				
				553	513	476	428	386	349	317	288	264	241	222	204				
18LH08	22	18	22400	876	843	812	784	758	717	680	641	604	571	540	512				
				577	534	496	462	427	387	351	320	292	267	246	226				
18LH09	24	18	24000	936	901	868	838	810	783	759	713	671	633	598	566				
				616	571	527	491	458	418	380	346	316	289	266	245				

Joist Designation	Approx. Wt. in Lbs. per Linear Ft.	Nominal Depth in Inches	SAFE LOAD** in Lbs. Between	25	26	27	28	29	30	31	32	33	34	35	36	37	38	39	40
			22–24	25	26	27	28	29	30	31	32	33	34	35	36	37	38	39	40
20LH02	13	20	11300	442	437	431	410	388	365	344	325	307	291	275	262	249	237	225	215
				306	303	298	274	250	228	208	190	174	160	147	136	126	117	108	101
20LH03	14	20	12000	469	463	458	452	434	414	395	372	352	333	316	299	283	269	255	243
				337	333	317	302	280	258	238	218	200	184	169	156	143	133	123	114
20LH04	16	20	14700	574	566	558	528	496	467	440	416	393	372	353	335	318	303	289	275
				428	406	386	352	320	291	265	243	223	205	189	174	161	149	139	129
20LH05	17	20	15800	616	609	602	595	571	544	513	484	458	434	411	390	371	353	336	321
				459	437	416	395	366	337	308	281	258	238	219	202	187	173	161	150
20LH06	19	20	21100	822	791	763	723	679	635	596	560	527	497	469	444	421	399	379	361
				606	561	521	477	427	386	351	320	292	267	246	226	209	192	178	165
20LH07	21	20	22500	878	845	814	786	760	711	667	627	590	556	526	497	471	447	425	404
				647	599	556	518	484	438	398	362	331	303	278	256	236	218	202	187
20LH08	22	20	23200	908	873	842	813	785	760	722	687	654	621	588	558	530	503	479	457
				669	619	575	536	500	468	428	395	365	336	309	285	262	242	225	209
20LH09	24	20	25400	990	953	918	886	856	828	802	778	755	712	673	636	603	572	544	517
				729	675	626	581	542	507	475	437	399	366	336	309	285	264	244	227
20LH10	27	20	27400	1068	1028	991	956	924	894	865	839	814	791	748	707	670	636	604	575
				786	724	673	626	585	545	510	479	448	411	377	346	320	296	274	254

Notes: Based on a maximum allowable tensile stress of 30,000 psi. The Steel Joist Institute publishes both Specifications and Load Tables; each of these contains standards which are to be used in conjunction with one another.

**To solve for the *safe uniform load* between spans shown, divide the safe load in pounds by the net span in feet + 0.67 ft. (The added 0.67 ft is necessary to obtain the proper span for which the tables were developed. To solve for the *live load* (which will produce approximately span/360 deflection) between the spans shown, multiply the live load of the shortest net span shown in the table by (shortest net span + 0.67 ft)^2, and divide by (actual net span + 0.67 ft)^2. The live load shall not exceed the safe uniform load.

Source: Reproduced by permission of the Steel Joist Institute.

TABLE A-2 STANDARD LOAD TABLE: LONGSPAN STEEL JOISTS, LH-SERIES (*cont.*)

CLEAR OPENING OR NET SPAN IN FEET

Joist Designation	Approx. Wt. in Lbs. per Linear Ft.	Depth in Inches	SAFE LOAD** in Lbs. Between 28–32	33	34	35	36	37	38	39	40	41	42	43	44	45	46	47	48
24LH03	14	24	11500	342	339	336	323	307	293	279	267	255	244	234	224	215	207	199	191
				235	226	218	204	188	175	162	152	141	132	124	116	109	102	96	90
24LH04	16	24	14100	419	398	379	360	343	327	312	298	285	273	262	251	241	231	222	214
				288	265	246	227	210	195	182	169	158	148	138	130	122	114	107	101
24LH05	17	24	15100	449	446	440	419	399	380	363	347	331	317	304	291	280	269	258	248
				308	297	285	264	244	226	210	196	182	171	160	150	141	132	124	117
24LH06	19	24	20300	604	579	555	530	504	480	457	437	417	399	381	364	348	334	320	307
				411	382	356	331	306	284	263	245	228	211	197	184	172	161	152	142
24LH07	21	24	22300	665	638	613	588	565	541	516	491	468	446	426	407	389	373	357	343
				452	421	393	367	343	320	297	276	257	239	223	208	195	182	171	161
24LH08	22	24	23800	707	677	649	622	597	572	545	520	497	475	455	435	417	400	384	369
				480	447	416	388	362	338	314	292	272	254	238	222	208	196	184	173
24LH09	24	24	28000	832	808	785	764	731	696	663	632	602	574	548	524	501	480	460	441
				562	530	501	460	424	393	363	337	313	292	272	254	238	223	209	196
24LH10	27	24	29600	882	856	832	809	788	768	737	702	668	637	608	582	556	533	511	490
				596	559	528	500	474	439	406	378	351	326	304	285	266	249	234	220
24LH11	29	24	31200	927	900	875	851	829	807	787	768	734	701	671	642	616	590	567	544
				624	588	555	525	498	472	449	418	388	361	337	315	294	276	259	243

Joist Designation	Approx. Wt. in Lbs. per Linear Ft.	Depth in Inches	SAFE LOAD** in Lbs. Between 33–40	41	42	43	44	45	46	47	48	49	50	51	52	53	54	55	56
28LH05	16	28	14000	337	323	310	297	286	275	265	255	245	237	228	220	213	206	199	193
				219	205	192	180	169	159	150	142	133	126	119	113	107	102	97	92
28LH06	19	28	18600	448	429	412	395	379	364	350	337	324	313	301	291	281	271	262	253
				289	270	253	238	223	209	197	186	175	166	156	148	140	133	126	120
28LH07	21	28	21000	505	484	464	445	427	410	394	379	365	352	339	327	316	305	295	285
				326	305	285	267	251	236	222	209	197	186	176	166	158	150	142	135
28LH08	21	28	22500	540	517	496	475	456	438	420	403	387	371	357	344	331	319	308	297
				348	325	305	285	268	252	236	222	209	196	185	175	165	156	148	140
28LH09	24	28	27700	667	639	612	586	563	540	519	499	481	463	446	430	415	401	387	374
				428	400	375	351	329	309	291	274	258	243	228	216	204	193	183	173
28LH10	27	28	30300	729	704	679	651	625	600	576	554	533	513	495	477	460	444	429	415
				466	439	414	388	364	342	322	303	285	269	255	241	228	215	204	193
28LH11	29	28	32500	780	762	736	711	682	655	629	605	582	561	540	521	502	485	468	453
				498	475	448	423	397	373	351	331	312	294	278	263	249	236	223	212
28LH12	33	28	35700	857	837	818	800	782	766	737	709	682	656	632	609	587	566	546	527
				545	520	496	476	454	435	408	383	361	340	321	303	285	270	256	243
28LH13	36	28	37200	895	874	854	835	816	799	782	766	751	722	694	668	643	620	598	577
				569	543	518	495	472	452	433	415	396	373	352	332	314	297	281	266

Joist Designation	Approx. Wt. in Lbs. per Linear Ft.	Depth in Inches	SAFE LOAD** in Lbs. Between 38–48	49	50	51	52	53	54	55	56	57	58	59	60	61	62	63	64
32LH06	18	32	16700	338	326	315	304	294	284	275	266	257	249	242	234	227	220	214	208
				211	199	189	179	169	161	153	145	138	131	125	119	114	108	104	99
32LH07	20	32	18800	379	366	353	341	329	318	308	298	288	279	271	262	254	247	240	233
				235	223	211	200	189	179	170	162	154	146	140	133	127	121	116	111
32LH08	21	32	20400	411	397	383	369	357	345	333	322	312	302	293	284	275	267	259	252
				255	242	229	216	204	194	184	175	167	159	151	144	137	131	125	120
32LH09	24	32	25600	516	498	480	463	447	432	418	404	391	379	367	356	345	335	325	315
				319	302	285	270	256	243	230	219	208	198	189	180	172	164	157	149
32LH10	26	32	28300	571	550	531	512	495	478	462	445	430	416	402	389	376	364	353	342
				352	332	315	297	282	267	254	240	228	217	206	196	186	178	169	162
32LH11	28	32	31000	625	602	580	560	541	522	505	488	473	458	443	429	416	403	390	378
				385	363	343	325	308	292	277	263	251	239	227	216	206	196	187	179
32LH12	33	32	36400	734	712	688	664	641	619	598	578	559	541	524	508	492	477	463	449
				450	428	406	384	364	345	327	311	295	281	267	255	243	232	221	211
32LH13	36	32	40600	817	801	785	771	742	715	690	666	643	621	600	581	562	544	527	511
				500	480	461	444	420	397	376	354	336	319	304	288	275	262	249	238
32LH14	37	32	41800	843	826	810	795	780	766	738	713	688	665	643	622	602	583	564	547
				515	495	476	458	440	417	395	374	355	337	321	304	290	276	264	251
32LH15	41	32	43200	870	853	837	821	805	791	776	763	750	725	701	678	656	635	616	597
				532	511	492	473	454	438	422	407	393	374	355	338	322	306	292	279

Joist Designation	Approx. Wt. in Lbs. per Linear Ft.	Depth in Inches	SAFE LOAD** in Lbs. Between 42–56	57	58	59	60	61	62	63	64	65	66	67	68	69	70	71	72
36LH07	20	36	16800	292	283	274	266	258	251	244	237	230	224	218	212	207	201	196	191
				177	168	160	153	146	140	134	128	122	117	112	107	103	99	95	91
36LH08	20	36	18500	321	311	302	293	284	276	268	260	253	246	239	233	227	221	215	209
				194	185	176	168	160	153	146	140	134	128	123	118	113	109	104	100
36LH09	23	36	23700	411	398	386	374	363	352	342	333	323	314	306	297	289	282	275	267
				247	235	224	214	204	195	186	179	171	163	157	150	144	138	133	127
36LH10	26	36	26100	454	440	426	413	401	389	378	367	357	347	338	328	320	311	303	295
				273	260	248	236	225	215	206	197	188	180	173	165	159	152	146	140
36LH11	28	36	28500	495	480	465	451	438	425	412	401	389	378	368	358	348	339	330	322
				297	283	269	257	246	234	224	214	205	196	188	180	173	166	159	153
36LH12	32	36	34100	593	575	557	540	523	508	493	478	464	450	437	424	412	400	389	378
				354	338	322	307	292	279	267	255	243	232	222	213	204	195	187	179
36LH13	36	36	40100	697	675	654	634	615	596	579	562	546	531	516	502	488	475	463	451
				415	395	376	359	342	327	312	298	285	273	262	251	240	231	222	213
36LH14	37	36	44200	768	755	729	706	683	661	641	621	602	584	567	551	535	520	505	492
				456	434	412	392	373	356	339	323	309	295	283	270	259	247	237	228
36LH15	41	36	46600	809	795	781	769	744	721	698	677	656	637	618	600	583	567	551	536
				480	464	448	434	413	394	375	358	342	327	312	299	286	274	263	252

Reproduced by permission of the Steel Joist Institute

TABLE A-2 STANDARD LOAD TABLE: LONGSPAN STEEL JOISTS, LH-SERIES (*cont.*)

Joist Designation	Approx. Wt. in Lbs. per Linear Ft.	Depth in Inches	SAFE LOAD** in Lbs. Between	CLEAR OPENING OR NET SPAN IN FEET															
			47–64	65	66	67	68	69	70	71	72	73	74	75	76	77	78	79	80
40LH08	20	40	16600	254/150	247/144	241/138	234/132	228/127	222/122	217/117	211/112	206/108	201/104	196/100	192/97	187/93	183/90	178/86	174/83
40LH09	23	40	21800	332/196	323/188	315/180	306/173	298/166	291/160	283/153	276/147	269/141	263/136	256/131	250/126	244/122	239/118	233/113	228/109
40LH10	25	40	24000	367/216	357/207	347/198	338/190	329/183	321/176	313/169	305/162	297/156	290/150	283/144	276/139	269/134	262/129	255/124	249/119
40LH11	27	40	26200	399/234	388/224	378/215	368/207	358/198	349/190	340/183	332/176	323/169	315/163	308/157	300/151	293/145	286/140	279/135	273/130
40LH12	32	40	31900	486/285	472/273	459/261	447/251	435/241	424/231	413/222	402/213	392/205	382/197	373/189	364/182	355/176	346/169	338/163	330/157
40LH13	36	40	37600	573/334	557/320	542/307	528/295	514/283	500/271	487/260	475/250	463/241	451/231	440/223	429/214	419/207	409/199	399/192	390/185
40LH14	37	40	43000	656/383	638/367	620/351	603/336	587/323	571/309	556/297	542/285	528/273	515/263	502/252	490/243	478/233	466/225	455/216	444/209
40LH15	41	40	48100	734/427	712/408	691/390	671/373	652/357	633/342	616/328	599/315	583/302	567/290	552/279	538/268	524/258	511/248	498/239	486/230
40LH16	47	40	53000	808/469	796/455	784/441	772/428	761/416	751/404	730/387	710/371	691/356	673/342	655/329	638/316	622/304	606/292	591/282	576/271

Joist Designation	Approx. Wt. in Lbs. per Linear Ft.	Depth in Inches	SAFE LOAD** in Lbs. Between																
			52–72	73	74	75	76	77	78	79	80	81	82	83	84	85	86	87	88
44LH09	22	44	20000	272/158	265/152	259/146	253/141	247/136	242/131	236/127	231/122	226/118	221/114	216/110	211/106	207/103	202/99	198/96	194/93
44LH10	25	44	22100	300/174	293/168	286/162	279/155	272/150	266/144	260/139	254/134	249/130	243/125	238/121	233/117	228/113	223/110	218/106	214/103
44LH11	27	44	23900	325/188	317/181	310/175	302/168	295/162	289/157	282/151	276/146	269/140	264/136	258/131	252/127	247/123	242/119	236/115	232/111
44LH12	31	44	29600	402/232	393/224	383/215	374/207	365/200	356/192	347/185	339/179	331/172	323/166	315/160	308/155	300/149	293/144	287/139	280/134
44LH13	35	44	35100	477/275	466/265	454/254	444/246	433/236	423/228	413/220	404/212	395/205	386/198	377/191	369/185	361/179	353/173	346/167	338/161
44LH14	36	44	40400	549/315	534/302	520/291	506/279	493/268	481/259	469/249	457/240	446/231	436/223	425/215	415/207	406/200	396/193	387/187	379/181
44LH15	41	44	47000	639/366	623/352	608/339	593/326	579/314	565/303	551/292	537/281	524/271	512/261	500/252	488/243	476/234	466/227	455/219	445/211
44LH16	47	44	54200	737/421	719/405	701/390	684/375	668/362	652/348	637/336	622/324	608/313	594/302	580/291	568/282	555/272	543/263	531/255	520/246
44LH17	54	44	58200	790/450	780/438	769/426	759/415	750/405	732/390	715/376	699/363	683/351	667/338	652/327	638/316	624/305	610/295	597/285	584/276

Joist Designation	Approx. Wt. in Lbs. per Linear Ft.	Depth in Inches	SAFE LOAD** in Lbs. Between																
			56–80	81	82	83	84	85	86	87	88	89	90	91	92	93	94	95	96
48LH10	25	48	20000	246/141	241/136	236/132	231/127	226/123	221/119	217/116	212/112	208/108	204/105	200/102	196/99	192/96	188/93	185/90	181/87
48LH11	27	48	21700	266/152	260/147	255/142	249/137	244/133	239/129	234/125	229/120	225/117	220/113	216/110	212/106	208/103	204/100	200/97	196/94
48LH12	31	48	27400	336/191	329/185	322/179	315/173	308/167	301/161	295/156	289/151	283/147	277/142	272/138	266/133	261/129	256/126	251/122	246/118
48LH13	35	48	32800	402/228	393/221	384/213	376/206	368/199	360/193	353/187	345/180	338/175	332/170	325/164	318/159	312/154	306/150	300/145	294/141
48LH14	36	48	38700	475/269	464/260	454/251	444/243	434/234	425/227	416/220	407/212	399/206	390/199	383/193	375/187	367/181	360/176	353/171	346/165
48LH15	41	48	44500	545/308	533/298	521/287	510/278	499/269	488/260	478/252	468/244	458/236	448/228	439/221	430/214	422/208	413/201	405/195	397/189
48LH16	47	48	51300	629/355	615/343	601/331	588/320	576/310	563/299	551/289	540/280	528/271	518/263	507/255	497/247	487/239	477/232	468/225	459/218
48LH17	54	48	57600	706/397	690/383	675/371	660/358	646/346	632/335	619/324	606/314	593/304	581/294	569/285	558/276	547/268	536/260	525/252	515/245

Reproduced by permission of the Steel Joist Institute

Index